NATIONAL GEOGRAPHIC

FIELD GUIDE TO THE
BIRDS
OF THE UNITED STATES AND CANADA
WEST

SECOND EDITION

FIELD GUIDE TO THE
BIRDS
OF THE UNITED STATES AND CANADA
WEST

SECOND EDITION

TED FLOYD

NATIONAL GEOGRAPHIC
WASHINGTON, D.C.

CONTENTS

- 6 Introduction
- 22 **Geese, Swans, and Ducks** | *Anatidae*
- 52 **New World Quail** | *Odontophoridae*
- 56 **Pheasants and Kin** | *Phasianidae*
- 68 **Grebes** | *Podicipedidae*
- 72 **Sandgrouses** | *Pteroclidae*
- 72 **Pigeons and Doves** | *Columbidae*
- 76 **Roadrunners and Cuckoos** | *Cuculidae*
- 80 **Goatsuckers** | *Caprimulgidae*
- 82 **Swifts** | *Apodidae*
- 86 **Hummingbirds** | *Trochilidae*
- 96 **Rails, Gallinules, and Coots** | *Rallidae*
- 100 **Cranes** | *Gruidae*
- 102 **Stilts and Avocets** | *Recurvirostridae*
- 102 **Oystercatchers** | *Haematopodidae*
- 104 **Plovers** | *Charadriidae*
- 110 **Sandpipers** | *Scolopacidae*
- 134 **Skuas and Jaegers** | *Stercorariidae*
- 138 **Auks, Murres, and Puffins** | *Alcidae*
- 148 **Gulls, Terns, and Skimmers** | *Laridae*
- 174 **Tropicbirds** | *Phaethontidae*
- 176 **Loons** | *Gaviidae*
- 180 **Albatrosses** | *Diomedeidae*
- 182 **Southern Storm-Petrels** | *Oceanitidae*
- 182 **Northern Storm-Petrels** | *Hydrobatidae*
- 186 **Petrels and Shearwaters** | *Procellariidae*
- 198 **Storks** | *Ciconiidae*
- 198 **Frigatebirds** | *Fregatidae*
- 200 **Boobies** | *Sulidae*
- 202 **Cormorants** | *Phalacrocoracidae*
- 204 **Pelicans** | *Pelecanidae*
- 206 **Bitterns, Herons, and Egrets** | *Ardeidae*
- 214 **Ibises** | *Threskiornithidae*
- 214 **New World Vultures** | *Cathartidae*
- 216 **Osprey** | *Pandionidae*
- 218 **Hawks, Eagles, and Kin** | *Accipitridae*
- 232 **Barn Owls** | *Tytonidae*
- 232 **Typical Owls** | *Strigidae*
- 242 **Trogons** | *Trogonidae*
- 242 **Kingfishers** | *Alcedinidae*
- 244 **Woodpeckers** | *Picidae*
- 258 **Caracaras and Falcons** | *Falconidae*
- 262 **Parakeets and Parrots** | *Psittacidae*
- 264 **Australasian Parrots and Lovebirds** | *Psittaculidae*
- 266 **Becards** | *Tityridae*
- 266 **Tyrant Flycatchers** | *Tyrannidae*
- 284 **Vireos** | *Vireonidae*
- 290 **Monarch Flycatchers** | *Monarchidae*
- 292 **Shrikes** | *Laniidae*
- 292 **Jays and Crows** | *Corvidae*
- 302 **Verdin** | *Remizidae*
- 302 **Chickadees and Titmice** | *Paridae*
- 308 **Larks** | *Alaudidae*
- 310 **Reed Warblers** | *Acrocephalidae*
- 310 **Swallows** | *Hirundinidae*
- 316 **Bushtits** | *Aegithalidae*
- 316 **Bush-Warblers** | *Cettiidae*
- 316 **Leaf Warblers** | *Phylloscopidae*
- 318 **Bulbuls** | *Pycnonotidae*
- 318 **Sylviid Warblers** | *Sylviidae*
- 320 **White-eyes** | *Zosteropidae*
- 320 **Laughingthrushes** | *Leiothrichidae*
- 322 **Kinglets** | *Regulidae*
- 324 **Waxwings** | *Bombycillidae*
- 324 **Silky-flycatchers** | *Ptiliogonatidae*
- 326 **Nuthatches** | *Sittidae*
- 326 **Treecreepers** | *Certhiidae*
- 328 **Gnatcatchers** | *Polioptilidae*
- 330 **Wrens** | *Troglodytidae*
- 336 **Thrashers and Mockingbirds** | *Mimidae*
- 340 **Starlings and Mynas** | *Sturnidae*
- 342 **Dippers** | *Cinclidae*
- 342 **Thrushes** | *Turdidae*
- 350 **Old World Flycatchers** | *Muscicapidae*
- 352 **Olive Warbler** | *Peucedramidae*
- 352 **Weavers** | *Ploceidae*
- 352 **Whydahs** | *Viduidae*
- 354 **Waxbills** | *Estrildidae*
- 356 **Old World Sparrows** | *Passeridae*
- 356 **Wagtails and Pipits** | *Motacillidae*
- 360 **Finches** | *Fringillidae*
- 380 **Longspurs and *Plectrophenax* Buntings** | *Calcariidae*
- 384 **Old World Buntings** | *Emberizidae*
- 386 **New World Sparrows** | *Passerellidae*
- 410 **Yellow-breasted Chat** | *Icteriidae*
- 410 **Blackbirds** | *Icteridae*
- 424 **Wood-Warblers** | *Parulidae*
- 444 **Cardinals and Kin** | *Cardinalidae*
- 452 **True Tanagers** | *Thraupidae*
- 454 **Appendix A** | Rare Birds in the West
- 473 **Appendix B** | Extinct and Likely Extinct Birds in the West
- 474 Glossary
- 479 Illustrations Credits
- 481 About the Author/Acknowledgments
- 483 Index

Opposite: A male Calliope Hummingbird visits the blossoms of a pride of Madeira shrub. Page 2: Steller's Jays

INTRODUCTION

Welcome to the second edition of the *National Geographic Field Guide to the Birds of the United States and Canada—West*. It has been more than 15 years since the first edition of this book was published. In that time, both the science of ornithology and the experience of birding have changed tremendously. This new edition reflects and responds to those changes.

Advances in genetic science are informing a new understanding of bird evolution and taxonomy, redrawing the relationships among species. Crowdsourced data are expanding our knowledge of bird status and distribution and driving new maps like those integral to this edition, which were created in partnership with the Cornell Lab of Ornithology. And careful field research is greatly refining our knowledge of avian behavior and ecology.

Smartphones are now ubiquitous in the birder's toolkit, offering access to massive databases with visual and auditory aids for identification and online platforms where individuals can record their finds. Today digital cameras, powerful and affordable, can be as important in bird identification as binoculars.

In this new era, where does a book—a traditional field guide like this one—fit in? Here you can flip through, hold up, compare, bookmark, and annotate descriptions and illustrations of all 717 species likely to be encountered in the western regions of the United States (including Hawaii) and Canada. You can read informative entries on every single species, learn of their relationships to others, and see telling characteristics pointed out in art on every spread.

By browsing through this book before you go birding, you can get a notion of what to look for. Thumbing through it afterward will help you confirm identifications and enrich your understanding of what you have observed. And of course this book is intended for quick reference in the field as well. The guiding principle behind this volume, whether you use it at home or in the field, is efficiency of presentation. Species accounts, art, and maps have been created to succinctly convey the essentials for accurate identification of all the bird species occurring regularly in the West.

Birding offers the chance to become immersed in the natural world, as in this lush Hawaiian forest.

INTRODUCTION

Field guides not only introduce us to new avian species but also teach us about their behavior and biology, such as molting, shown here by this male Western Tanager.

ORGANIZATION AND TAXONOMY

This book includes all the birds of the western United States and Canada (see pages 13 and 14 for maps defining the region), focusing especially on those most likely to be observed. The accounts that form the bulk of the book, pages 22–453, represent the 717 species that regularly occur in the region. Longer accounts represent the 595 more widespread and common species. Shorter accounts add another 122 species; while these reliably occur within the region, they are either highly localized or regionally rare. Special sections in the back of the book list 254 species found less than annually in the West; 37 species presumed to be extinct in the wild, including 33 endemic to Hawaii, are listed on page 473.

This book's contents are organized taxonomically, driven by what we know today about the evolutionary relationships among bird species. Biologists classify birds in a hierarchical, or tiered, system that groups them according to those relationships. All birds belong to the class Aves, which is divided into about 40 orders. Those orders are in turn subdivided into about 250 families; the families are further organized in about 2,200 genera (plural of genus); and, within those genera, there are more than 11,000 species of birds worldwide. Many species are further divided into subspecies, or groups of closely related yet morphologically distinct individuals from different geographic regions. Species-level identifications are the chief concern for most birders, but getting the ID right is greatly assisted in many instances by placing the bird within the correct grouping—especially its family and often its genus.

The order of species in this book follows the checklist of the American Ornithological Society (AOS), the definitive authority for bird taxonomy in the Americas, as of July 2023. Birds in the same family appear together, sometimes on a single spread and sometimes on a suite of pages. An indication of the family or families on each two-page spread, citing both common and scientific names, appears in the upper corner of each right-hand page. A visual index of bird families found in this book appears inside the front and back covers.

Every bird species has both a common name and a scientific name, the latter usually derived from Latin or Greek and recognized universally by scientists working in all languages. The scientific name indicates the bird's genus (first word, capitalized and italicized) and species (second word, lowercase and italicized). The scientific name of the American Crow, for example, is *Corvus brachyrhynchos*. Species in the same genus are each other's closest relatives. The tradition among ornithologists and birders is to capitalize the common name, as we do here.

In 2023, the AOS announced its intention to assign new common names to all bird species in the area under its jurisdiction that are currently named after people. By removing these names, many with disturbing historical associations, the AOS hopes to create a more welcoming and inclusive space for all those who care about birds. The process of assigning new names is a careful and deliberate one, and no changes had been made when we went to press.

INTRODUCTION

ABOUT THIS BOOK

The accounts of all bird species in this book are organized on facing pages, with the left-hand page describing the species and the right-hand page providing annotated art. The species most likely to be observed are represented by longer accounts; the others, present but less common, are represented by shorter accounts. Whether full-length or shorter, each account and its accompanying art capture the key identifying characteristics of the bird described.

ON THE LEFT

Species accounts are grouped by family, starting with a short description of the characteristics of each family. Every account begins with the bird's common and scientific names, the typical size of an adult, and its four-letter banding code—an abbreviation widely used as shorthand in the birding community. For species currently listed by the International Union for Conservation of Nature (IUCN) as critically endangered or endangered, we have included a CR or EN symbol, respectively; national, state, and provincial authorities were consulted for the status of subspecies. A range map (mainland species) or map icon (Hawaii species) accompanies every full-length account (see pages 14–16).

The longer accounts are divided into sections that address four questions important for field identification: What does the bird look like? What does it sound like? What are its key behaviors? Where and when does it occur?

Appearance covers physical features, such as size, shape, color, and pattern, with emphasis on field marks—the visual details key to field ID.

Vocalizations describes what the bird sounds like, both shorter calls and more complex songs. Nonvocal sounds (drumming, bill snapping, etc.) are noted when relevant to identification.

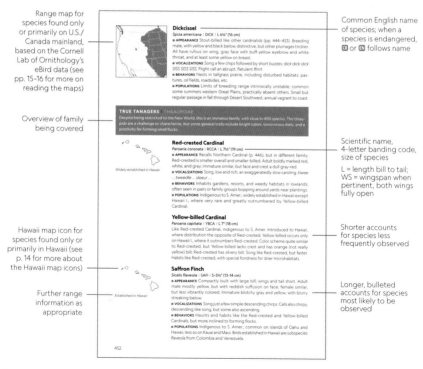

INTRODUCTION

Behaviors provides information about foraging methods, favored foods, and especially habitat. Even at a glance, what a bird is doing—and where it is doing it—can help to confirm an ID.

Populations details where and when birds can be found, with particular emphasis on geographic variation and seasonal movements. This section may also mention subspecies and occurrence patterns not captured in the range maps. Season, geography, and habitat can be as important for identification as appearance, vocalizations, and behaviors.

The shorter accounts describe the key features of a bird, but without maps or map icons, since these species have either a very limited range or an irregular pattern of occurrence.

ON THE RIGHT

Right-hand pages match the facing full-length and shorter accounts with expert color illustrations of the species. The artists whose work is collected in this field guide are esteemed for their knowledge about birds and their behavior. Based on that knowledge, they have represented species in postures and plumages likely to be found in ordinary field encounters. Illustrations such as these are somewhat idealized, and individual birds in the field will not always look exactly like the art: Birds in real life look different from season to season and even minute to minute. Light and shadow hugely influence what we see (and don't see) in the field, and birds' feathers and bare parts vary greatly with ordinary wear and tear.

Annotations alongside the illustrations draw attention to certain important details, especially those that characterize a species or distinguish it from close look-alikes. Labels in bold signal differences in sex, life stage, or molt within a species. When multiple subspecies are pictured, they are indicated in bold (for common name or geographic region) and italic (for scientific name).

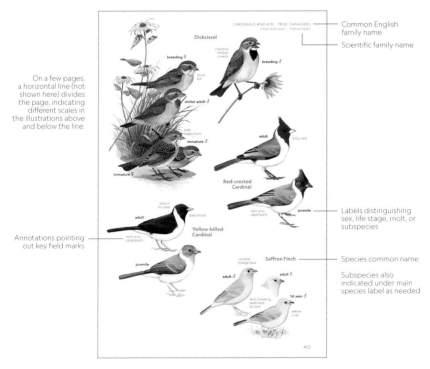

INTRODUCTION

PARTS OF A BIRD

Every detail of a bird's anatomy has a technical name, but even serious field ornithologists often use casual, descriptive language to describe the parts of a bird. The language in this book's species accounts tends toward the conversational, with some technical terms used for clarity. The diagrams here help define many of those terms, and they appear in the glossary (pages 474-478) as well.

There is one case in which some amount of terminological precision is essential, however, and that is for the feathers involved in flight. A bird's wing comprises two main tracts of flight feathers: an outer tract, or primaries, and an inner tract, or secondaries. The innermost secondaries may be called tertials. Together, these three groups of feathers are called the remiges (singular, remex). All the flight feathers of a bird's tail are called rectrices (singular, rectrix). The feathers covering the remiges and the rectrices are called coverts, and they often differ in color and pattern from the flight feathers.

Field identification also requires noticing a bird's bare parts—the eyes, bill, and feet. The color of the bare parts is, in some instances, as important for identification as the color of the feathers. The size and shape of the bill and feet, as well as the size and position of the eyes, can also be important factors in identifying a bird.

Birds molt—or grow new feathers—once, twice, or sometimes even more times each year, and the appearance of a bird's feathers can, and often does, change drastically through the seasons. Bare parts can change in color through the year as well. All in all, a bird's total appearance—the bare parts plus all the feather tracts—is an integrated whole, reflecting the bird's age and sex and the overall effects of diet, season, and climate, not to mention intrinsic variation among individuals.

Here we offer annotated diagrams of a sparrow—representing the passerines, or songbirds, the largest group of birds in the West and indeed all the world—in addition to representative diagrams of three other familiar birds: a hummingbird, a shorebird, and a gull.

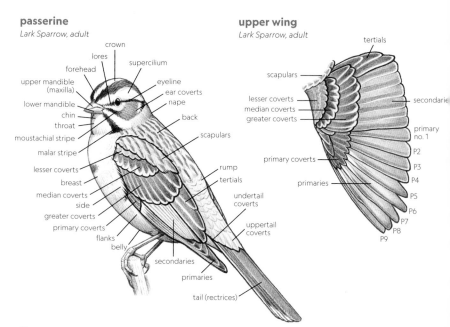

passerine
Lark Sparrow, adult

upper wing
Lark Sparrow, adult

INTRODUCTION

hummingbird
Rufous Hummingbird, adult male

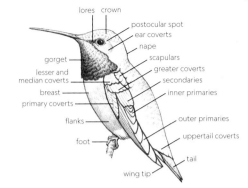

shorebird
Semipalmated Sandpiper, juvenile

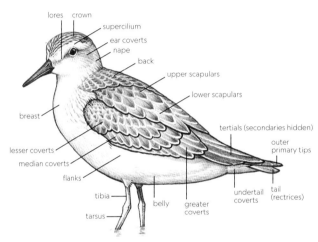

gull
Herring Gull, breeding adult

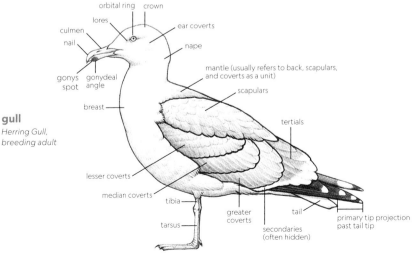

INTRODUCTION

Protected by a sea of cactus spines, Curve-billed Thrashers, widespread across the Desert Southwest, prefer nesting on cholla cacti.

DEFINING THE WEST

The region of the United States and Canada we call "the West" is biogeographically varied—more so than the region we call "the East." Roger Tory Peterson himself originally doubted a unified bird guide could even be created for the West, writing that "although the plan worked out well for eastern North America, it would be almost impossible to do the same thing for the West, where the situation was, it seemed to me, much more complicated." And yes, the situation in the West still is, as Peterson put it, considerably more complicated, including distinctly different biological and geographical regions with distinctly different populations of birds.

THE PACIFIC REGION

The Pacific region extends from Seattle to San Diego coastally, and from Spokane to the Salton Sea inland. The avian diversity in the three Pacific states—Washington, Oregon, and California—is staggering: Washington and Oregon have lists of well over 500 species; California approaches 700. Baja California, comprising two Mexican states, is biogeographically aligned with this region but not included in this book.

Biodiversity in the region arises in part from the complex interactions of ocean dynamics, steep mountains, and great climatic variability. Consider that Inyo County, California, is home to both the summit of Mount Whitney and the depths of Death Valley. Characteristic birds of the Pacific region vary tremendously: Sooty Grouse in coastal forests, Black-footed Albatrosses offshore, White-headed Woodpeckers in the mountains, Wrentits in chaparral, and Costa's Hummingbirds in the desert.

THE INTERIOR WEST

Within the United States, one often hears the distinction between east or west of the Rockies, but that proposed boundary line is biologically deceptive. The Interior West—roughly the region east of the Cascade-Sierra axis and west

INTRODUCTION

of the east flank of the Rockies, is highly distinctive. Arid and mountainous, this region is neither eastern nor Pacific, but in fact a third major zone of biological diversity. The breeding ranges of three closely related vireos nicely illustrate the three regions, with Blue-headed Vireos in the East, Cassin's Vireos mostly in the Pacific region, and Plumbeous Vireos only in the Interior West.

Distinct in many ways but included in the Interior West are the U.S.-Mexico borderlands, especially the regions encompassing southeastern Arizona, southwestern New Mexico, and the Rio Grande drainage of West Texas.

WESTERN CANADA

In the same way that the West region of the U.S. lower 48 comprises two major biogeographic regions, so too does western Canada, but with a twist: The Interior West biome barely touches western Canada, yet the biome of the U.S. East does. The great Boreal Shield, stretching across Canada's central Prairie Provinces, links the avifauna of the Canadian Rockies with that of Appalachia!

Meanwhile, the Pacific Slope bioregion extends along the entire coastal plain and adjoining slopes of British Columbia. Eastern and Pacific Slope populations come into contact here. Even more extraordinary is a biogeographic trifecta near Prince George, British Columbia, where one can see Yellow-bellied Sapsuckers from the East, Red-breasted Sapsuckers from the Pacific Slope, and Red-naped Sapsuckers from the Interior West.

North of the boreal forest in Canada lies a transitional taiga zone, and north of that the true Arctic tundra, treated below.

ALASKA

The map of Alaska, superimposed over the U.S. lower 48, stretches from San Francisco, California, to Jacksonville, Florida, and north to Duluth, Minnesota. It is hardly surprising, then, that the state encompasses such biogeographic variety. Full-on Arctic tundra is found in the state chiefly north of the Brooks Range; here one finds nesting Bluethroats, Eastern Yellow Wagtails, and Northern Wheatears, species with Asian affiliations. Alaska's vast boreal/taiga zone supplies breeding habitat to species that winter in South America and migrate mostly through eastern North America. And the rainforest of the Alaska Panhandle, one of the wettest places on the planet, is home to forest nesters like Marbled Murrelets, Pacific Wrens, and Varied Thrushes.

For many birders, it is the Bering Sea region that is the most exciting of all in Alaska. Spanning thousands of miles of coastline, including some of the most fabled birding destinations on Earth, this stretch of the highly productive Northern Pacific ecosystem is a challenge to treat in a field guide. The avifauna here takes on a distinctively Asian character. Almost all species occurring annually in this region are included in the main part of this book, with the rarer species (called casuals and accidentals), plus a few that are annual but restricted mostly to the infrequently visited outer Aleutians, described in a separate section.

INTRODUCTION

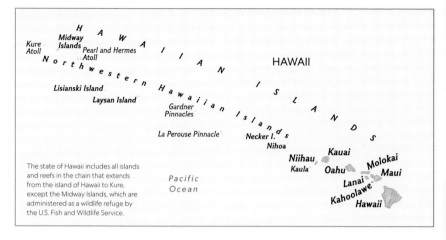

The state of Hawaii includes all islands and reefs in the chain that extends from the island of Hawaii to Kure, except the Midway Islands, which are administered as a wildlife refuge by the U.S. Fish and Wildlife Service.

HAWAII

In October 2016, members of the American Birding Association (ABA) voted to expand the ABA Checklist Area to include Hawaii. Following that decision, and answering the curiosity of birders today, the rich avifauna of the Hawaiian archipelago is included in this book. In a departure from many other print field guides, Hawaiian species appear in the main text alongside the other species of the West, within the families to which they belong. This approach provides context, taxonomically and ecologically, for everything from endemic solitaires to regionally rare seabirds to wintering sandpipers.

The Hawaiian Islands are a hotbed for endemism, referring to the proliferation of species found nowhere else—especially the Hawaiian honeycreepers, with bright colors, haunting songs, and unusual bills. Hawaii is also notable for harboring many deliberately introduced species. Avian conservation challenges are more urgent in the Hawaiian Islands region than anywhere else in the West—perhaps anywhere else in the world. Introduced rats and hunting were major threats from the arrival of Polynesians more than a millennium ago all the way into the 20th century. Today mosquito-borne malaria, vectored by nonindigenous birds and amplified by anthropogenic climate change, is the greatest peril.

The ranges of many birds found in Hawaii are highly local, and maps at the scale created for this book cannot reflect that level of detail. Species found only or primarily in Hawaii (within the scope of this book) are accompanied by one of two map icons of the Hawaiian Islands with annotations below summarizing their occurrence. Species found in both Hawaii and the mainland West are represented by a range map showing the species' mainland distribution with annotations below noting Hawaiian appearances.

Main Hawaiian Islands

All Hawaiian Islands

We refer to the whole archipelago as the *Hawaiian Is.*; *Hawaii* denotes the state (which excludes Midway); *Hawaii I.* indicates the Big Island. The term *main islands* refers generally to the larger southeastern islands. For birds found only there, a map icon shows the main islands. The small islands from Kure to Niihau are the *Northwestern Hawaiian Is.*; for species that occur beyond the main islands, a map icon shows the whole island chain.

INTRODUCTION

NEW APPROACH TO RANGE MAPS

To capture the most current knowledge about bird distribution throughout the West, National Geographic has partnered with the Cornell Lab of Ornithology to present data-driven range maps with key innovations for understanding how bird populations occur in space and time. These maps are made possible because of data from eBird, a collaborative enterprise with dozens of partner organizations, thousands of regional experts, and more than a million birders whose observations represent one of the most valuable resources for birding and conservation science in the digital age. These contributions from birders include more than a billion bird observations, photos, and sound recordings from around the world.

The Cornell Lab has used the past 15 years of observations from 400,000 eBird users to power the statistical analyses for its eBird Status and Trends data products and maps, including those created for this book. Experts have reviewed our maps against current knowledge, fine-tuning them to create the most accurate and up-to-date range maps in any field guide for this region.

Different from those found in other field guides, including previous editions of this and other National Geographic birding field guides, these maps use a new and conceptually powerful approach to mapping bird ranges. Rather than depicting the time of year when a bird can be found in a certain location, the maps in this book show the distributions of birds according to the phases of their life cycle: their nonbreeding, migration, and breeding seasons.

A given species breeds at the same time every year, but the timing of breeding varies from one species to another. Many birds found in our region breed in the Northern Hemisphere during our spring and summer. After breeding, migratory birds typically travel to their

PHASES OF BIRD LIFE CYCLE: BREEDING, MIGRATING, NONBREEDING, MIGRATING

The maps in this book use these colors to represent key phases in each species' life cycle.

Watching birds interact, such as these two adult male Mountain Bluebirds, can help birders better understand avian relationships with each other and their environment.

INTRODUCTION

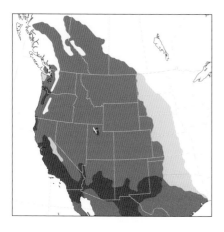

Sample Range Map
Cinnamon Teal

For most species in this book, like the one shown here, the annual cycle corresponds to Northern Hemisphere seasonality, with breeding during our spring and summer, migration primarily in spring and fall, and the nonbreeding season during winter.

■ Year-round
■ Breeding season
■ Nonbreeding season
▨ Pre- and postbreeding migration season

nonbreeding grounds; they then migrate back to their breeding grounds the following year. Other bird species remain in the same region year-round.

For pelagic, or open ocean, species that breed in the Southern Hemisphere—several of which can be found in the West—seasonal cycles are reversed from our point of view here in the Northern Hemisphere. These birds breed far to our south during our winter (the austral summer), and they are found in our area during their nonbreeding season, our summer.

A species' annual cycle is depicted on the maps with four colors. *Purple* indicates where a species occurs year-round, with breeding usually occurring spring to summer and nonbreeding fall to winter—although some individuals may remain as nonbreeders throughout the year. *Red* indicates where a species is normally present only during its breeding season. *Blue* indicates that a species is normally present in that area only during its nonbreeding season. *Yellow* typically shows where a species occurs on migration between its breeding and nonbreeding seasons, but yellow can also indicate other movements, including dispersal of juveniles or molt migration.

Keep in mind that not all individual birds will breed, even when present in the area indicated during the breeding season. Additionally, migration timing and distances vary among species, and not all species migrate. In general, these maps depict breeding and nonbreeding seasons as periods during which bird populations are relatively stationary, and migrations as times when birds are generally on the move. But at times, migratory birds can be on the move within their breeding (red) and nonbreeding (blue) seasons; conversely, many bird species engage in prolonged sojourns, called stopovers, within their migratory (yellow) seasons.

While lines on a printed map stay permanent, within and across years, species' ranges may expand, contract, or change in other ways. For seabirds in particular, ranges often shift annually as ocean conditions change due to warming waters and food availability. In many years, "irruptive" land birds stray far from their normal ranges, often driven by regional failures in food supplies, such as conifer seeds. "Vagrants," or individual birds that wander far from the normal limits of their occurrence, may be found hundreds or even thousands of miles out of range. Be sure to consult the accompanying text for additional nuances of seasonal and geographic occurrence, using a mix of caution and open-mindedness when you see a bird apparently outside its expected range.

You can explore the latest sightings and contribute your own data to eBird to help shape the next generation of maps for field guides, research, and conservation by downloading the eBird app or visiting *eBird.org*.

INTRODUCTION

NOTICING THE DETAILS

Any bird, whether it is a tiny Calliope Hummingbird or a titanic American White Pelican, has feathers and bare parts (eyes, feet, and bill). All extant species in the West can fly most of the year, almost all can walk, many swim, and all make sounds. Add all those things up—color and pattern, behavior, calls and songs—and you are well on your way to putting the right name on the bird.

Learning to identify birds requires precision in noticing the parts of a bird—for example, distinguishing between the coverts and flight feathers (remiges) of a bird's wings, or distinguishing between the "eyebrow" (technically, the supercilium) and the "eyeline" (or transocular) of a warbler or sparrow. The same goes for calls and songs: Listen carefully and learn to distinguish between high-pitched and low-pitched, rising and falling, whistled (pure-tone) and buzzy (modulated).

Also keep in mind that birds, like humans, differ from individual to individual. Birds often display reliable and recognizable differences between the sexes, among different age classes, and through the annual cycle. Most juvenile birds, especially among smaller species, are more likely to be seen in summer and early

Small details distinguish male and female Varied Thrushes. Males have a black breastband and face mask whereas these marks are duskier on females.

autumn than in winter or spring. Adults in fresh plumage are more likely to be seen in fall, right after the annual molt, than in summer, when many are quite bedraggled.

Birds also display situational differences having to do with behavior, health, and, importantly, the bird's location relative to the observer. A bird puffed up in cold weather looks quite different

There's nothing else in our area quite like an American Dipper (left), which can be found deftly plucking its small aquatic prey from swift currents, but the Spotted Owl (right) can be easily confused with the similar Barred Owl unless details of its appearance and vocalizations are known.

17

INTRODUCTION

It can be challenging to identify birds when multiple species are present, such as this flock on Tern Island in the Northwestern Hawaiian Islands. Knowing which species to expect in a given habitat helps narrow down the possibilities.

from one of the same species singing from a summer perch. A bird infected with feather mites might not resemble others of its species. Even the context within which a bird is spotted—out in the open in bright sunlight, as opposed to in the shadows of the forest understory, for instance—may influence an individual's appearance in ways that a field guide simply cannot cover.

THE BIG PICTURE

Adding up all these details in the service of identifying birds in the wild is a gratifying mental exercise. A "mostly blue bird with orange and white patches below" could be a Lazuli Bunting or a Western Bluebird or even a Belted Kingfisher. A birder would rarely if ever confuse them, though, despite their superficial similarities. For one thing, their body structures are different, but even without that knowledge, these birds' behaviors alone distinguish them. The kingfisher waits on a streamside branch, then dives into the water for fish. Differences between the bunting and bluebird are subtler but nonetheless important to note. The bunting sings from high perches at broadleaf forest edges, whereas the bluebird flits about snags in conifer groves. All these distinctive behaviors say so much more about these species than the shared color of their feathers.

Getting good at bird identification requires embracing the idea that the whole is greater than the sum of its parts. Feathers, vocalizations, and body structure are all a part of it, but behavior and habitat are critically important as well, along with knowledge of when and where certain species occur. All these features are dynamic and interactive, coalescing in the living creature—and in our human recognition and appreciation of that creature in its environment.

A BIRDER'S TOOLS

It is possible—indeed it can be wonderfully rewarding—to go out in the field and simply watch and wonder at birds without any equipment at all. But a few items of hardware can contribute valuably to the experience, especially if the emphasis is on identification.

Most birders use binoculars to watch birds. There is, unsurprisingly, a direct relationship between the price of a binocular and how good it is. The best are technological marvels, exquisitely bright and clear, and they are very expensive. Regardless of the price of the binocular you use, give thought to some basic considerations of performance and ergonomics. Weight (compact vs. full-size) can make a big difference if you are in the field for any amount of time. Lower magnifications, all things considered, deliver brighter images than higher magnifications—less is more. And the size and feel of the binoculars in your own hands, although personal and subjective, importantly affect the experience of birding. Research what you buy. Shop around, talk to friends, and, ideally, test a pair in the field (not just in a store) before purchasing.

Binoculars were until recently considered obligatory for birding, but that is changing. Today's digital cameras are small, powerful, and far more affordable than the best binoculars. They are superb for identifying birds in the field, and they preserve the experience as digital images—something binoculars, at least given present technology, cannot do. Just as with binoculars, shop around and ask questions before making a purchase; the variety of digital cameras out there is staggering, and some are much more suited to use by birders than others.

Another game changer has been the smartphone. New birding apps are launched frequently, and many are free. The Merlin Bird ID app, from the Cornell Lab of Ornithology, can instantly identify birds based on a description, photo, or sound. Powered by AI and data from eBird, Merlin can help identify thousands of bird species from around the world. Seek—an app powered by iNaturalist, another community science initiative—uses visual recognition software

Encountering a new species thrills all birders, regardless of their experience level or age.

INTRODUCTION

In an expansive landscape like the open ocean, binoculars bring telling details into view, making them an essential tool for all birders.

to identify not only birds but also non-avian animals, plants, other organisms, and even "sign," like nests, tracks, and scat. Point your phone's camera at a bird (or a wildflower, or even a slime mold), and it will suggest an ID within seconds.

Keep in mind that bird ID apps are not foolproof. "Trust, but verify," as the saying goes. The AI software that powers app-based IDs, although getting better all the time, is fallible. Confirm the output from Merlin or Seek with your own eyes and ears, and consult with human experts and, of course, this field guide. If the app makes a mistake, you may choose to report the mistake and thus help improve the performance of the app.

Apps and digital cameras, along with more traditional binoculars and print field guides, are means to an end. That end is to enjoy and identify birds in the field and, ultimately, to understand and care for them.

THE BIRDING COMMUNITY

Today more than ever before, the study and enjoyment of wild birds in the field is a shared experience. New birders often find their way into the extensive online communities organized around eBird and iNaturalist, as well as social media groups focused on towns and cities, states and provinces. These are superb starting points, but they cannot—and they were never intended to—substitute for the in-person experience of birding with others.

Essentially everywhere in the United States and Canada, local birdwatching clubs welcome those with an interest in wild birds and their habitats. Outings and programs are often free, and annual memberships are generally affordable. Beginners and experts alike can join in their shared wonder at birds and nature, whether attending a talk given by a visitor or enjoying a walk led by a knowledgeable local. You can encounter other birders at your local park or preserve, and many will be eager to share information about how to become more involved with the local birding scene.

Community birding, whether online or in person, has the potential to make a difference for bird conservation. For starters, every upload to eBird or iNaturalist adds to databases that inform conservation science about the current state of birds in your region. Bird clubs and ornithological societies frequently sponsor

A hallmark of the birding community is a willingness to help others. Seasoned birders are always ready to share their knowledge.

INTRODUCTION

Birding allows us to appreciate how birds adapt to changing seasons, whether they migrate or stay put in harsh winters.

projects such as nest box monitoring, migration counts, and habitat restoration. The Christmas Bird Count—a survey of bird populations in the Western Hemisphere in late December and early January, sponsored by the National Audubon Society—is perhaps the most famous long-term ecological survey in history, powered entirely by the birding community.

As you spend time birding with other people, you begin to respect the importance of responsible birding. An essential resource in this regard is the American Birding Association's Code of Birding Ethics: *aba.org/ethics*.

BIRDING TODAY AND TOMORROW

Why birds? Why do we go birding? Those are questions that almost any birder gets asked, whether they've been birding for six weeks or 60 years.

An undeniable part of the reason is to "get back to nature," to reconnect with the rhythms of the outdoors—the dawn chorus we've known since childhood, the predictable seasonal movements of hawks on fall migration and Tundra Swans in the spring, the simple pleasure of songbirds in the quiet winter woods or Sanderlings chasing the ocean waves. Experiencing nature has such intrinsic value that, for most of us, it requires no further justification. But it is also worthwhile to reflect on the other benefits of time spent outdoors.

Especially in the post-pandemic era, there has been considerable interest in the health and wellness benefits of spending time in nature and of birding in particular. Birding brings us into contact with the world around us—and with others who share our delight and wonder at the beauty of birdsong, the vibrant colors and intricate patterns of birds' feathers, and the pageantry of bird migration.

Birding challenges us. Watching and listening to birds; studying their ecology, behaviors, and populations; and pondering birds' roles in the broader environment are mentally and emotionally stimulating. Birding leads us along paths of science and conservation, activism and environmentalism, art and poetry, and more. Time spent in the field with birds keeps us sharp.

Perhaps the grandest motivation of all for the birder is the promise of discovery and new knowledge. We enjoy birdwatching because we are seekers of surprise and wonder; we go birding with open minds, receptive to the possibility of new ways of engaging with nature. *National Geographic Field Guide to the Birds of the United States and Canada—West* is a traditional field guide, comprising bird art and species accounts, bound together between two covers. But it is intended for a contemporary audience, eager to chart a new course for nature study, and dedicated to bettering the welfare of the human and nonhuman inhabitants of the world around us.

GEESE, SWANS, AND DUCKS | ANATIDAE

Collectively known as waterfowl, the birds in this large family were heavily hunted by inhabitants of the West for millennia. The biggest threat facing them today is habitat loss. All are habitual swimmers, and many are excellent divers. Three subfamilies are indigenous to the West: the whistling-ducks (Dendrocygninae), scarce here; the closely related geese and swans (Anserinae, pp. 22–29); and all the rest in the duck subfamily Anatinae (pp. 30–51).

Black-bellied Whistling-Duck
Dendrocygna autumnalis | BBWD | L 21" (53 cm)

With their large size, lumbering wingbeats, and relatively long necks, whistling-ducks are somewhat gooselike in structure. Black-bellied is widespread in the Neotropics, north to southeastern U.S.; recent records in West mostly from Ariz., where uncommon but regular in small numbers. Adult, with coral bill and black belly, distinctive. Gray-billed juvenile, with more muted colors overall, can suggest juvenile Fulvous Whistling-Duck; Black-bellied at all ages has extensive white in wings, prominent in flight, and dark uppertail coverts. Call, often heard dusk and dawn, is squeaky and wheezy ("whistling"), rushed and frantic: *peep! wee-dee wee-dee wip.*

Fulvous Whistling-Duck
Dendrocygna bicolor | FUWD | L 20" (51 cm)

Formerly regular in Calif., but has become very rare in West in 21st century; accidental to Hawaii. A whistling-duck in the West, especially Ariz., is much more likely to be Black-bellied. Fulvous in all plumages told from Black-bellied by entirely dark wings and broad white band at base of tail; female Northern Pintail (p. 38), also long-necked, gray-necked, and relatively large, may suggest Fulvous Whistling-Duck, but even bright pintails are colder and grayer. Haunts and habits of Fulvous generally like those of Black-bellied: loafs by day near shorelines, more active dusk and dawn. "Whistles" like Black-bellied, but call is shorter: usually a two-note *pip-wee*, often repeated slowly.

Emperor Goose
Anser canagicus | EMGO | L 26" (66 cm)

Accidental to Hawaii

- **APPEARANCE** Stocky, short-necked, and stub-billed. All are dark-bodied with contrasting white tail; feet orange-yellow. Adult has gray body, finely barred; head and hindneck white, often stained yellowish in summer. Juvenile mostly dark-headed but shows finely barred body and white tail of adult.
- **VOCALIZATIONS** Honk relatively high and wimpy, sometimes wavering: *hrenk, hrweannk,* etc.
- **BEHAVIORS** On coastal breeding grounds, mills about sloughs and estuaries, also to nearby grassy tundra; winterers to U.S. West Coast often find their way to flocks of more common geese.
- **POPULATIONS** Breeds eastern Russia and western Alas., with major concentration in Yukon-Kuskokwim Delta near Bethel; most winter in coastal Alas., but also regular in very small numbers south along and near coast to Calif.; accidental to Hawaii. Unlike most other goose species, does not usually feed in agricultural fields. Declined sharply in latter half of 20th century but is recovering now.

GEESE, SWANS, AND DUCKS
ANATIDAE

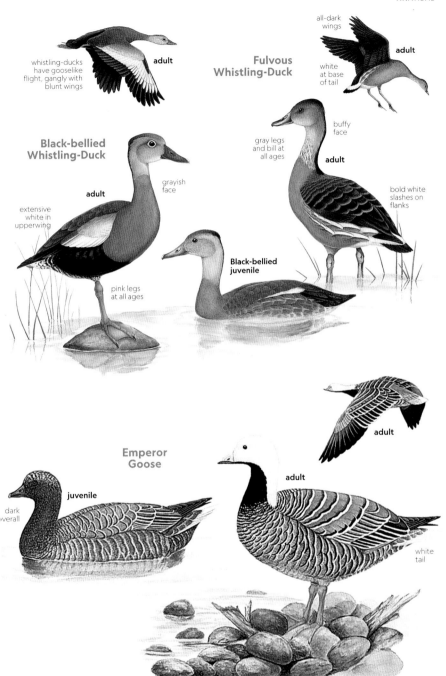

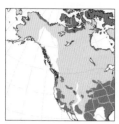

Annual to Hawaii

Snow Goose
Anser caerulescens | SNGO | 26-33" (66-84 cm)
- **APPEARANCE** Medium to large goose; all plumages have dark primaries. Two color morphs: Adults of both have pink legs and long, pink bill with odd "grinning patch." Most adults in West white except for black primaries and, on some birds, pale rusty suffusion on face. Dark morph ("Blue Goose") largely slate brown, but with wings paler and head mostly white. Juveniles darker overall than corresponding adult morphs.
- **VOCALIZATIONS** Calls higher, less growling than Canada Goose's (p. 26): *clee, cleek*, etc., on one pitch, monosyllabic but often wavering.
- **BEHAVIORS** Often in single-species flocks, but stray individuals find their way to assemblages of other goose species. Quiet while grazing or swimming, but flocks in flight clamorous.
- **POPULATIONS** Increasing; an emerging threat to fragile Arctic ecosystems where species breeds—and feeds voraciously. Dark morph ("blue") rare in winter west of the Rockies. Very rare but annual to Hawaii.

Ross's Goose
Anser rossii | ROGO | L 23" (58 cm)
- **APPEARANCE** Plumages and morphs correspond to those of Snow Goose, but body shape different: Ross's smaller overall, smaller-headed, and stub-billed. Bill lacks "grinning patch"; instead, has blue-green splotches at base. Adult white morph even more immaculate than adult Snow; juvenile correspondingly less dusky than juvenile Snow.
- **VOCALIZATIONS** Calls softer and more nasal than Snow Goose's; in big mixed flocks, drowned out by louder, richer calls of noisier species.
- **BEHAVIORS** Mixes with other geese but feeds differently: Small bill causes it to specialize on tender shoots above ground; does not "root" like larger geese.
- **POPULATIONS** Like Snow Goose, increasing rapidly; sometimes hybridizes with that species. Dark-morph ("blue") Ross's very rare, and many reports likely refer to "blue" Snow Geese or to hybrids with dark-morph Snow Geese.

Annual to Hawaii

Greater White-fronted Goose
Anser albifrons | GWFG | L 28" (71 cm)
- **APPEARANCE** Same heft as Snow Goose; muddy gray-brown in all plumages. The white "front" refers to prominent feathering at base of bill of adult; an old name, "Specklebelly," refers to splotchy black on underparts of adult. Colorful bill and especially feet stand out at all ages. Juvenile lacks white front, has plain belly. Most in West are "Tundra" subspecies; darker "Tule Goose" subspecies, wintering largely in northern Central Valley of Calif., is larger and longer-billed. Beware similarity with domestic geese, which sometimes escape from captivity; compare especially with Graylag Goose, *Anser anser* (also pictured opposite), of Old World, common in captivity in West.
- **VOCALIZATIONS** Call of widespread "Tundra" a ringing, trisyllabic *klee-wee leek*, or disyllabic *klee leek*, high and squeaky; "Tule" also has trisyllabic call, but deeper and more honking.
- **BEHAVIORS** Away from breeding grounds, roosts in flocks on or near water; "Tundra" forages widely, often in farm fields, with "Tule" foraging mostly on aquatic vegetation.
- **POPULATIONS** Sometimes hybridizes with other geese, especially Canada (p. 26). "Tule Goose," distinctive in ecology and morphology, may deserve full-species rank. "Tundra" very rare but annual to Hawaii.

Annual to Hawaii

Brant
Branta bernicla | BRAN | L 25" (64 cm)
■ **APPEARANCE** Slightly built goose, mostly dark in all plumages. Head, neck, and breast of adult black with intricate white collar ("sash" or "bow tie"); dark extends to belly, but flanks and vent lighter. Juvenile like adult, but smudgier, with white collar reduced or nearly absent. Most in West are subspecies *nigricans* ("Black Brant"), but gray-bellied birds, paler below, regular in winter in Puget Sound region.
■ **VOCALIZATIONS** Usually heard in chorus of hundreds; low, growling *ccrruuk, rrrook*, etc. More laid-back than urgent-sounding calls of other geese.
■ **BEHAVIORS** Tied to coastal habitats year-round; on migration and in winter, feeds heavily on eelgrass, genus *Zostera*. Gregarious; vagrants inland typically find their way to flocks of other goose species.
■ **POPULATIONS** Winter range has shifted considerably; climate change, die-off of eelgrass, and novel diets likely all at play. Most winter in Mexico. Annual in small numbers to Hawaii.

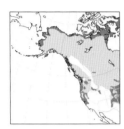
Annual to Hawaii

Cackling Goose
Branta hutchinsii | CACG | L 23-33" (58-84 cm)
■ **APPEARANCE** Smaller, more compact, and shorter-necked than Canada Goose, with stubbier bill that meets forehead at relatively steep angle. Like Canada, has black neck and head with white "chinstrap." Taxonomy complex: *minima*, dark and diminutive, scarcely the size of Mallard (p. 36); nominate *hutchinsii*, pale and relatively large, overlaps in size with smaller Canadas. Subspecies *leucopareia* ("Aleutian") often has white border ("collar") between black breast and paler belly, but note that any Cackling or Canada may be thus marked. Juveniles and adults similar.
■ **VOCALIZATIONS** "Cackling" is a misnomer; honks like Canada, but calls average higher; *heek, heenk, heek-a-leenk*.
■ **BEHAVIORS** "Cacklers" in winter join big, mixed-species flocks, but they don't intermingle indiscriminately; even within big, rowdy flocks, they keep close company with conspecifics.
■ **POPULATIONS** Formally split from Canada Goose in 2004; some details of taxonomy still uncertain. Hybridization likely extensive where "Cacklers" and Canadas co-occur as breeders. Annual in small numbers but increasing in Hawaii.

Casual to Hawaii

Canada Goose
Branta canadensis | CANG | L 30-43" (76-109 cm)
■ **APPEARANCE** Large, long-billed, and long-necked; black neck and head with white "chinstrap" distinguishes it from all geese except closely related Cackling. Most are told from smaller Cackling by head structure: Long bill of Canada slopes smoothly with forehead (stubby bill of Cackling forms sharp angle with forehead). Mustard-colored hatchlings transition quickly to adult size and plumage.
■ **VOCALIZATIONS** Honest-to-goodness *honk*, often doubled or trebled: *ka-lonk, honk-a-lonk*. Feeding flocks murmur quietly; hatchlings wheeze softly.
■ **BEHAVIORS** Adaptable and aggressive; the most common goose—and one of the most conspicuous animals—in many urban and commercial areas.
■ **POPULATIONS** Formerly an icon of bird migration; well-intentioned but misguided captive breeding efforts in mid-20th century resulted in non-migratory flocks, flourishing today. Small numbers of feral origin present year-round in Hawaii; natural vagrants aren't even annual.

Endemic; breeds on Kauai, Maui, Hawaii

Hawaiian Goose
Branta sandvicensis | HAGO | L 22-26" (56-66 cm)

- **APPEARANCE** Midsize goose with short bill and steep forehead. Neck of adult gray-buff with black streaks; cream-colored cheek contrasts with dark crown. Juveniles and adults similar.
- **VOCALIZATIONS** Honk less strident than other geese; has endearingly sad quality, *uhhhhh* and *uhhnnn*. Indigenous name, Nēnē, in wide use, is onomatopoetic.
- **BEHAVIORS** Usually seen on land; forages in fields and on lawns like other geese, but also browses fruits from shrubs.
- **POPULATIONS** Endemic to, and state bird of, Hawaii. Nearly went extinct in 20th century, but is recovering, with breeding on islands of Kauai, Maui, and Hawaii. Some movement among islands.

Mute Swan
Cygnus olor | MUSW | L 60" (152 cm)

Old World species, long established in East; not considered established in West, but adults, pairs, and occasional family groups widely noted in parks, at farm ponds, etc. Longer-tailed than other swans; holds neck in broad arc. Bill of adult deep orange with black knob. Juvenile dusky overall; bill dark, lacks knob of adult. Escaped or released Black Swans, *Cygnus atratus* (also pictured opposite), of Australia, occasionally seen in West, especially Hawaii (resorts, parks, etc.).

Trumpeter Swan
Cygnus buccinator | TRUS | L 60" (152 cm)

- **APPEARANCE** Very similar to smaller Tundra Swan. Note precise differences in bill shape: Base of Trumpeter's bill cuts a straight line with white feathers on sides of face; bill and feathers abut in pointy V-shape in space between eyes. Bill longer, more smoothly sloping than on Tundra. Most adults have all-black bill, lacking yellow patch of Tundra. Juvenile has dusky gray-white plumage with dusky pink bill.
- **VOCALIZATIONS** Double honk, *hunk-unk*, muffled and nasal.
- **BEHAVIORS** Herbivorous like other swans. Feeds mostly on lakes, with tail straight up and rest of body submerged, but flocks sometimes forage in agricultural fields, especially Puget Sound region. Nests on ponds and marshes, often wooded.
- **POPULATIONS** Following severe losses in 20th century, is recovering rapidly; widespread in winter in very small numbers in Interior West.

Accidental to Hawaii

Tundra Swan
Cygnus columbianus | TUSW | L 52" (132 cm)

- **APPEARANCE** Base of bill subtly different from larger Trumpeter Swan's: Border with white facial feathers not as straight as on Trumpeter, cuts across forehead in shallower and smoother arc; bill of Tundra shorter overall, not as straight. Most adult Tundras have a bit of yellow at base of bill near eye. Juveniles of both species have dirty-pink bills; use bill structure, cautiously, to separate species.
- **VOCALIZATIONS** Nasal call higher than Trumpeter's, not usually doubled. N. Amer. subspecies formerly called "Whistling Swan" for the baying of distant flocks in chorus.
- **BEHAVIORS** Nests on tundra, generally well separated from breeding habitat of Trumpeter, but the two winter widely, often on the same waterbodies.
- **POPULATIONS** Eurasian subspecies ("Bewick's Swan"), with more yellow on bill, very rare but annual coastally and well inland. Accidental to Hawaii.

Egyptian Goose
Alopochen aegyptiaca | EGGO | L 25-29" (64-74 cm)
African species, rapidly establishing in U.S., especially Fla. and Tex. Despite the name, is more closely related to ducks than geese. Largest numbers in West centered around Los Angeles, but widely noted and perhaps establishing elsewhere. Large; the size of a small goose. Oddly colored in beige, sienna, black, and white, with red "lipstick" and dark "eye shadow" contributing to the bird's overall weirdness. White coverts and dark remiges striking in flight; in good light, note dark green secondaries and orange tertials.

Common Shelduck
Tadorna tadorna | COMS | L 25" (64 cm)
Widespread Palearctic duck related to Egyptian Goose. Popular in captivity, with escapes occasionally seen in wild; unlike Egyptian Goose, never seems to become established in the wild. Adult male has a peculiar mix of components: Red bill has knob like Mute Swan (p. 28); green head, white breast, and orange below recall Northern Shoveler (p. 32); and overall proportions gooselike. Compare also with Ruddy Shelduck, *T. ferruginea* (also pictured opposite), closely related but mostly marmalade-colored.

Muscovy Duck
Cairina moschata | MUDU | L 26-33" (66-84 cm)
Large Neotropical duck whose indigenous range barely reaches South Tex.; feral populations well established in Fla., with non-established singles and small flocks widely noted in the wild, including in the West. Birds of feral stock typically show some to much white; bill usually red, and most show red facial skin around eyes and bill. Frequently hybridizes with other waterfowl, especially Mallard (p. 36), producing "frankenducks" that may create momentary bewilderment at park ponds.

Wood Duck
Aix sponsa | WODU | L 18½" (47 cm)
- **APPEARANCE** Fairly small duck; long-tailed, long-necked, and block-headed. Female's eye surrounded by broad white patch; looks surprised. Adult male in breeding plumage unmistakable; eclipse male (summer only) resembles female, but eye and bill red.
- **VOCALIZATIONS** Female, a piercing *whoo-eek*, uttered as bird flushes; male, a drawn-out, feeble hiss, like air being expelled from bike tire.
- **BEHAVIORS** Shy; despite its brilliance, has a knack for avoiding detection in the wooded waterways it favors. In flight, frequently glances down, as if to see what's going on.
- **POPULATIONS** Nest box supplementation has aided the species' recovery in recent decades. Does not usually congregate in dense flocks.

Mandarin Duck
Aix galericulata | MAND | L 16" (41 cm)
East Asian counterpart of larger Wood Duck, well established in western Europe and widely noted in West. Most in the wild here are recent escapes from captivity, but small populations can persist for decades; highest numbers around Los Angeles and (formerly) Bay Area of Calif., with decent outposts around Salt Lake City and elsewhere. Breeding male unique, but female similar to female Wood Duck; female Mandarin has thinner eye ring and fine "bridle" extending well behind eye.

GEESE, SWANS, AND DUCKS
ANATIDAE

Casual to Hawaii

Blue-winged Teal
Spatula discors | BWTE | L 15½" (39 cm)
- **APPEARANCE** Fairly small duck; smoothly contoured and rides low to the water's surface. Breeding male sports white swatch on dark purple-gray head. Female like Green-winged (p. 38) and especially Cinnamon: Green-winged is smaller-bodied, slimmer-billed, pale at rear, and often shows a bit of green when standing or swimming; Cinnamon is larger-billed with blank face lacking Blue-winged's dark eyeline and white at base of bill.
- **VOCALIZATIONS** Male, especially when courting, gives squeaky chatter, *peek! peek! peek!* Female's note a nasal *kvenk*.
- **BEHAVIORS** Like other *Spatula* ducks, favors shallow wetlands with emergent vegetation; feeds at or immediately below water's surface, rarely if ever diving.
- **POPULATIONS** Eastern counterpart of closely related Cinnamon Teal; although widespread in West, Blue-winged is decidedly scarce within much of Cinnamon's range. Unlike Cinnamon, Blue-winged returns relatively late in spring. Not quite annual to Hawaii; has bred there.

Casual to Hawaii

Cinnamon Teal
Spatula cyanoptera | CITE | L 16" (41 cm)
- **APPEARANCE** Bill larger and wider than that of Blue-winged. Breeding male's gingerbread plumage distinctive. Female Blue-winged, with patterned face and smaller bill, is darker overall than female Cinnamon; female Green-winged smaller, a bit darker, and notably smaller-billed. The name *cyanoptera* means "blue-winged," signifying the chalk-blue secondary coverts of both species.
- **VOCALIZATIONS** Male call unlike that of male Blue-winged: a rapid rapping, *ta'ta'ta'ta*. Female's nasal *kvenk* like that of female Blue-winged.
- **BEHAVIORS** Habits and habitats much like Blue-winged, with which it often co-occurs, but Cinnamon accepting of more saline environments.
- **POPULATIONS** Hybridizes with Blue-winged, with drakes intermediate in plumage, typically showing splotchy white facial crescent (Blue-winged trait) and dusky chestnut tones overall (Cinnamon trait); females difficult and maybe impossible to separate. Unlike late-arriving Blue-winged, Cinnamon is among the earliest of all spring migrants. Casual to Hawaii.

Annual to Hawaii

Northern Shoveler
Spatula clypeata | NSHO | L 19" (48 cm)
- **APPEARANCE** Larger and longer than teal. Bill broad, very long, and peculiarly shaped; note the odd, baleenlike structure, an adaptation for filter feeding. Adult male has green head, white breast, and rusty belly and flanks. Female plumage like female Mallard (p. 36), and first-winter males show variable white facial crescent suggesting Blue-winged, but bill very different in all plumages.
- **VOCALIZATIONS** Common call disyllabic, repeated slowly: *shook shook ... shook shook ... shook shook*.
- **BEHAVIORS** Forages and roosts widely, but typically in shallow water. Gathers in large masses (more than 100 birds) that bunch up tight and rotate in slow-moving circles, pulling up food in the water below; shovelers also pair off in twos, pinwheeling endlessly.
- **POPULATIONS** Widespread, including Hawaii in winter, but local abundance constrained in part by availability of small crustaceans, which the birds strain through their amazing bills.

GEESE, SWANS, AND DUCKS
ANATIDAE

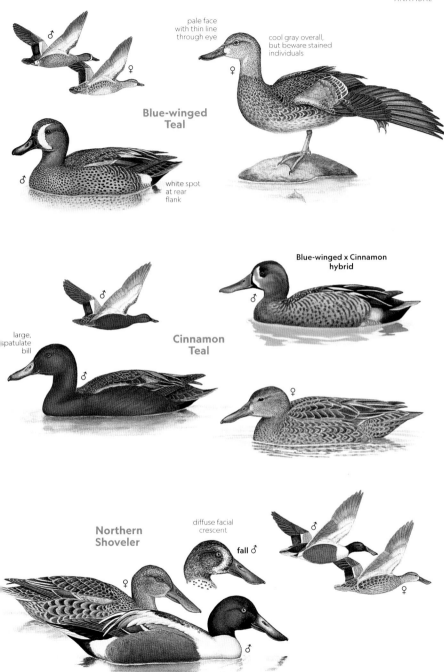

Blue-winged Teal — pale face with thin line through eye; cool gray overall, but beware stained individuals; white spot at rear flank

Cinnamon Teal — large, spatulate bill; Blue-winged x Cinnamon hybrid

Northern Shoveler — diffuse facial crescent; fall ♂

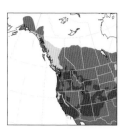
Casual to Hawaii

Gadwall
Mareca strepera | GADW | L 20" (51 cm)
- **APPEARANCE** A midsize duck; blocky overall, with forehead steep and bill relatively slight. Adult male gray with a black "butt"; close up, the gray breast is exquisitely vermiculated; note also rich chestnut on secondary coverts. Female mostly gray-brown, but with whitish belly; good mark is bill pattern, dark with thin orange stripe. Both sexes flash square of white on secondaries in flight.
- **VOCALIZATIONS** Female quack, far-carrying and nasal, *geb;* male gives a deep, short whistle.
- **BEHAVIORS** A habitat generalist, occurring in freshwater marshes, in shallow saltwater lagoons, and on deep humanmade reservoirs. Although generally common where it occurs, this relatively unflamboyant duck is easily overlooked among showier species.
- **POPULATIONS** Beneficiary of habitat protections across N. Amer. range; following steep losses earlier, numbers have been rebounding for 50+ years. Has expanded breeding range in northwestern lower 48 states. Nearly annual in winter to Hawaii.

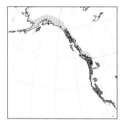
Annual to Hawaii

Eurasian Wigeon
Mareca penelope | EUWI | L 20" (51 cm)
- **APPEARANCE** The two wigeons are slighter-billed and a bit less hefty overall than congeneric Gadwall. Breeding male Eurasian, with mostly gray body and rusty head, could be confused with Redhead (p. 38), but note creamy yellow forehead and rosy-gray breast of Eurasian. Female wigeons harder to differentiate; Eurasian has warmer tones on head and cooler tones on body, the opposite of American. Wings of both sexes uniform dusky gray below; on American, white underwing coverts contrast with darker gray remiges.
- **VOCALIZATIONS** Call most commonly heard in West has timbre of American, but simpler: *wheee-oooh,* rising then falling.
- **BEHAVIORS** Habits in nonbreeding season much like those of American, with which it frequently consorts, often around ponds in parks and golf courses. A lone Eurasian amid Americans is often at the periphery of, or slightly away from, the main flock.
- **POPULATIONS** Old World counterpart of American Wigeon. Regular in very small numbers coastally in winter, except Salish Sea region, where fairly common. Annual to Hawaii, although outnumbered there by American.

Annual to Hawaii

American Wigeon
Mareca americana | AMWI | L 19" (48 cm)
- **APPEARANCE** Adult male has gray head, finely stippled, topped off by gleaming white crown; broad green crescent extends behind eye. Female told from female Gadwall by blue-gray bill with fine black edging; both wigeons have notably small bills. Female and eclipse (summer) male very similar to female of uncommon Eurasian Wigeon; that species is warmer-headed, with darker underwing.
- **VOCALIZATIONS** Males vocal; give frenzied, trisyllabic *whee WHEE whew,* often in chorus.
- **BEHAVIORS** Although perfectly capable of swimming, this common duck has a particular affinity for grassy stretches, where it grazes in midsize flocks.
- **POPULATIONS** Has recovered impressively since bottoming out in mid-20th century; breeding range expanding and numbers increasing in northern Canada and Alas. Annual in winter to Hawaii.

Laysan Duck 🆑

Anas laysanensis | LAYD | L 16" (41 cm)

Rare, critically endangered, and local; the species is heavily managed and is found today only on the Northwestern Hawaiian Is., hundreds of miles from where Hawaiian Ducks occur. A small and dark duck, typically active at night; eye of adult encircled by a broad and often "messy" white ring of feathers around the eye, reduced in juveniles. Vocalizations Mallard-like, but faster and buzzier.

Hawaiian Duck 🆔

Endemic; found on main islands

Anas wyvilliana | HAWD | L 20" (51 cm)

- **APPEARANCE** Like a smallish female Mallard. All plumages have plain lead-gray bill; many males show thin eye ring, never as thick as Laysan Duck's.
- **VOCALIZATIONS** A rather quiet duck; calls like Mallard's, but more subdued.
- **BEHAVIORS** Skittish and secretive; a "Hawaiian Duck" in a park, especially one receiving handouts, is likely a Mallard derivative. In dramatic nuptial flight, multiple birds fly in small circles 100 feet above ground.
- **POPULATIONS** Hawaiian endemic; found on main islands, but mostly Kauai. Endangered; former impacts were hunting, habitat loss, and biological introductions, but hybridization with Mallards the greatest current threat.

Mallard

Introduced and vagrant to Hawaii

Anas platyrhynchos | MALL | L 23" (58 cm)

- **APPEARANCE** A familiar, large duck with broad bill and round head. Breeding male sports glistening green head, white neck ring, chestnut breast. Female mostly brown year-round, with bill marbled orange and black. Late summer (eclipse) adult male resembles female, but with solid dusky yellow bill.
- **VOCALIZATIONS** Female, not male, gives the universally recognized, honest-to-goodness *quack!* Male's utterance raspier, less exuberant. In rarely observed courtship "song," male grunts, gasps, and whistles.
- **BEHAVIORS** Aggressive and adaptable; takes well to duck ponds in our largest metropolises, but equally at home in large marshes in remote mountain valleys.
- **POPULATIONS** Interbreeds with closely related species, eroding their genetic distinctiveness. Rare natural vagrant to Hawaii, but also introduced and established there.

Mexican Duck

Anas diazi | MEDU | L 22" (56 cm)

- **APPEARANCE** In all plumages, suggests female Mallard. Most of the year in the West, a "female Mallard" with a uniform yellowish bill is likely a male Mexican Duck; check for brownish tail (whitish on Mallard) lacking curled feathers of male Mallard. Pale head of drake Mexican contrasts with darker breast. Female Mexican has drab yellow-orange bill (splotchy orange and black on female Mallard). Field ID difficult in summer when males of both species are in disheveled eclipse plumage, resembling female Mallard.
- **VOCALIZATIONS** Similar or identical to Mallard's; calls may average slightly buzzier and more nasal than Mallard's.
- **BEHAVIORS** A bit "wilder" overall than Mallard, and more likely to be found in natural settings like playas and riverbanks; that said, frequently mingles with Mallards at park ponds in towns and cities.
- **POPULATIONS** In West, expanding north and slowly increasing. Many show introgression with Mallards.

GEESE, SWANS, AND DUCKS
ANATIDAE

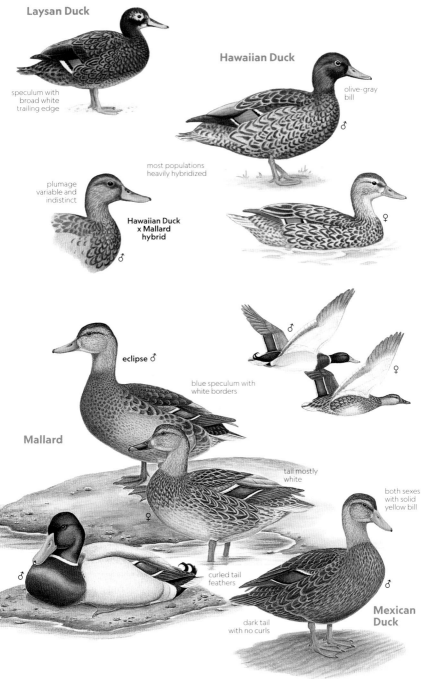

Annual to Hawaii

Northern Pintail
Anas acuta | NOPI | L 20-26" (51-66 cm)

■ **APPEARANCE** The "pointy duck"; in flight, both sexes appear long-winged, long-tailed, long-necked, and pointy-billed. Most of year, male instantly recognized by central tail streamers; note also thin white stripe running up chocolate head, white breast, and gray body. Female uniform gray-brown, with long neck, thin bill, and relatively long tail.

■ **VOCALIZATIONS** Male call a rising hiss transitioning to a musical hoot, then trailing off, *ssssSSLEEERPppsss*. Female cackles quietly.

■ **BEHAVIORS** Nests in freshwater marshes; in winter, often grazes in agricultural districts.

■ **POPULATIONS** Spring migration and breeding season early, with "fall" migration briskly underway by late summer. Regular in winter to Hawaii.

Annual to Hawaii

Green-winged Teal
Anas crecca | GWTE | L 14½" (37 cm)

■ **APPEARANCE** Our smallest dabbling duck, with bill gray and thin; green secondaries prominent in flight. Male has chestnut head with broad green patch behind eye. Female at rest told by small size, pale rear; often, a bit of the green wing pokes through.

■ **VOCALIZATIONS** Male's froglike *pleep* enlivens marshes; females mutter nasally, *kwenk* or *kwunk*.

■ **BEHAVIORS** Nests in forested waterways. Tight flocks feed energetically on lakeshores, sometimes alongside dowitchers (p. 126).

■ **POPULATIONS** Hardy; winters well inland. Eurasian subspecies breeds Aleutians and is regular but scarce in winter to Hawaii and West Coast.

Casual to Hawaii

Canvasback
Aythya valisineria | CANV | L 21" (53 cm)

■ **APPEARANCE** Shape distinctive: long overall, with sloping forehead and long bill. Rides close to the water. Most of year, male has red-brown head and gray-white ("canvas") back. Female separated from female Redhead by paler tones overall, all-dark bill, and especially body shape.

■ **VOCALIZATIONS** Rarely heard in winter, but courtship song arresting: a mirthful, musical cackle, *he-lee-we-loo*.

■ **BEHAVIORS** Like other *Aythya* ducks, dives for food. Nests in permanent freshwater marshes with deep water; winters on lakes and reservoirs.

■ **POPULATIONS** Breeding range shifting north into northwestern Canada and interior Alas.; has been undergoing dietary diversification in recent decades. Casual in winter to Hawaii.

Accidental to Hawaii

Redhead
Aythya americana | REDH | L 19" (48 cm)

■ **APPEARANCE** More compact than Canvasback. Bill color tripartite: A thin white stripe separates the black tip from the broad blue base; pattern more muted on female. Female darker than female Canvasback, but much variation; body shape always different.

■ **VOCALIZATIONS** Courting males in chorus astonishing: *whee* and *whew* notes, soft but far-carrying; given by many drakes at once, quietly frenzied.

■ **BEHAVIORS** Ecologically intermediate between dabblers (pp. 32-39) and other *Aythya* ducks: Especially on breeding grounds, forages in shallow marshes; mixes dabbling and diving, and loafs with dabblers.

■ **POPULATIONS** Like Canvasback, breeding range expanding north into northwestern Canada and interior Alas. Accidental to Hawaii.

Annual to Hawaii

Ring-necked Duck
Aythya collaris | RNDU | L 17" (43 cm)
- **APPEARANCE** Head obviously peaked; bill prominently marked. Male has faint purplish neck ring. Adult male has white wedge at base of breast, appearing vertical on swimming bird. Female sports white eye ring and postocular stripe ("bridle"), more obvious than on female Canvasback and Redhead (both on p. 38).
- **VOCALIZATIONS** Flushing, gives annoyed, muffled quacks, *runk ... runk ....*
- **BEHAVIORS** Nests in boreal bogs and marshes. In winter, more accepting than other *Aythya* ducks of smaller waterbodies, often around woodlands and farm country.
- **POPULATIONS** Breeding range on continent slowly expanding, especially northward. Annual in winter to Hawaii.

Tufted Duck
Aythya fuligula | TUDU | L 17" (43 cm)
Widespread Palearctic breeder. Uncommon but regular migrant in Bering Sea region, much rarer elsewhere in West, with most detections near coast, mainly winter and often with scaup; casual to Hawaii. Adult male usually adorned with very long crest. Dark-headed female typically shows at least a hint of the male's crest; dark blue bill has broad black tip. Both sexes show extensive white in wing in flight.

Annual to Hawaii

Greater Scaup
Aythya marila | GRSC | L 18" (46 cm)
- **APPEARANCE** Best distinction from slightly smaller Lesser Scaup is head shape: shallowly rounded in Greater, with peak above or in front of eye. Bill broader than Lesser's, with larger black tip; white in wing extends well out onto primaries. Adult male head glossed greenish; sides and back cleaner white than on Lesser. Female plumage very similar to female Lesser's, but averages more white on face. First-winter males of both scaup species intermediate between adult males and females.
- **VOCALIZATIONS** Rarely heard in winter; flushing birds sometimes give a rough, snarling sound. Courting male utters muffled hoots and whistles.
- **BEHAVIORS** Where the two scaup species winter together coastally, Greater is more inclined to marine and brackish conditions, but there is great overlap in these proclivities.
- **POPULATIONS** Possible anywhere in Interior West fall to spring, but almost always outnumbered there by Lesser; also annual to Hawaii, where Lesser is more numerous.

Annual to Hawaii

Lesser Scaup
Aythya affinis | LESC | L 16½" (42 cm)
- **APPEARANCE** Compare carefully with similar Greater Scaup. Head of Lesser peaked, with high point behind eye. Bill slighter than Greater's, with less black at tip; male's head glossed purplish, female's face with less white than Greater's. White in wing restricted mostly to secondaries, but getting diagnostic view of flying bird is tricky.
- **VOCALIZATIONS** Mostly silent in winter; when flushing, sometimes gives weak, wavering growl. Courting male makes popping and whooping sounds: *whoopa* and *whee-up.*
- **BEHAVIORS** Away from breeding grounds, forms tight flocks ("rafts") on lakes and rivers.
- **POPULATIONS** Spring migration fairly late for a duck, with fall movements southward also late. Regular in small numbers in winter to Hawaii.

GEESE, SWANS, AND DUCKS
ANATIDAE

Steller's Eider
Polysticta stelleri | STEI | L 17" (43 cm)
- **APPEARANCE** Despite the name, distinct from other eiders; closest relative is apparently the extinct Labrador Duck (p. 473). Smaller than other eiders. Male boldly patterned in black, white, and buff; other plumages dark brown with broad, blue-gray bill.
- **VOCALIZATIONS** On breeding grounds, harsh and buzzy *frrrp* or *zzzrrr*; usually silent in winter.
- **BEHAVIORS** Nests near freshwater ponds, moving to salt water in winter. Feeds by both dabbling and diving.
- **POPULATIONS** Winters mostly Aleutians, but vagrates south coastally to Northern Calif. Numbers declining and range contracting.

Spectacled Eider
Somateria fischeri | SPEI | L 21" (53 cm)
- **APPEARANCE** Feathering extends farther down bill than on other *Somateria* eiders, making Spectacled appear small-billed. Huge "goggles" striking on adult male, more muted in other plumages. Female barred rusty and black; her forehead is dark.
- **VOCALIZATIONS** Female guarding nest protests with low grunts; usually silent in winter.
- **BEHAVIORS** Diet generalist on breeding grounds; in winter, narrows diet to mollusks, especially clams, scarfed up from sea floor.
- **POPULATIONS** Winters in Bering Sea polynyas; rare in winter away from open ocean.

King Eider
Somateria spectabilis | KIEI | L 22" (56 cm)
- **APPEARANCE** About the same heft as Spectacled Eider. Head compact; bill distinctively shaped in all plumages. Parti-colored drake unmistakable; young and nonbreeding males dark-backed with orangish bill. Border between black bill and brown face of female not as pointed as on female Common; body feathers of female King have black V marks.
- **VOCALIZATIONS** Adult male on breeding grounds gives low, booming hoots, sometimes heard at sea toward the end of the winter.
- **BEHAVIORS** Nests in drier microhabitats than other eiders; in winter often feeds near ice floes and ledges.
- **POPULATIONS** In West restricted essentially to Alas. in winter, but scattered records well south along coast; accidental inland.

Common Eider
Somateria mollissima | COEI | L 24" (61 cm)
- **APPEARANCE** Largest duck in Northern Hemisphere. Bill long and sloping; in side view, feathering on face extends far onto bill in sharp point. Black cap of male contrasts with huge white cheek; western subspecies, *v-nigrum*, named for thin black V on chin of drake. Bill of male yellow-orange; bill of female dark, but structure like that of male. Female barred like female Spectacled, but ground color colder. Young and nonbreeding males suggest female, but with colorful bill and usually some white on breast.
- **VOCALIZATIONS** Male's call, heard mostly on breeding grounds, *oooOOOooh*, rising and falling.
- **BEHAVIORS** Closely linked to marine habitats throughout entire annual cycle. Nests colonially; winters at sea, both close to shore and well offshore.
- **POPULATIONS** Range expanding in East and in Europe, but evidently not in West; Alas. breeders negatively affected by introduced predators.

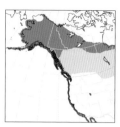
Accidental to Hawaii

Surf Scoter
Melanitta perspicillata | SUSC | L 20" (51 cm)

■ **APPEARANCE** Scoters are sea ducks that combine dark plumage with brightly colored, oddly shaped bills. This species and Black Scoter are dark-winged, useful for quick ID at a distance. Black and white head of adult male Surf distinctive; note also colorful bill. Female shows two white splotches on face, a vertical one at base of bill and a rounder one behind and below eye.

■ **VOCALIZATIONS** Relatively quiet, even on breeding grounds; courting males give rapid popping, *puh puh puh puh puh puh* ....

■ **BEHAVIORS** Nests on taiga, sometimes in broken forest away from water's edge. Winters in nearshore marine habitats, where it feeds on bottom-dwelling invertebrates.

■ **POPULATIONS** Sexes segregate to some extent in winter and while molting. Rare but regular in winter well inland; accidental to Hawaii.

White-winged Scoter
Melanitta deglandi | WWSC | L 21" (53 cm)

■ **APPEARANCE** Larger than Surf and Black Scoters; bold white wing patch prominent in flight, variably concealed on swimming bird. Adult male sports broad white "teardrop"; female has two white blobs on dark face, more rounded than corresponding white patches on face of female Surf. See Stejneger's Scoter.

■ **VOCALIZATIONS** Like Surf Scoter, fairly quiet, even in summer. Gruff croaks and rough whistles occasionally heard in winter.

■ **BEHAVIORS** Nests in thick ground cover near freshwater boreal lakes, migrating to coasts in winter. Feeds mostly on mollusks in winter, but more broadly carnivorous when breeding.

■ **POPULATIONS** In long-term decline; formerly bred farther south. Like Black and Surf Scoters, rare but widespread inland annually.

Stejneger's Scoter
Melanitta stejnegeri | STSC | L 21" (53 cm)

Closely related to White-winged Scoter; elevated to full-species status in 2019. Compared to White-winged, adult male Stejneger's has blacker flanks and darker orange bill with larger knob. Stejneger's breeds widely in northern Asia; in recent years, has been annual in small numbers to Bering Sea region, doubtless reflecting observer interest in a suddenly "countable" bird; has occurred as far south as central coastal Calif. and as far inland as Mont.

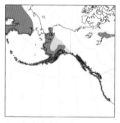
Accidental to Hawaii

Black Scoter
Melanitta americana | BLSC | L 19" (48 cm)

■ **APPEARANCE** Our smallest and darkest scoter; dark-winged like Surf Scoter, but remiges from below paler gray than coverts. Adult male appears all-black at rest, except for bright yellow on bill. Extensive pale gray on face of adult female separates her from other scoters; an unexpected point of confusion is Ruddy Duck (p. 50) in winter, similar in overall plumage.

■ **VOCALIZATIONS** Unlike other scoters, vocal even in winter. Gives drawn-out, sighing whistle on different pitches, often doubled: *ooooo eeeee*.

■ **BEHAVIORS** Nests in dense grass on tundra; winters coastally where benthic mollusks are plentiful.

■ **POPULATIONS** Heaviest concentration in winter from Wash. to southern Alas.; widespread but rare inland. Accidental to Hawaii. Males migrate earlier in spring than females.

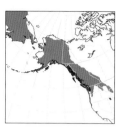

Accidental to Hawaii

Harlequin Duck
Histrionicus histrionicus | HARD | L 16½" (42 cm)
- **APPEARANCE** Small, round-bodied, and short-billed. Adult male histrionically spectacular. Female, brown-bodied with white spots on face, suggests female Surf Scoter (p. 44), but bill structure different.
- **VOCALIZATIONS** Squeaky, monosyllabic utterances, *jee* and *jeep*, often in rapid series; when seas are calm, close-up birds often audible.
- **BEHAVIORS** Habitat requirements may be the most unusual of any waterfowl species in West: Nests on ledges near turbulent streams and rivers; keeps close to sea rocks and jetties in winter. Scarce but often present for weeks to months at a time at popular coastal birding hotspots where easily seen.
- **POPULATIONS** Has never been a common species. Formerly bred in Colo. and the Sierra Nevada; breeding range in Wyo. may be expanding. Accidental to Hawaii.

Accidental to Hawaii

Long-tailed Duck
Clangula hyemalis | LTDU | L 16–22" (41–56 cm)
- **APPEARANCE** Small-bodied, short-billed, and long-tailed. Variable, often extensive, white above; underwings dark. Annual variation complex; has three molts per year, a very unusual strategy. Male sports extremely long tail most of year; upperparts mostly white in winter, extensively black in summer. Female lacks central tail streamers of male, but nevertheless pointy-tailed; note white plumage overall, becoming darker spring into summer. Both sexes in transitional plumages throughout much of year. Bill of male mostly pinkish year-round, grayish year-round on female.
- **VOCALIZATIONS** Sonorous and far-carrying, a nasal hoot followed by two or three more: *ow! ... ow-uh-WOW! (owl om-e-let);* flocks yodel melodiously round the clock, even in the dead of winter.
- **BEHAVIORS** Nests in freshwater habitats in Arctic; moves to coasts in winter, where it feeds on benthic invertebrates in cold waters, often near edge of pack ice.
- **POPULATIONS** Appears to be in steep decline in West, with several former molting and staging grounds seemingly gone. Coastal wintering range mostly Wash. northward; rare but regular far inland. Accidental to Hawaii.

Annual to Hawaii

Bufflehead
Bucephala albeola | BUFF | L 13½" (34 cm)
- **APPEARANCE** Smallest diving duck in West. Forehead steep; bill small and dark. Male has huge white patch on black head, iridescent in bright light; female sports broad white swath across dark chocolate head.
- **VOCALIZATIONS** Quiet even on breeding grounds; flushing birds in winter sometimes give harsh, flatulent quacks.
- **BEHAVIORS** Nests around freshwater ponds in boreal forests, often with hardwoods like aspens. Winters mostly coastally, favoring more protected sites than other bay ducks: inlets, lagoons, and sheltered bays. Away from breeding grounds, often consorts with Hooded Mergansers (p. 48). At all seasons, "Buffleheads rock," tilting slowly from side to side as they fly past.
- **POPULATIONS** Increasing; historically depended on nest holes drilled by large woodpeckers, especially Northern Flicker (p. 256), but in recent years has taken to human-proffered nest boxes. Widespread in small numbers inland in winter. Annual to Hawaii.

GEESE, SWANS, AND DUCKS
ANATIDAE

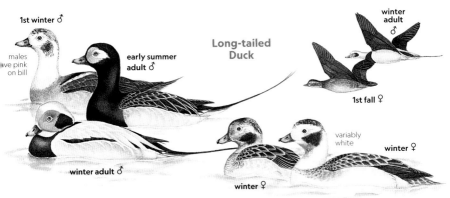

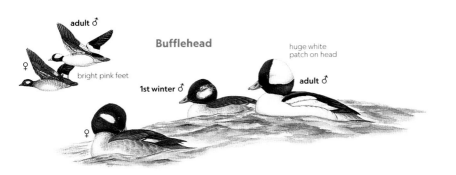

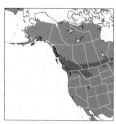
Casual to Hawaii

Common Goldeneye
Bucephala clangula | COGO | L 18½" (47 cm)

■ **APPEARANCE** Midsize diving duck with sloping bill and forehead. Adult male's glossy green head marked with large, white, oval blob. Female brown-headed; her black bill is variably tipped with yellow. Male in first winter brown-headed like female, but starting to show white facial blob of adult male.

■ **VOCALIZATIONS** Male courtship song a mechanical buzz followed by shrill *peent!* Also known as "whistlers," both goldeneye species make a loud trilling sound with their wings in flight.

■ **BEHAVIORS** Breeds around freshwater wetlands with standing deadwood for nest holes; also accepts next boxes. Hardy; winters coastally, but also well inland where there is open water. Courting begins midwinter, far from breeding grounds.

■ **POPULATIONS** Fall migration late; spring migration fairly early. Nonbreeders sometimes noted in summer far from breeding range; casual to Hawaii. Hybridizes with Barrow's Goldeneye; see below.

Barrow's Goldeneye
Bucephala islandica | BAGO | 18" (46 cm)

■ **APPEARANCE** Forehead steep; forms relatively sharp angle with relatively short bill. Adult male's head glossed purplish, and white blob on face crescent-shaped; mostly black scapulars make swimming male appear darker-sided than Common. Female's bill mostly or entirely deep yellow. Plumage of first-winter male like that of first-winter male Common, but steep forehead of Barrow's often obvious. Hybridizes with Common: Drakes intermediate in extent of white on face and flanks; hybrids in other plumages difficult to recognize.

■ **VOCALIZATIONS** As in Common, nonvocal wing whistle loud and far-carrying. Otherwise fairly quiet; courting male not as noisy as courting male Common Goldeneye, and display less elaborate.

■ **BEHAVIORS** Like Common, nests in tree cavities near standing water, on average a bit more alkaline than preferred by Common. Winter flocks keep to themselves or mix only casually with Common.

■ **POPULATIONS** Uncommon; in regions where both species winter, Common often outnumbers Barrow's. Western breeders completely isolated from eastern breeders (mostly Que.).

Casual to Hawaii

Hooded Merganser
Lophodytes cucullatus | HOME | L 18" (46 cm)

■ **APPEARANCE** Smallest and most distinctive merganser; like other mergansers, has serrated bill for capturing and manipulating prey. Adult male's extraordinary bonnet can be raised or lowered in less than a second; at quick glance, male could be mistaken for male Bufflehead (p. 46). Female's crest tawny; rest of body colder brown. First-winter male has black face with staring yellow eye.

■ **VOCALIZATIONS** Gives soft grunts and croaks around nest. Mostly silent on migration and in winter, but nonvocal wing whistle distinctive year-round: a rapid trill like a ground cricket.

■ **BEHAVIORS** Cavity nester; accepts both natural and artificial nest holes. Consumes diverse animal matter, especially crayfish.

■ **POPULATIONS** Breeding range expanding, probably in response to nest box supplementation. In winter, spreads out across landscape singly or in small flocks. Casual to Hawaii.

Accidental to Hawaii

Common Merganser
Mergus merganser | COME | L 25" (64 cm)

- **APPEARANCE** The largest merganser, muscular and smoothly contoured. In all plumages, bill and forehead structure distinctive: Broad-based bill slopes gradually toward long, shallow forehead. Males on the wing look like flying bowling pins, with white bodies and all-dark heads; breast white, sometimes tinged rosy. Female's orange-brown head sharply demarcated from pure-white breast; head color on female Red-breasted blends more gradually with paler breast.
- **VOCALIZATIONS** Flushing and sometimes swimming, female issues muffled quacks, in slow succession: *woohf ... woohf ... woohf ....*
- **BEHAVIORS** Breeds around lakes and rivers with nearby tree holes and bird boxes for nesting; also nests on cliffs near water. Hardy; winters anywhere there is deep water with fish. More inclined to freshwater than Red-breasted.
- **POPULATIONS** Migrates quite early in spring, fairly late in fall; sex segregation, especially in late summer, can be pronounced, with males dispersing farther to molt than females. Accidental to Hawaii.

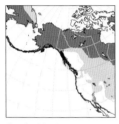
Casual to Hawaii

Red-breasted Merganser
Mergus serrator | RBME | L 23" (58 cm)

- **APPEARANCE** A bit smaller than Common Merganser. Shaggy-headed in all plumages; thin-based bill forms sharp angle with steep forehead, distinct from sloping profile of Common. Adult male's rusty breast and dark green head separated by broad white band. Some females closely resemble Common, but bill and forehead structure usually a reliable point of distinction.
- **VOCALIZATIONS** Female's quack a bit higher, more nasal than female Common's; heard less frequently in winter than Common.
- **BEHAVIORS** The most seafaring merganser. Breeds on tundra and in boreal forest, often coastally; usually nests on ground, unlike other mergansers. Most winter coastally, but small to occasionally large flocks sometimes occur well inland.
- **POPULATIONS** Like Common, exhibits marked sex segregation during parts of the annual cycle. Spring migration averages later than that of Common. Casual to Hawaii.

Accidental to Hawaii

Ruddy Duck
Oxyura jamaicensis | RUDU | L 15" (38 cm)

- **APPEARANCE** An odd duck. Small; body shape, proportioned like a rubber ducky, with tail cocked, distinctive. Breeding male deep rufous with black cap, bright white face, intense blue bill; in winter, retains white face but bill black and body cold brown. Female has long dark line across pale dusky face.
- **VOCALIZATIONS** Silent in winter, but courting male's song notable: a loud rapping, followed by a harsh buzz, *tik-tik-tik-tik-tik-rrrrrr.*
- **BEHAVIORS** In winter, loafs in small to medium "rafts" on lakes and bays; individuals and whole flocks can seem entirely inert for long periods of time. Temperamentally the opposite in summer, when courtship is highly animated.
- **POPULATIONS** Unlike most other waterfowl in West, acquires breeding plumage in spring and performs courtship in summer. Introduced to Europe, but has since been largely eradicated because it competes with endangered White-headed Duck, *O. leucocephala.* Accidental to Hawaii.

NEW WORLD QUAIL | ODONTOPHORIDAE

Small and plump, often with elegant crests and plumes, these upland gamebirds were until recently placed within the large family Phasianidae (pp. 56–67). Most New World quail gather in tight flocks called coveys, and their most characteristic vocalizations are far-carrying whistles. Note that the well-known Common Quail of Europe, *Coturnix coturnix,* is not in this family—and that these New World quail in fact have some representatives in Africa!

Mountain Quail
Oreortyx pictus | MOUQ | L 11" (28 cm)

■ **APPEARANCE** Our largest quail, with two plumes, long and straight, sticking out of the head. Adults of both sexes gray-brown overall with bold chestnut and white markings on the face and flanks; young birds smaller and scruffier with shorter plumes.

■ **VOCALIZATIONS** Most commonly heard call a rich, wavering, far-carrying whistle or hoot: *eeeerrk* or *eee-yeerk*. At a distance, can be confused with Gambel's and California Quails (both on p. 54). Like all quail, cackles softly and continually when feeding in coveys.

■ **BEHAVIORS** As its name suggests, a denizen of high country, especially where slopes are steep and covered in dense undergrowth.

■ **POPULATIONS** Undergoes limited altitudinal migration downslope before winter. Introductions and translocations cloud status in some mountain ranges.

Northern Bobwhite
Colinus virginianus | NOBO | L 9¾" (25 cm)

Two populations in West, both with problematic, peripheral statuses here. All are rotund and rufescent, with small head and short bill. Familiar nominate eastern subspecies, with bold white and chestnut (males) or buff and brown (females) on face, reaches limit of range in western Great Plains, where populations frequently augmented by deliberate releases. Mostly Mexican "Masked Bobwhite" was formerly resident in Southeast Ariz., but extirpated there in 20th century; males distinctive, with black face and rich cinnamon below, but females like female of eastern subspecies. Recent sightings of "Masked" probably refer to released birds and their progeny. Well-known *bob ... white!* song, given by all populations, usually preceded by much softer introductory note, inaudible at a distance.

Montezuma Quail
Cyrtonyx montezumae | MONQ | L 8¾" (22 cm)

■ **APPEARANCE** Rotund, even for a quail; appears nearly tailless and neckless. Also known as "Harlequin Quail" for the male's intricately patterned face; female's odd head shape calls to mind a food service worker's hairnet. Juvenile like female, but often notably smaller.

■ **VOCALIZATIONS** Haunting song, a loud, long whistle, descending slightly and weakly modulated: *eeeeuuuurrrrrr*.

■ **BEHAVIORS** Wary and secretive, usually stumbled upon by accident; found in savanna-like habitats with scattered shrubs and trees, especially oaks, with extensive bunchgrass. Probably most common in foothills and adjoining valley floors, but ranges well upslope too.

■ **POPULATIONS** Because it is so secretive and cryptic, details of disjunct range in U.S. are imperfectly known. Practically sedentary, but may undergo weak seasonal movements altitudinally.

NEW WORLD QUAIL
ODONTOPHORIDAE

Scaled Quail
Callipepla squamata | SCQU | L 10" (25 cm)

■ **APPEARANCE** Popularly known as "Cottontop" for the male's puffy white crest, which can be smooshed down or impressively erect; crest reduced and browner on female. On both sexes, breast feathers leaden with fine black edges (the "scales").

■ **VOCALIZATIONS** Song comprises two notes, one right after the other, the pairings in rapid succession: *chipchurr-chipchurr-chipchurr ... (Pecos, Pecos...)*. Also gives an explosive, startling *rreurk!* Flocks murmur and cackle quietly but continually.

■ **BEHAVIORS** Typically seen scooting down a dirt road or across the desert, singly or in frantic groups. When startled, coveys flush a short distance with much fuss. Habituates to feeding stations but is never unwary.

■ **POPULATIONS** Intrinsically variable; numbers crash regionally following heavy snowfall, but also rise in response to good growing seasons. Probably in long-term decline owing to overgrazing; hybridizes with Gambel's Quail where ranges overlap.

California Quail
Callipepla californica | CAQU | L 10" (25 cm)

■ **APPEARANCE** Identical in overall heft and general build to Scaled and Gambel's Quails. Both sexes adorned with curlicue plume, prominent on male, sticking out of head. Very similar to Gambel's, but ranges on mainland barely touch; belly of California scaly (belly of Gambel's plain or weakly streaked).

■ **VOCALIZATIONS** Classic call a rich, ringing *r'REER! reer (Chicago!)*. Male advertising call a steady *reeeer;* can be confused at a distance with Mountain Quail's (p. 52).

Introduced to Hawaii

■ **BEHAVIORS** Arid-adapted like Gambel's Quail, but more generalist, favoring cooler climes; an old name was "Valley Quail," distinguishing it from Mountain Quail. Tends feeders and water features, sometimes in towns and cities.

■ **POPULATIONS** Widely introduced, including to Hawaii, where most common on islands of Hawaii and Maui. Within continental range, frequent releases for hunting cloud understanding of status; hybridizes with Gambel's Quail where ranges overlap.

Gambel's Quail
Callipepla gambelii | GAQU | L 10" (25 cm)

■ **APPEARANCE** Warm desert counterpart of California Quail; the two species' natural ranges on mainland barely overlap. A bit paler and brighter overall than California; pale belly of Gambel's unmarked or weakly streaked (belly scaled and darker on California). Hybridizes with California Quail in Mojave Desert and, more commonly, with Scaled Quail in Chihuahuan Desert.

■ **VOCALIZATIONS** Similar to California Quail's. *Chicago!* call not as rich and low. Male advertising call of Gambel's not as steady: *rrEEEeer,* building in intensity, then dropping off.

Introduced to Hawaii

■ **BEHAVIORS** Where range overlaps with Scaled, more likely to be found in dense washes and canyons, with Scaled out in shrubby grasslands. Habituates to feeders, becoming beloved attractions at park and refuge headquarters.

■ **POPULATIONS** As with California Quail, introductions complicate understanding of original range on mainland. Introduced to Hawaii; present strongholds there are Lanai and Kahoolawe, near Maui.

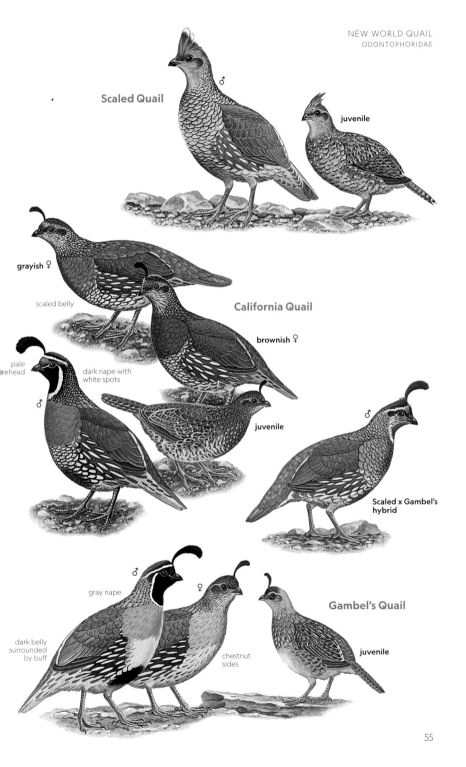

PHEASANTS AND KIN | PHASIANIDAE

Together with the New World quail (pp. 52–55), these are the western representatives of the order Galliformes, the upland fowl. Like New World quail, they feed mostly on the ground on vegetable matter. Species in the family Phasianidae vary in size, but most are larger than New World quail; they engage in some of the most remarkable courtship displays of any birds.

Introduced to Hawaii

Wild Turkey
Meleagris gallopavo | WITU | L 37–46" | (94–117 cm)

- **APPEARANCE** Huge; our largest upland gamebird. Appears all-dark in poor light but glistens bronze and vinaceous in the sunshine. Male has naked reddish head; female's unfeathered head is gray. Many adults, especially older males, sport "beard" of feathers hanging from breast.
- **VOCALIZATIONS** Male's well-known gobble audible at great distances; flocks cackle quietly, sometimes erupt into short shrieks.
- **BEHAVIORS** Usually in medium-size flocks. Until recently, avoided humans and limited activities to forests and forest edges, but of late has invaded suburbs and even cities. Capable of decent flight, but usually walks or runs from danger.
- **POPULATIONS** Birds from Great Basin westward, including Hawaii, are introduced. Extensive management has led to mixture of forms, varying especially in tail color. Formerly distinctive "Merriam's Wild Turkey," indigenous to southwestern U.S., unfortunately has been homogenized through human agency.

Ruffed Grouse
Bonasa umbellus | RUGR | L 17" (43 cm)

- **APPEARANCE** Named for black neck feathers, puffed out by male in display. Both sexes have modest crest and dark terminal band on tail. Otherwise cryptic, matching the grays and browns of tree trunks and forest floor.
- **VOCALIZATIONS** Nonvocal drumming of male caused by beating of wings. In most of range, accelerates and intensifies: *bup ... bup ... bup bup bu'bu'bu'B'B'B'B'*.
- **BEHAVIORS** A forest gamebird; reaches highest densities in second-growth aspen woods. Splits time between foraging on ground and gleaning buds and catkins up in trees.
- **POPULATIONS** Color varies regionally, with many in Pacific Northwest more reddish brown than elsewhere. Drumbeats of brownish *brunnescens* subspecies of southwestern B.C. not as smoothly accelerating as other subspecies.

Spruce Grouse
Canachites canadensis | SPGR | L 16" (41 cm)

- **APPEARANCE** Compared to Ruffed Grouse, relatively uncrested and slightly smaller overall. Male Spruce shows much more black below than Ruffed; red patch above eye hard to see except in display. In most of West, both sexes have dark tail with pale red-brown band, the inverse of Ruffed. Tail of male "Franklin's Grouse" all-dark with white spots at base.
- **VOCALIZATIONS** Male's display, like that of Ruffed, created by beating of wings; "Franklin's" ends display with loud clap.
- **BEHAVIORS** Range in West generally overlaps Ruffed Grouse, but Spruce is more inclined to conifers. Unwary and easily approached, but also cryptic and easily overlooked.
- **POPULATIONS** "Franklin's," from central B.C. southward, treated by some authorities as separate species.

Willow Ptarmigan
Lagopus lagopus | WIPT | L 15" (38 cm)

■ APPEARANCE The largest ptarmigan. Like congeners, has three plumages per year: mostly white in winter; cryptically brown (female) and deep chestnut (male) in summer; and intermediate in spring and fall, with spring males often strikingly bicolored. Black tail (even in winter) separates Willow from White-tailed; female year-round very similar to smaller Rock, but Rock has proportionately even smaller bill.

■ VOCALIZATIONS Calls impressively humanlike: short utterances like *Oh? Oh!* and longer series, *Go back! Go back! Go BACK!*

■ BEHAVIORS All three species of ptarmigan are fantastically adapted to life on the tundra. In addition to morphological adaptations like cryptic plumage, feathered tarsi, and rotund bodies (for retaining heat), they burrow in snowbanks for shelter and walk across snowfields on "snowshoes" of feathered talons.

■ POPULATIONS Textbook example of a vertebrate with natural oscillations in abundance, typically 10–11 years; causes not well understood, might even involve sunspot cycles! Withdraws from northern portion of range in winter; sometimes seen flying over open water, even at sea.

Rock Ptarmigan
Lagopus muta | ROPT | L 14" (36 cm)

■ APPEARANCE Small bill, even for a ptarmigan, is an adaptation for extreme Arctic lifestyle. Male in winter plumage nearly pure white; differs from Willow and White-tailed by black line through eye; male's spring and summer plumages grayer overall than Willow's. Female Rock told from female White-tailed by black tail; from Willow, with caution, by proportionately smaller bill and smaller size overall.

■ VOCALIZATIONS Courting male gives unmusical rattle or rapping, typically in short bouts of flight.

■ BEHAVIORS Perhaps the most ecologically extreme bird in the Northern Hemisphere. Occurs even farther north than Willow; where ranges overlap, Rock prefers drier, rockier, sparser habitats.

■ POPULATIONS Annual molt schedules vary complexly across range; details still being worked out. Male "Evermann's" of outer Aleutians notably dark. Migration poorly understood; some move considerable distances over land and even sea.

White-tailed Ptarmigan
Lagopus leucura | WTPT | L 12½" (32 cm)

■ APPEARANCE Smallest ptarmigan; both sexes have white tail year-round. Cold gray-brown plumage in warmer months similar to Rock Ptarmigan; Willow warmer red-brown at that time of year. Red "comb" above eye of breeding male on average less prominent on White-tailed than other ptarmigan.

■ VOCALIZATIONS Cackles and murmurs while walking around on tundra, sometimes erupting into louder notes, squeakier than calls of Willow.

■ BEHAVIORS Much like other ptarmigan species. Highest-elevation breeders reach 14,000 ft (4,270 m) in southern Rockies; some move downslope a bit after breeding.

■ POPULATIONS Range disjunct in lower 48 states; introduced in several mountain ranges, indigenous to many others. Climate change an emerging threat to all ptarmigan, with rapid loss of snowpack in southern Rockies particularly worrisome for this species.

Greater Sage-Grouse
Centrocercus urophasianus | GRSG | L 22-28" (56-71 cm)

- **APPEARANCE** Heavier than even a pheasant; tail long and pointed. Both sexes cold gray-black with black belly; flying away, appears uniformly dark above. Compare with Gunnison Sage-Grouse.
- **VOCALIZATIONS** Lekking males emit vocal and nonvocal popping and whooshing sounds, many with liquid quality; far-carrying but not especially loud.
- **BEHAVIORS** All phases of annual cycle tightly associated with sagebrush-steppe landscapes: Males lek in clearings in late winter; hens and chicks forage on sage in summer; and extended families flock together starting in fall.
- **POPULATIONS** Declining range-wide due to ongoing deterioration of sagebrush habitats caused by overgrazing; disturbance from oil field infrastructure a new threat.

Gunnison Sage-Grouse EN
Centrocercus minimus | GUSG | L 18-22" (46-56 cm)

Described to Western science only in 2000, the first such *nova avis* in the U.S. and Canada in 100 years! Occurs only in a few places in remote southwestern Colo. and southeastern Utah; total population less than 5,000 and declining. Range does not overlap with Greater Sage-Grouse. In addition to smaller size, Gunnison differs from Greater by shorter tail feathers with white bands, longer crest (filoplumes) in displaying male, and lower, steadier hooting at leks.

Dusky Grouse
Dendragapus obscurus | DUGR | L 20" (51 cm)

- **APPEARANCE** Large grouse; dark gray-brown overall. Similar to Sooty Grouse, but ranges barely overlap. A curious distinction is the number of tail feathers: 20 on Dusky, 18 on Sooty. Male Dusky displays with air sac more purplish and tail feathers more square-tipped than Sooty's, but differences slight and variable. Female averages paler than female Sooty.
- **VOCALIZATIONS** Displaying male gives soft hoots, hard to hear. In family groups, hens cackle softly and young whistle endearingly.
- **BEHAVIORS** Male displays on ground or fallen log, unlike Sooty. Usually found in forests, but wanders well into open shrublands where confusion with other grouse possible. In winter subsists mostly on conifer needles.
- **POPULATIONS** After breeding, some migrate counterintuitively upslope for winter feeding on conifer needles.

Sooty Grouse
Dendragapus fuliginosus | SOGR | L 20" (51 cm)

- **APPEARANCE** Nearly identical to Dusky Grouse, with differences in tail structure as noted for that species. Color of air sac of displaying male varies geographically: yellowish in south of range, but red-purple (like Dusky) farther north.
- **VOCALIZATIONS** Powerful hoots audible at great distance, weirdly different from Dusky's; usually hoots from trees (Dusky hoots on or near ground).
- **BEHAVIORS** More arboreal than Dusky, but note that any Dusky (even young) can get well up into trees.
- **POPULATIONS** Like Dusky, undergoes seasonal movements that reflect availability of conifer needles for feeding. The two species were formerly treated as one; they hybridize in parts of Wash. and B.C.

Sharp-tailed Grouse
Tympanuchus phasianellus | STGR | L 17" (43 cm)
- **APPEARANCE** Similar in size and shape to closely related prairie-chickens; tail, pale and pointed, distinctive in all plumages. Spotted and checkered below; prairie-chickens are mostly barred beneath. Beware similarity with female Ring-necked Pheasant (p. 64); tail of female pheasant is longer and lacks extensive white of Sharp-tailed.
- **VOCALIZATIONS** Males at leks (communal courtship grounds) combine loud, nonvocal stomping of feet and rattling of tail, interspersed with shrill, yapping calls.
- **BEHAVIORS** Lekking behavior highly animated, with males capering about, often leaping high. More accepting of agricultural districts than prairie-chickens; sometimes comes to feeding stations.
- **POPULATIONS** Fairly common across much of extensive range in northern prairie biome, but declining at western and especially southern peripheries. Southern subspecies, including *columbianus* ("Columbian Sharp-tailed Grouse"), are smaller and paler; gone from much of former range, *columbianus* is of particular concern. Hybridizes very rarely with Greater Sage-Grouse (p. 60) and less infrequently with Greater Prairie-Chicken.

Greater Prairie-Chicken
Tympanuchus cupido | GRPC | L 17" (43 cm)
- **APPEARANCE** About the same heft as Sharp-tailed Grouse, a bit larger than Lesser Prairie-Chicken. Both sexes prominently barred, especially below; tail short, rounded. Similar Sharp-tailed Grouse and female Ring-necked Pheasant (p. 64) have pointed tails, lack barring. Compare with Lesser Prairie-Chicken.
- **VOCALIZATIONS** "Booming" of males in display a long, low moan, unsteady and wavering. Notably ventriloquial, especially in morning mist of the featureless prairie; the effect of hearing them is arresting.
- **BEHAVIORS** Occurs year-round in tallgrass prairie with light grazing; less likely than Sharp-tailed Grouse to be found around woodland edges and in croplands.
- **POPULATIONS** Range greatly contracted from historic levels, with numbers continuing to decline overall. Responds well to protection of tallgrass prairie, however, and has shown some local increases in well-managed landscapes.

Lesser Prairie-Chicken
Tympanuchus pallidicinctus | LEPC | L 16" (41 cm)
- **APPEARANCE** Very similar to Greater Prairie-Chicken, but essentially no range overlap in West. Lesser is smaller and slighter overall, with finer barring. Air sacs of male Lesser are reddish; yellow-orange in Greater.
- **VOCALIZATIONS** Male in courtship gives short, clipped notes, *poo* and *p'doo*, in rapid series. Lacks the long, low "booming" of Greater.
- **BEHAVIORS** Favors shorter grass than Greater, especially where sandy soils support low-stature sagebrush with stunted oaks. Sometimes wanders into agricultural habitats, especially after breeding season.
- **POPULATIONS** Has always had a small range; in West, nearly gone from Colo., doing a bit better on eastern plains of N. Mex. Hybridizes with Greater in Kans., but hybrids not expected in our area. Introduction efforts, even in Hawaii, have been unsuccessful.

PHEASANTS AND KIN
PHASIANIDAE

Gray Partridge
Perdix perdix | GRAP | L 12½" (32 cm)

- **APPEARANCE** Small for a phasianid; suggests a large quail (pp. 52–55). Adults gray and russet; orangish face contrasts with gray underparts, finely barred. Male has dark chestnut belly patch; female pale-bellied.
- **VOCALIZATIONS** Calls, squeaky and creaking, include a three-syllable call, the middle note shortest: *kreeh d' kriih,* like a rusty gate swinging open. Flushing flocks boisterous.
- **BEHAVIORS** Found in agricultural country, especially around hedgerows and other plantings, where cold-weather crops predominate. Usually seen in coveys.
- **POPULATIONS** Eurasian species that became established in N. Amer. in early 20th century; has declined since then.

Ring-necked Pheasant
Phasianus colchicus | RNEP | L 21–33" (53–84 cm)

- **APPEARANCE** Large and slender with notably long tail. Dapper male unmistakable, but females suggest indigenous grouse; note long tail of female and warm buff-tan tones overall.
- **VOCALIZATIONS** Male in display gives double honk followed by rushing of wings. The two honking notes carry far, but the wing fluttering does not.
- **BEHAVIORS** Often seen singly along roadsides in farm country. Runs impressively fast; flushes in a startling explosion of sound and feather.
- **POPULATIONS** Indigenous to Eurasia. Canadian and U.S. birds derived originally from stock introduced long ago to western Europe from Asia. Constant reintroductions by game agencies cloud understanding of whether the species is truly self-sustaining here, and stock from diverse sources—a mix of domestic variants and well-differentiated subspecies—add further complication. Dark individuals of uncertain ancestry occasionally seen in Hawaii, where Kalij Pheasant also occurs.

Introduced to Hawaii

Kalij Pheasant
Lophura leucomelanos | KAPH | L 20–28" (51–71 cm)

Indigenous to South Asia. Nominate Nepalese subspecies introduced in 1960s to Hawaii, where most extensively established on Hawaii I. Male, prominently crested, has extensive red facial skin like male Ring-necked but is otherwise blue and blue-gray. Female shares crest and red facial skin of male but is dusky brown overall. Male display involves rushing of wings like Ring-necked, but without the loud double honk; instead, rattles shrilly.

Well established on main islands

Indian Peafowl
Pavo cristatus | INPE | L 40" (102 cm) ♀; 60–80" (152–203 cm) ♂

- **APPEARANCE** Huge, approaching heft of Wild Turkey (p. 56), but more slender. Male impossible to misidentify; female, although somewhat more demurely attired, nevertheless striking, especially with unique parasol.
- **VOCALIZATIONS** Caterwauling of male, with tormented quality, audible at tremendous distances.
- **BEHAVIORS** Feeds on ground during day. In evening, flies clamberingly into treetops and even rooftops to roost.
- **POPULATIONS** Indigenous to Indian subcontinent. First introduced here in 1860 in Hawaii, where now well established. Introduction to Calif. followed soon thereafter; status there and elsewhere in West complicated. In many places, small groups persist for a few years, then die out; however, may be considered recently established in Southern Calif.

Gray Francolin
Francolinus pondicerianus | GRAF | L 13" (33 cm)
Indigenous to India and Middle East; well established in Hawaii. Similar in build to a large quail (pp. 52–55). Clad overall in grays and browns, barred below and above; face orangey, with pale throat set off from gray breast by black "chinstrap." Varied calls have squeaky, raspy, frantic quality. Found in open habitat with cover: shrublands, unkempt pastures, etc.

Black Francolin
Francolinus francolinus | BLFR | L 13–14" (33–36 cm)
Similar in size and structure to Gray Francolin. Like that species, indigenous to Middle East and South Asia; well established in Hawaii. Male rich and dark overall, with white cheek and chestnut nape. Female paler overall, but with extensive black barring and much rufous on nape. Call a weak buzz, often in quick series of two to four notes. Habitats and distribution like those of Gray, but Black is more of a recluse.

Red Junglefowl
Gallus gallus | REJU | L 17" (43 cm)
The familiar chicken, indigenous to Southeast Asia, seen and heard in semiwild settings around human habitation in the continental West and especially Hawaii. Junglefowl at higher elevations on Kauai are descendants of birds established there centuries ago by Polynesians; these wild type birds are deep red-orange overall (male) or duskier orange-gray (female). Otherwise, domestic variants of all sorts are widely noted.

Himalayan Snowcock
Tetraogallus himalayensis | HISN | L 28" (71 cm)
Like a massively oversize Chukar. Indigenous to Central Asia; established in rugged terrain of Ruby Mts., Nev., where easily overlooked amid scree and boulders. Gray overall; pale head and breast marked with darker bands. In flight, flashes extensive white in wings. Wailing call impressively curlew-like.

Chukar
Alectoris chukar | CHUK | L 14" (36 cm)
■ **APPEARANCE** Rotund; suggests large quail (pp. 52–55). Gray-brown overall with red bill and black-barred flanks; buff-cream face bordered by thick black band. In flight, shows rufous in tail.
■ **VOCALIZATIONS** Calls raucous, *chaa* or *chaak*, often given in quick succession in long series.
■ **BEHAVIORS** Abounds in rocky, sparsely vegetated foothills and lower slopes, especially where invasive grasses dominate.
■ **POPULATIONS** Indigenous to Central Asia and Middle East, well established in U.S. in western interior and at higher elevations in Hawaii. Escapes from game farms, typically single birds, possible anywhere.

Introduced to Hawaii

Erckel's Francolin
Pternistis erckelii | ERFR | L 16" (41 cm)
Indigenous to Ethiopia and Eritrea; widely established in Hawaii. Compact; solidly built. Heavily streaked, with rusty crown and white throat; legs sturdy, yellow. More catholic in habitat preferences than Gray and Black Francolins, getting well into forests. Monosyllabic calls nasal and barking, often in quick series.

PHEASANTS AND KIN
PHASIANIDAE

Gray Francolin
black "chinstrap"

Black Francolin ♂
rufous on lower nape ♀

Red Junglefowl ♂
typically with cocked tail ♀

Himalayan Snowcock ♂
rufous on lower nape
dark belly

Chukar
adults
red bill
juvenile

Erckel's Francolin
white throat
heavily streaked

GREBES | PODICIPEDIDAE

Although typically seen swimming, often in the company of other waterfowl, grebes are not ducks. Instead, they are related to flamingos and pigeons, in an assemblage known as the Mirandornithes ("miraculous birds"). Grebes are superb divers; many have strange calls, bizarre courtship displays, and remarkable physiologies.

Casual to Hawaii

Pied-billed Grebe
Podilymbus podiceps | PBGR | L 13½" (34 cm)

- **APPEARANCE** Small, with bulbous bill. Gray-brown overall with puffy white rear. Bill of breeding adult pale blue-gray with black ring; nonbreeder has plain, straw-colored bill. Young birds transition slowly from strikingly patterned fledglings to weakly striped juveniles.
- **VOCALIZATIONS** Song a long series of clucking or gulping notes, decelerating like Yellow-billed Cuckoo's (p. 78). Adults issue monosyllabic grunts and shrieks year-round; dependent young pipe urgently.
- **BEHAVIORS** Submerges by simply sinking; like a submarine, instantly reduces buoyancy by expelling air from the spaces among its feathers.
- **POPULATIONS** Winters well inland where there is open water. Despite seeming a weak flier, capable of extensive dispersal; casual to Hawaii.

Accidental to Hawaii

Horned Grebe
Podiceps auritus | HOGR | L 13½" (34 cm)

- **APPEARANCE** Larger and less buoyant than Eared Grebe, which rides higher on water than Horned. Breeding adult has black head with extensive yellow "horns" behind eye; neck dark rufous. Horned in winter shows more contrast on face than Eared. In all plumages, the relatively stout bill has a bulbous white tip.
- **VOCALIZATIONS** On breeding grounds, gives a gull-like squeal, wavering, *uh-reee*. Silent in winter.
- **BEHAVIORS** Nests in freshwater wetlands, where aggressively territorial. Winters mainly in marine settings, from protected inlets out onto the ocean just beyond the shore.
- **POPULATIONS** Scarce but widespread in winter in Interior West, where generally less rare than Eared. Regular in small numbers to Bering Sea region; accidental to Hawaii. Eared molts later in spring than Horned; thus, by the time most Horned Grebes are in spiffy breeding plumage, many Eared Grebes are very "messy," still molting.

Accidental to Hawaii

Eared Grebe
Podiceps nigricollis | EAGR | L 12½" (32 cm)

- **APPEARANCE** Smaller, more buoyant than Horned; all-dark bill of Eared fine and upturned. Neck black in breeding plumage; the "ears," analogous to "horns" of Horned, are a tuft of yellow-orange feathers behind the eye. In gray winter plumage, markings on face blurry; Horned shows sharper contrast.
- **VOCALIZATIONS** On breeding grounds, a slightly rising whistle ending in a stutter: *wheeeeee-ippuh*, often in slow series.
- **BEHAVIORS** A physiological marvel: During the course of the annual cycle, combines hyperphagy (extreme eating) with rapid molt and drastic reduction in flight muscle mass. Most are flightless more than half the year, but they also perform impressive long-distance migrations.
- **POPULATIONS** Widespread breeder in interior; stages in early autumn at a few large inland lakes (Great Salt Lake, Mono Lake, etc.), then spreads out for winter. Accidental to Hawaii.

GREBES
PODICIPEDIDAE

Pied-billed Grebe

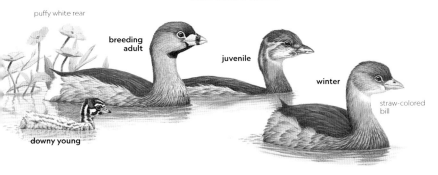

Horned Grebe

Eared Grebe

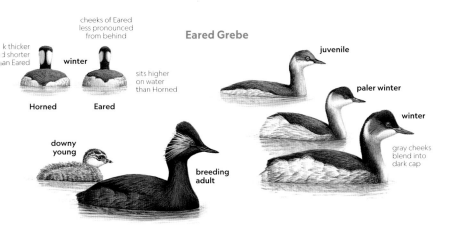

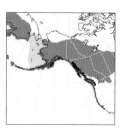

Accidental to Hawaii

Red-necked Grebe
Podiceps grisegena | RNGR | L 20" (51 cm)

- **APPEARANCE** Close in size and gross structure to *Aechmophorus* grebes, but more closely related to Horned and Eared (both on p. 68). Breeding adult has blackish cap, pale gray face, and rufous neck; bill is long and straw yellow. In winter, especially when hunched, can be confused with smaller grebes, including diminutive Eared, but even the dingiest Red-necked shows yellow at base of long bill.
- **VOCALIZATIONS** Animated on breeding grounds, with displaying birds mixing nasal wailing and prolonged chatter; usually silent on migration and in winter.
- **BEHAVIORS** A shape-shifter, especially in winter; a lone bird on the water can appear blobbily amorphous one moment, elongate and alert the next.
- **POPULATIONS** Most winter coastally, including both inshore and just offshore, but also widespread in tiny numbers well inland where there is open water on deep reservoirs. Accidental to Hawaii.

Western Grebe
Aechmophorus occidentalis | WEGR | L 25" (64 cm)

- **APPEARANCE** At all seasons, the two "swan-necked grebes," genus *Aechmophorus,* are slim overall with slender, daggerlike bills. Western is the darker of the two, being darker-sided, with black on cap extending below eye; but beware winter Clark's, appearing duskier than in summer. Year-round, bill color is a good mark: dusky yellow-olive on Western, brighter school bus yellow on Clark's.
- **VOCALIZATIONS** Advertising call, heard year-round, day and night, a shrill, sudden, strongly disyllabic *reeeh-rreeeek*. Short piping notes in rapid series given on breeding grounds.
- **BEHAVIORS** Mated pairs perform astonishing courtship ritual of synchronized swimming, involving wild rushing and lunging, and culminating in an extraordinary "weed ceremony."
- **POPULATIONS** Range in West overlaps broadly with that of Clark's year-round; both concentrate in winter near coasts, with Clark's less likely to get out to sea.

Clark's Grebe
Aechmophorus clarkii | CLGR | L 25" (64 cm)

- **APPEARANCE** Brighter, whiter counterpart of Western Grebe; formerly treated as white phase or white morph of that species. Garnet eye stares from white face of breeders; on Western, black cap extends below eye. Clark's averages whiter-sided than Western, but lounging grebes of either species can show much white. Black and white of face turns to grayish mush on many Clark's and Westerns in winter, so focus on bill color: bright yellow-orange on Clark's, even on dingy winter birds.
- **VOCALIZATIONS** Key difference with Western is the advertising call, a drawn-out, monosyllabic *rreeeeek* in Clark's. Sometimes slurred, *reee-eek*, but rarely as strongly disyllabic as in Western.
- **BEHAVIORS** Nests in wetlands, freshwater or alkaline, with standing vegetation, often in company of Westerns; remarkably, one short vocal element in the two species' elaborate courtship displays is enough to prevent interspecific pairing in most instances—but hybrids sometimes reported.
- **POPULATIONS** Mixed-species pairs uncommon but routinely noted on breeding grounds; hybrid progeny, especially in winter, difficult to recognize.

GREBES
PODICIPEDIDAE

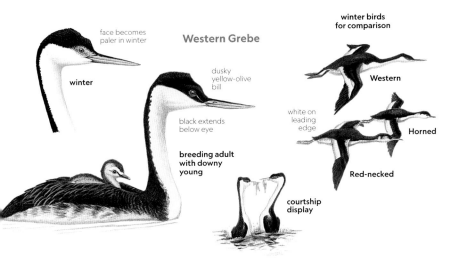

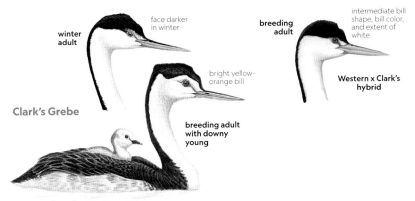

SANDGROUSES | PTEROCLIDAE

In their own order, Pterocliformes, the sandgrouses are birds of shrublands and grasslands in Africa and Asia. Previously treated as relatives of quail and grouse (pp. 52–67), they are now thought to be more closely related to pigeons and doves (pp. 72–77).

Chestnut-bellied Sandgrouse
Pterocles exustus | CBSA | L 12" (30 cm)

Pigeon-like species indigenous to Africa, India, and Middle East; established Hawaii I. Male warm butterscotch with diffuse black bars; tail quite long. Female has denser black barring. Seen in small flocks in open dry country along roadsides, typically in flight or on the ground.

PIGEONS AND DOVES | COLUMBIDAE

Worldwide in distribution, the columbids are plump and small-headed, slight-billed and short-legged. They sing simple, cooing songs, and many are strong fliers. Species called pigeons average larger than those called doves, but there is no real biological distinction between the two.

Rock Pigeon
Columba livia | ROPI | L 12½" (32 cm)

- **APPEARANCE** The familiar city pigeon. Wild type light blue-gray with black crescents across wings; neck iridescent green and lavender; variants may be pure white, nearly black, or quite red. All show tuft of white feathering at base of upper mandible; underwing linings gleam white in flight.
- **VOCALIZATIONS** Low-pitched, burry cooing soft and ventriloquial.
- **BEHAVIORS** Often in flocks on wires and rooftops and under viaducts. Flourishes in cities, yet many are surprisingly wild, nesting on remote cliffs.
- **POPULATIONS** Indigenous to Eurasia, but established in Americas long ago and genetically differentiable from Old World counterparts.

Introduced to Hawaii

Band-tailed Pigeon
Patagioenas fasciata | BTPI | L 14½" (37 cm)

- **APPEARANCE** Similar in build and general appearance to Rock Pigeon but longer-tailed. Adult told by white band across nape; juvenile paler, lacks white nape band.
- **VOCALIZATIONS** Song a simple low hoot, often doubled: *hooo hoooo*.
- **BEHAVIORS** A woodland species; seen perched unobtrusively in treetops or in powerful flight above the canopy. Tends feeders.
- **POPULATIONS** Paler and smaller in Interior West than in Pacific coast region.

Eurasian Collared-Dove
Streptopelia decaocto | EUCD | L 12½" (32 cm)

- **APPEARANCE** Big and bulky, with ample tail. Eponymous collar absent from juveniles, but all have broadly white-cornered tail. African Collared-Dove, *S. roseogrisea* (also pictured opposite), an occasional escape from captivity, has paler wings and undertail coverts.
- **VOCALIZATIONS** Tripartite song has accent on second syllable: *hoo HOO hoo*, repeated slowly or quickly. In flight, gives harsh growl.
- **BEHAVIORS** Found anywhere there is a bit of infrastructure: roads, railyards, farms, suburbs.
- **POPULATIONS** Indigenous to Old World; has dramatically invaded western N. Amer. in recent decades. Frequently intergrades with domestic strains, resulting in paler progeny.

Spotted Dove
Streptopelia chinensis | SPDO | L 12" (30 cm)
In same genus as Eurasian Collared-Dove (p. 72) and, like that species, introduced from Eurasia. Declining and very scarce in Southern Calif., where formerly much more common; widespread and numerous in Hawaii, mostly lowlands. Big and broad-tailed like collared-dove. Adult has fine white spotting on black nape; nape of juvenile plain, concolor with face and neck. Plumage warmer, more pinkish brown than collared-dove. Song similar to collared-dove's, but harsher and not as syncopated: *cuh cuh cruuuuuhh*, repeated slowly.

Zebra Dove
Geopelia striata | ZEBD | L 8" (20 cm)
Small, slender dove; indigenous to Southeast Asia. Introduced widely to Pacific Ocean islands; widespread and often abundant in Hawaii. Seen in flocks of 100+, but also in much smaller numbers. With long tail and barred plumage, suggests a mainland *Columbina* dove; unlikely to be confused with larger and plainer Mourning (p. 76) and Spotted Doves, also in Hawaii. Note bluish tones, especially around eyes. Song a rapid series of short popping coos, a simple *poo p'poooo* or longer *p'doo b'doo dooo*.

Inca Dove
Columbina inca | INDO | L 8¼" (21 cm)
- **APPEARANCE** Small and long-tailed. Adult heavily scalloped, juvenile even more so. In flight, wings flash rufous, tail boldly white-edged. Juvenile Mourning Dove (p. 76), smaller than adult and variably scalloped, can be mistaken for Inca.
- **VOCALIZATIONS** Song a pair of weakly descending coos, the first harsher than the second, repeated endlessly, even in the hottest parts of the day: *whirlpool ... whirlpool ... whirlpool ....* Also a wing rattle in flight.
- **BEHAVIORS** Has adapted well to human haunts; readily tends lawns, feeders, birdbaths, etc. Although generally slow-moving and inconspicuous, this bird often wanders onto lawns and perches on wires.
- **POPULATIONS** Species expanded north in late 20th century, but that trend has stalled and, in some places, reversed.

Common Ground Dove
Columbina passerina | CGDO | L 6½" (17 cm)
- **APPEARANCE** Our smallest dove, plump and bobtailed. Like Inca Dove, flashes extensive rufous in wings; tail much shorter than Inca's, with limited white. Fine scaling on head and breast; Inca has coarser, more extensive scalloping. Bill reddish; dark on Inca. Gray-brown wings have black blobs, suggesting Mourning. Similar Ruddy Ground Dove (p. 455) is casual to Desert Southwest, especially Ariz. Ruddy told from Common by dark bill (reddish on Common), plain (not scaled) plumage, and grayer face and crown.
- **VOCALIZATIONS** Song a weakly rising, slightly slurred *ooo-wup*, repeated steadily; recalls song of Sora (p. 96). Lacks wing rattle of Inca Dove.
- **BEHAVIORS** Favors agricultural districts with brush and trees; also gets into suburban and urban habitats. Like Inca Dove, easily overlooked although not particularly shy.
- **POPULATIONS** Expanded north in Calif. in 20th century with rise of agriculture, especially avocado farming, but has since withdrawn somewhat. Generally uncommon in West; a few stray well north every fall.

PIGEONS AND DOVES
COLUMBIDAE

White-winged Dove
Zenaida asiatica | WWDO | L 11½" (29 cm)

- **APPEARANCE** Smaller than Eurasian Collared-Dove (p. 72), larger than Mourning Dove. White on upperwings always visible; appears as long crescent on perched bird, bright white patch in flight. Collared-dove in flight, with dark primaries, gray-fawn back, and paler gray coverts, is similar, but contrast stronger on White-winged. Tail from below also similar to collared-dove: black with broad white band at tip, but shorter.
- **VOCALIZATIONS** Song a tetrasyllabic, syncopated *ooo ooo uh ooooo (Who cooks for you?)*; often runs out to 5–10 syllables.
- **BEHAVIORS** Originally a bird of bosques and well-vegetated washes, but has taken well to ranches, towns, even big cities. Along tree-lined streets in desert cities, among the most conspicuous species.
- **POPULATIONS** Expanding north, especially east of our area. Migration complex: Some move south in winter, but many also stay put, and some actually move north after breeding.

Introduced to Hawaii

Mourning Dove
Zenaida macroura | MODO | L 12" (30 cm)

- **APPEARANCE** Slender overall. All but central feathers of long, thin tail tipped white. Wings monochrome gray-tan, except for some irregular black blobs. Young fledge well before attaining adult size; with their scalloped plumage, can suggest smaller species like Inca Dove (p. 74).
- **VOCALIZATIONS** Familiar "mourning" heard much of the year, from first light well into dusk: *ooOOOOh oooh oooh oooh*. In flight, wings whistle airily.
- **BEHAVIORS** Found in many habitats, from austere deserts to major metropolises, but eschews dense forest.
- **POPULATIONS** Competition from expanding Eurasian Collared-Dove (p. 72) worries some, but there is little evidence that the two species interact significantly. Most withdraw from northern Great Plains. Introduced to Hawaii, where widespread but uncommon.

ROADRUNNERS AND CUCKOOS | CUCULIDAE
Numbering about 150 species worldwide, the family Cuculidae is represented in the West by the distinctive Greater Roadrunner and by several cuckoos; the Groove-billed Ani (p. 455) is barely annual. All are in the order Cuculiformes. Our species are slender and lanky, with long tails; they are often associated with dense vegetation.

Greater Roadrunner
Geococcyx californianus | GRRO | L 23" (58 cm)

- **APPEARANCE** Huge and crested. Densely streaked overall, with oily greenish sheen to tail and red, white, and blue "racing stripe" behind eye.
- **VOCALIZATIONS** Does not say *"Meep! Meep!"* Instead, issues a series of low, sonorous coos, descending; faint but far-carrying, heard on still mornings, especially late winter. Also makes snapping sounds with bill.
- **BEHAVIORS** Basks in the morning sun, but is wary and fleet of foot as soon as the day starts to warm up. Trots across shrublands at an impressive clip, frequently stopping to cock its tail and raise its crest.
- **POPULATIONS** Nonmigratory and largely sedentary; at edge of range, numbers crash following harsh winters. Urbanization and high-yield agriculture have driven the species from parts of its former range in Calif.

PIGEONS AND DOVES | ROADRUNNERS AND CUCKOOS
COLUMBIDAE | CUCULIDAE

White-winged Dove

black flight feathers

gray rump

square-cornered tail with broad white border

Mourning Dove

♀

♂

long, intricately patterned tail

juvenile

scaly like Inca Dove

short flights, often sailing to ground before running

Greater Roadrunner

long, iridescent tail with white tips

Common Cuckoo
Cuculus canorus | COCU | L 13" (33 cm)

The cuckoo of cuckoo-clock fame, perhaps the best known harbinger of spring across Eurasia. Wanders to Bering Sea region spring to summer in very small numbers; a few records elsewhere in Alas., plus one-offs to Southern Calif. and Hawaiian Is. Barred plumage and direct flight call to mind a small raptor. Male gray overall; female varies from mostly gray with rufous highlights (gray morph) to heavily rufous (hepatic morph). Familiar "cuckoo" song, delivered slowly, not likely to be heard from birds in our area.

Yellow-billed Cuckoo
Coccyzus americanus | YBCU | L 12" (30 cm)

Accidental to Hawaiian Is.

- **APPEARANCE** Slender overall; tail is long, bill decurved. Yellow bill of adult matches yellow orbital ring. All plumages brownish above and white below, with prominent white spots on undertail; wings flash rufous in flight. Juvenile has dark bill like Black-billed, but usually shows extensive rufous in wings and white in tail like adult Yellow-billed.
- **VOCALIZATIONS** Two main songs: (1) slow, steady series of rather harsh, slightly descending *whoo!* notes; (2) rapid series of clipped *cu!* notes, decelerating to stuttering, clucking notes: *cowlp-cowlp-cowlp.*
- **BEHAVIORS** Seems awkward and hesitant when probing treetop foliage for caterpillars, but is a powerful flier, blasting through clearings with aplomb and migrating every autumn to S. Amer. Sometimes lays eggs in other bird species' nests, a practice called brood parasitism; Common Cuckoo is notorious for this behavior, but it is less common in *Coccyzus* cuckoos.
- **POPULATIONS** Geographic variation fraught. Birds of Desert Southwest, called *occidentalis* but not considered a valid subspecies, are declining breeders in cottonwood forests around waterways; they return extremely late in spring. Breeders farther north in western U.S. represent nominate eastern *americanus*, accepting of more diverse forest types. Accidental to Hawaiian Is.

Black-billed Cuckoo
Coccyzus erythropthalmus | BBCU | L 12" (30 cm)

- **APPEARANCE** Shaped like Yellow-billed, but even slimmer; bill thinner. Dark bill and red orbital ring often hard to see; more useful marks are yellowish wash on breast and reduced white on undertail. Flushing, adult shows little rufous in wing, but some juveniles do, inviting confusion with similar juvenile Yellow-billed, but note smaller white tips to tail feathers.
- **VOCALIZATIONS** Song a series of rapidly uttered *cu* notes: *cu'cu'cu'... cu'cu'cu'... cu'cu'cu'....* Surprisingly, can sound like some songs of unrelated Least Bittern (p. 206).
- **BEHAVIORS** Occasional brood parasite, less so than Yellow-billed. Both of our *Coccyzus* cuckoos are lovers of hairy caterpillars, Black-billed especially so; it is particularly adept at "gutting" caterpillars, avoiding consumption of the most toxic parts of its prey.
- **POPULATIONS** Breeds every year in our region, but erratically and peripherally so, probably to foothills of Canadian Rockies. Status in West in any given year may reflect changing availability of caterpillar prey farther east. Secretive; likely somewhat overlooked here.

> **GOATSUCKERS** | CAPRIMULGIDAE
> These nocturnal insectivores get their odd name from an old legend that they raided farmyards at night to steal milk from goats. They are related to swifts and hummingbirds (pp. 82–95). All have tiny bills but very wide mouths. They are cryptically colored to evade detection by day, and their songs are decidedly strange.

Lesser Nighthawk
Chordeiles acutipennis | LENI | L 8½" (22 cm)
- **APPEARANCE** A bit smaller than Common Nighthawk, and buffier overall. Lesser in flight shows shorter, more rounded wings than Common, appearing more batlike. White bar closer to wing tips; the dark wing tip of Lesser forms an equilateral triangle, the longer wing tip of Common an acute isosceles triangle.
- **VOCALIZATIONS** Song a low, slow, toadlike trill, suggesting tremolo of Eastern Screech-Owl (p. 234), but droning on for 10–30+ seconds. In flight, gives mirthful, maniacal yips and yaps, run together.
- **BEHAVIORS** Widespread in hot lowlands, nesting in open desert; forages, sometimes in large flocks, above waterways. Forages and displays close to the ground; Common often much higher up.
- **POPULATIONS** Migrates earlier in spring and generally later in fall than Common. Strays annually well north of range, with records all the way to Yukon and Alas.

Accidental to Hawaii

Common Nighthawk
Chordeiles minor | CONI | L 9½" (24 cm)
- **APPEARANCE** A bit larger than Lesser Nighthawk, and proportionately longer-winged. Both sexes have prominent white bar across pointed wings, farther from wing tip than on Lesser. Male has white throat and broad white tail band, female does not.
- **VOCALIZATIONS** Nasal call, given in flight, a startling *peeent!* Or *peeezhr!* In display, plunge-dives almost to ground level, pulling up at the last split second, creating "sonic boom" with wings.
- **BEHAVIORS** This is the "Bullbat" that whips high above the lights of ball fields, shopping centers, and even busy downtowns continent-wide; flies erratically, "shifting gears" as it goes. Like all nighthawks, may be discovered roosting by day; assumes horizontal posture on ground, rooftops, and tree boughs.
- **POPULATIONS** Across much of West, a very late spring migrant; also an early "fall" migrant, departing midsummer from many areas. Has declined in recent years; culprits include loss of aerial prey and nest predation by urbanizing crows. Accidental to Hawaii.

Common Poorwill
Phalaenoptilus nuttallii | COPO | L 7¾" (20 cm)
- **APPEARANCE** Small and gray. Flushing, appears bobtailed with short, rounded wings. Often rests on back roads at dusk, where its orange eyeshine gleams in cars' headlights. Sexes vary slightly in tail pattern.
- **VOCALIZATIONS** A sad, haunting *poor-will-ip*, proclaimed repeatedly from shrubby badlands and wooded canyons. Terminal *ip* note much softer, inaudible at a distance.
- **BEHAVIORS** This smallest "will" engages in prolonged periods of torpor, when metabolic activity nearly ceases. Other birds are capable of brief bouts of torpor, but the poorwill takes it to an entirely different level.
- **POPULATIONS** Movements after nesting poorly understood; poorwills in torpor in winter reclusive and, unsurprisingly, hard to find.

GOATSUCKERS
CAPRIMULGIDAE

Buff-collared Nightjar
Antrostomus ridgwayi | BCNI | L 8¾" (22 cm)

■ **APPEARANCE** Intermediate in various respects between Common Poorwill (p. 80) and Mexican Whip-poor-will: larger and browner than the former, with more white in tail of male; smaller and grayer than the latter, with less white in tail of male. Buff collar diagnostic, but beware tricks of light, especially under artificial illumination.

■ **VOCALIZATIONS** 12–15 sharp *pik* notes, accelerating and rising in pitch slightly, then a drop at the end: *plee-pick!*

■ **BEHAVIORS** Found in desert washes with mesquite and acacia. In rare encounters with birds actually seen at night, may be noted perched attentively on snag or outcropping watching for moths.

■ **POPULATIONS** At northern limit of range in Ariz. and, previously, N. Mex. Virtually unrecorded fall to winter, so presumed to migrate after breeding.

Mexican Whip-poor-will
Antrostomus arizonae | MWPW | L 9¾" (25 cm)

■ **APPEARANCE** Large nightjar; about the heft of Common Nighthawk (p. 80), but more compactly built. Longer-tailed than Buff-collared Nightjar; lacks discrete buff ring around the neck of that species, but vague buff collar of whip-poor-will can throw off overzealous searchers of the rarer nightjar.

■ **VOCALIZATIONS** Chanting song has cadence similar to that of its famous eastern counterpart (p. 467), but song burrier; also, each "whip-poor-will" (actually *whip! poor-a-will!*) song element preceded by soft *ip* note.

■ **BEHAVIORS** Habitat a strong, but not foolproof, point of ID: Mexican Whip-poor-will haunts mid-elevation pinewoods, often in the company of Common Nighthawks; other western goatsuckers typically at lower elevations.

■ **POPULATIONS** U.S. breeders migratory. Increased detection in recent years in Calif., Nev., and Colo. may reflect actual range expansion, better searching by birders, or both.

SWIFTS | APODIDAE
The most aerial of all birds, swifts land only at their nest sites, completely hidden from human view. In size and foraging behavior, they bear superficial resemblance to swallows (pp. 310–315), but a close relationship with hummingbirds (pp. 86–95) has long been recognized.

Black Swift
Cypseloides niger | BLSW | L 7¼" (18 cm)

■ **APPEARANCE** Largest and darkest swift in West, but any swift can appear dark and large (or small) at a distance. Tail slightly forked. Juveniles, seen on the wing for only a few weeks in our area, stippled white below.

■ **VOCALIZATIONS** Not a noisy swift, but sometimes gives down-slurred chips, steadily repeated, when departing nest colony.

■ **BEHAVIORS** Nests only in bridal veil falls with special combinations of slope, aspect, and vegetation type; *always* nests at such falls, *never* elsewhere. An unbelievable aerialist, flying extremely fast, foraging at dizzying heights; winters Brazil, where it never lands. The definitive monograph (2010) of this bird is titled, simply, *The Coolest Bird*.

■ **POPULATIONS** Winter range only recently discovered, via geolocator recoveries. Nest sites inaccessible and therefore secure, but declining prey base of airborne arthropods may bode ill for this rare species.

GOATSUCKERS | SWIFTS
CAPRIMULGIDAE | APODIDAE

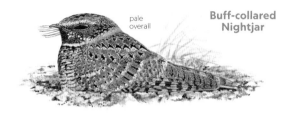

Buff-collared Nightjar

pale overall

less white in tail than on Mexican Whip-poor-will

in both of these species, females have buffier tails than males

Mexican Whip-poor-will

tail long

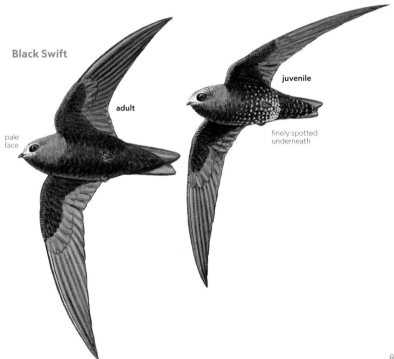

Black Swift

adult

pale face

juvenile

finely spotted underneath

83

Chimney Swift
Chaetura pelagica | CHSW | L 5¼" (13 cm)
- **APPEARANCE** Small swift; more compact than Black (p. 82) and White-throated Swifts. Tubular body and long, arc-shaped wings, swept back, separate this species and all other swifts from swallows (pp. 310–315). Wingbeats of Chimney and Vaux's "twinkle" in a manner very different from the languid wingbeats of swallows (wingbeats of larger swifts also twinkle, but not as much); also glides for short periods of time.
- **VOCALIZATIONS** Sharply down-slurred notes in rapid chatter.
- **BEHAVIORS** Feeds anywhere there is open sky, where it gleans aerial plankton. Nests almost exclusively in human-built structures.
- **POPULATIONS** Basically an eastern swift, yet breeds in cities and towns right up to foothills of Rockies. Status elsewhere in West poorly understood: Field ID of *Chaetura* swifts on migration west of Continental Divide in the Four Corners states fraught; Chimney has summered and even bred in Calif. Declining range-wide due to diminishing prey base of airborne arthropods. Local distribution affected in part by availability of chimneys.

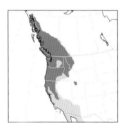

Vaux's Swift
Chaetura vauxi | VASW | L 4¾" (12 cm)
- **APPEARANCE** Western counterpart of Chimney Swift; on smaller Vaux's, rump contrasts more with rest of upperparts, and throat and breast are paler.
- **VOCALIZATIONS** Calls average higher than Chimney's, with chirpy, twangy quality.
- **BEHAVIORS** Routinely nests in natural cavities, unlike Chimney; but like that species, often roosts in artificial structures. May be seen anywhere there is open sky; less tied to towns and cities than Chimney.
- **POPULATIONS** Declining; hit by double whammy of snag removal in forests and decreasing popularity of stone chimneys in new homes.

Mariana Swiftlet 🆔
Aerodramus bartschi | MASW | L 4¼" (11 cm)

Perhaps the strangest avian introduction in the Hawaiian Is.; indigenous to Mariana Is., where endangered. Breeds in a single artificial structure on Oahu; most sightings are from the 'Aiea Loop Trail outside Honolulu. The only swift in the region; a bit smaller than a *Chaetura* swift and darker overall, with forked tail. Distinctive call notes short and scratchy, often in rapid succession; the effect of multiple birds is like typewriters going off in an old newsroom.

White-throated Swift
Aeronautes saxatalis | WTSW | L 6½" (17 cm)
- **APPEARANCE** Midway in size between larger Black Swift and smaller Chimney and Vaux's Swifts; sleek overall and long-tailed. Striking black-and-white plumage, if seen well, differentiates White-throated from all other swifts in region, but any fast-moving swift may appear uniformly colored. Compare with Violet-green Swallow (p. 310), with which it often co-occurs.
- **VOCALIZATIONS** Call, given on the wing, a shrill chatter, the whole series descending slightly in pitch.
- **BEHAVIORS** Nests out of sight on sheer walls of canyons in arid country: foothills, buttes, rimrocks, etc.
- **POPULATIONS** Breeding range has expanded, due in part to acceptance of artificial structures for nesting. Nonbreeding range shifting north in response to warmer winters.

SWIFTS
APODIDAE

Chimney Swift

soaring

rump same color as back

inner primary bulge

Vaux's Swift

pale rump

short-bodied and stubby-tailed

flutters like a bat

Mariana Swiftlet

long, wide forked tail

White-throated Swift

tail often closed in a single point

white flank patches visible from above

HUMMINGBIRDS | TROCHILIDAE

Tiny and hyperactive, hummingbirds are unmistakable with their extremely long bills, brilliant colors, and unique wingbeats. They are staggeringly diverse in northern South America, but only two dozen species have ever been recorded north of Mexico. Southeast Arizona is the epicenter of hummingbird variety in our region, although several species are widespread in the West. Swifts (pp. 82–85) are hummingbirds' closest relatives; together, they make up the order Apodiformes, alluding to their tiny feet.

Rivoli's Hummingbird
Eugenes fulgens | RIHU | L 5¼" (13 cm)

- **APPEARANCE** A magnificent hummingbird, proportionately long-billed and quite large overall. Adult male clad in blacks and dark greens, with paler green gorget and all-dark tail; forecrown purple, but often appears blackish. Adult female washed pale gray-green below, with a bit of white on tail corners; prominent white spot behind eye often continues as blurry white line extending rearward. Juvenile like adult female, but scaly; young male begins to acquire adult traits (purple forecrown, green gorget) by early autumn.
- **VOCALIZATIONS** Call a bright, sharply down-slurred *cheep*, often repeated slowly and steadily.
- **BEHAVIORS** Nests in pine-oak forests, often near streams. Visits feeders, but many go about their business far from prying human eyes.
- **POPULATIONS** U.S. breeders migratory, present late Mar. to late Sept.; a few winter. Not just an Ariz. specialty; breeds regularly well into N. Mex. and sparingly to mountains of West Tex. Accidental to Calif.

Plain-capped Starthroat
Heliomaster constantii | PCST | L 5" (13 cm)

Mid. Amer. hummingbird that breeds regularly almost to Ariz.; nearly annual here, with most records in or near wooded foothills. Rivoli's Hummingbird, Blue-throated Mountain-gem, and this species are the U.S. representatives of the "mountain-gem" clade, Lampornithini, our largest hummers. The starthroat is dull bronze-green overall with bold white marks on face and rump, and very long bill; gorget red in good light, but often appears dull black. Down-slurred call like Rivoli's, but not as bright.

Blue-throated Mountain-gem
Lampornis clemenciae | BTMG | L 5" (13 cm)

- **APPEARANCE** Shorter-billed and more solidly built than Rivoli's. All have dark "ears" (auriculars) bordered above and below with white stripes; tail, large and dark, always has bold white tips. Adult male has sky blue gorget; gray-throated female told by large size, short bill, and well-marked face.
- **VOCALIZATIONS** Call very different from Rivoli's; a short, rising, whistled *wheent*.
- **BEHAVIORS** Wingbeats slower, more fluttery than other western hummers, especially on landing. Tends feeders, but also spends much time in woods; many nest on human structures like bridges and outbuildings.
- **POPULATIONS** Like Rivoli's, at northern limit of range here; most withdraw in winter, although a few remain at lower elevations in Chiricahuas. Regular in summer in very small numbers in Big Bend N.P. and likely breeder in N. Mex.

Lucifer Hummingbird
Calothorax lucifer | LUHU | L 3¾" (10 cm)

■ **APPEARANCE** Small and compactly built, but with bill long and decurved. Long tail of adult male forked, but usually appears narrow and pointed (another name for species is "Lucifer Sheartail"); purple gorget of male elongated. Female and young shorter-tailed, with plain gray throat; note also buff underparts and white stripe on face of female.

■ **VOCALIZATIONS** Call a rapid twitter of sharply down-slurred notes, like a weak Chimney Swift (p. 84).

■ **BEHAVIORS** Desert hummingbird with decided affinity for century plants (genus *Agave*). Choreography of male flight display, performed for female on nest, notably complex.

■ **POPULATIONS** Most in U.S. are found in Big Bend N.P., Mar. to Sept., where a regular breeder. Uncommon in N. Mex. Bootheel; rare but annual to Ariz.

Ruby-throated Hummingbird
Archilochus colubris | RTHU | L 3½" (9 cm)

■ **APPEARANCE** Both *Archilochus* hummingbirds are mostly gray and green, usually lacking buff; compare with genera *Calypte* and *Selasphorus* (pp. 90–93). Ruby-throated is the brighter, greener eastern counterpart of Black-chinned. Adult male Ruby-throated has extensive ruby on gorget; females of the two species very similar, but note brighter green of Ruby-throated, especially on crown. Both sexes a bit shorter-billed than Black-chinned. In good view of perched bird, primaries of Ruby-throated thin and pointed, and longer tail of Ruby-throated extends beyond folded wings.

■ **VOCALIZATIONS** Down-slurred call note, often given in twos and threes, a muffled *choorf* or *chuh*. Also a weak twittering, especially in chases around feeders.

■ **BEHAVIORS** Hovering at flowers or feeders, tends to hold tail straight down and nearly motionless; Black-chinned usually pumps tail up and down while feeding.

■ **POPULATIONS** Barely overlaps with similar Black-chinned, but ranges of both expanding, and hybrids increasingly detected. Very rare vagrant throughout much of West.

Black-chinned Hummingbird
Archilochus alexandri | BCHU | L 3½" (9 cm)

■ **APPEARANCE** Looks washed-out in most views. Marginally longer-billed than Ruby-throated; primaries rounder and broader on Black-chinned, extend nearly to tail tip on perched bird. Purple swatch on adult male's gorget hard to see; except when lit, gorget appears dusky gray-black all over. Gray-green upperparts of female not as bright as on Ruby-throated; crown and face, in particular, appear dull compared to Ruby-throated. Some bright female Ruby-throateds have weak golden tint; female Black-chinned colder green-gray overall.

■ **VOCALIZATIONS** Calls given while feeding and fighting like those of Ruby-throated. Adult male in flight gives dull wing buzz; in arcing display, bottoms out with an endearing tinkling sound.

■ **BEHAVIORS** As hummingbirds go, relatively demure and low-drama. Widespread in canyon country, dry foothills, and residential districts.

■ **POPULATIONS** Breeding range expanding north and east; hybridizing with Ruby-throated in Okla. and perhaps Tex., and hybrids may be noted on migration in our area.

Anna's Hummingbird
Calypte anna | ANHU | L 3¾" (10 cm)
- **APPEARANCE** Medium-stature hummingbird with broad tail and sturdy, straight bill; body mostly green-gray like genus *Archilochus* (p. 88), lacking buff-orange washes of genus *Selasphorus* (pp. 90–93). Adult male has rosy gorget and extensive rosy on rest of head; adult female and immature male have reduced rosy on throat. Most show white "teardrop" behind eye, often prominent on females and immatures. In good photos of perched bird, note very short greater coverts.
- **VOCALIZATIONS** Male song fast, high, and wheezy: *shweewee shreeyee sweeyee* .... Male in display gives piercing *pleek*, recently proven to be created by tail. Both sexes give sharply down-slurred chip calls and variable twittering.
- **BEHAVIORS** Aggressive and conspicuous, even for a hummingbird; abounds in human-modified landscapes, but also flourishes in shrublands coastally and in foothills. Holds tail steady when feeding, unlike Costa's.
- **POPULATIONS** In southern part of range, nests very early, starting in Dec., to take advantage of winter rains. Range expanding rapidly, both breeding and wintering; recently established in decent numbers north in winter to Ketchikan, Alas.

Costa's Hummingbird
Calypte costae | COHU | L 3¼" (8 cm)
- **APPEARANCE** Smaller and more compact than closely related Anna's. Long gorget and much of head of adult male rich purple in good light, but appears black much of the time. Female, especially immature, similar to Black-chinned (p. 88): Costa's shorter-billed; in good photos of perched bird, primary tips of Costa's evenly spaced (inner primaries of Black-chinned bunched up).
- **VOCALIZATIONS** Vocal array complex; sings wheezy song like Anna's, but also gives a rising and falling song with an explosive middle portion too high for many humans to hear. Calls include snappy *tyit*, sharply rising, not falling like many other hummers' calls.
- **BEHAVIORS** A desert hummingbird, but note that Anna's and especially Black-chinned routinely get into such habitat. Like Black-chinned, pumps tail when feeding; Anna's tends not to.
- **POPULATIONS** Movements complex; most depart breeding grounds by late spring, reappearing in fall. Sometimes hybridizes with Anna's.

Calliope Hummingbird
Selasphorus calliope | CAHU | L 3" (8 cm)
- **APPEARANCE** Smallest bird in West and one of the tiniest on Earth. Shape distinctive: bill short and thin; tail short and slight; wings relatively long. Gorget of adult male comprises red-purple rays, sometimes splayed out wildly. Female and young told by pale but extensive buffy wash below, limited rufous in tail, and body shape.
- **VOCALIZATIONS** Befitting its small size, a rather quiet hummer. Call note a soft *tyip*, sharply down-slurred. Song of male includes a scratchy buzz, often interspersed with call notes.
- **BEHAVIORS** Nests in open woodland and around edges of meadows and bogs. Owing to small size and reticence, easily overlooked, but certainly capable of holding its own in disputes at feeders with bigger hummers.
- **POPULATIONS** Fall migration, beginning midsummer, more easterly than spring migration; regular in fall to western Great Plains.

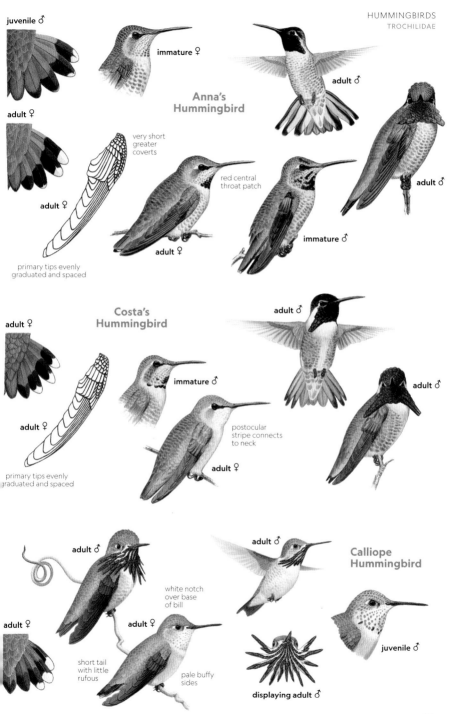

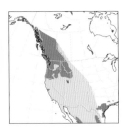

Rufous Hummingbird
Selasphorus rufus | RUHU | L 3½" (9 cm)

■ **APPEARANCE** Medium-small hummingbird of average proportions; very similar to Allen's. Adult male Rufous fiery orange above; male Allen's has mostly green back. Males hovering at feeders can be identified from photographic analysis of exact shape of splayed-out tail feathers: Male Allen's has thin tail feathers; male Rufous has broader tail feathers, with diagnostic notch in next-to-central tail feather. Females and immatures difficult to ID, but tail shape mirrors that of males: Outer tail feathers of female Rufous relatively broad, those of female Allen's narrow.

■ **VOCALIZATIONS** In aggression, a rapid, pentasyllabic *zee! zickety-zoo!* Male in display gives exceedingly harsh *KSHEESH*. When feeding, dull, down-slurred *slik* notes.

■ **BEHAVIORS** Even among hummingbirds, notably pugilistic. Chases off other hummers and even hawks and bears. Flight display of male J-shaped.

■ **POPULATIONS** Very early "fall" migrant, with the first southbound males on the move by June. An elliptical migrant: Southbound, occurs much farther east, regularly to western Great Plains, than northbound.

Allen's Hummingbird
Selasphorus sasin | ALHU | L 3¾" (10 cm)

■ **APPEARANCE** Nearly identical in build to Rufous. Adult male Allen's usually green-backed; adult male Rufous orange-backed, but some have green flecks on back, and a few have extensive green on back. Female and young Allen's and Rufous very hard to ID in the field, but differ in structure of individual tail feathers; see Rufous. Moreover, hybrids, practically impossible to identify, are not rare.

■ **VOCALIZATIONS** Nearly identical to Rufous; chase call may average faster.

■ **BEHAVIORS** Like Rufous, highly aggressive. Flight display of male U-shaped.

■ **POPULATIONS** Nominate *sasin* breeds in and around coastal fog belt; arrives from Mexican wintering grounds by early Jan., departs May to early July. Nonmigratory *sedentarius*, formerly restricted to Channel Is., has established on the mainland and is expanding across Southern Calif.

Broad-tailed Hummingbird
Selasphorus platycercus | BTHU | L 3¾" (10 cm)

■ **APPEARANCE** Built like Rufous and Allen's, but bigger-tailed and slightly larger overall. Adult male, mostly green above, has brilliant rosy gorget. Female told from female Rufous and Allen's by less orange in tail and only faint rufous below; compare also with smaller, shorter-tailed, slighter-billed female Calliope (p. 90).

■ **VOCALIZATIONS** Nonvocal wing trill, once learned, is one of the most characteristic sounds of mountain forests in interior. Many hummingbirds have wing trills, but Broad-tailed is notably low, loud, and ringing. Otherwise, chips and sputters like other hummers.

■ **BEHAVIORS** Adult males enliven mountain meadows and forest edges with frequent display flights: high dives preceded by impressive hovering.

■ **POPULATIONS** A hummer of the Interior West; sightings west of the Cascade-Sierra axis require careful documentation. Spring migrants, especially males, have been returning earlier in recent years.

Broad-billed Hummingbird
Cynanthus latirostris | BBIH | L 3¾" (10 cm)

■ APPEARANCE Medium-stature hummingbird with ample tail, shallowly forked; bill long and variably reddish. Adult male, with bill mostly red, is green overall with blue bib; belly and undertail coverts white. Female told by dusky gray below, white stripe behind eye, and at least some red on bill; compare with White-eared Hummingbird. Immature male intermediate in plumage between female and adult male.

■ VOCALIZATIONS Call a husky *ch'ch'* like Ruby-crowned Kinglet's (p. 322); also a faster chatter. In display, male gives a zinging buzz, created by wings.

■ BEHAVIORS Found mostly around foothills and lower-altitude canyons, especially where sycamores and cottonwoods predominate; also in urban habitats with flowers and feeders. Male display a shallow U or pendulum; may engage in some amount of communal courtship called lekking, best known in the West in some grouse (see pp. 60–63).

■ POPULATIONS At northern limit of range in West and mostly migratory here. Males arrive mid-Mar., females a bit later; depart around mid-Aug. Increasing and now present year-round in metro Tucson; a few winter elsewhere, especially around warmer canyon bottoms.

White-eared Hummingbird
Basilinna leucotis | WEHU | L 3¾" (10 cm)

Mid. Amer. hummingbird whose range barely reaches Ariz., where annual but rare in summer. Similar in size, shape, and general color scheme to much more common Broad-billed Hummingbird; bill reddish like Broad-billed's but shorter. Head of adult male blackish with bold white stripe behind eye. Black "ears" (auriculars) of female White-eared contrast with bold white stripe behind eye; female Broad-billed not as boldly marked. Song of male an endlessly repeated snappy and rapping note, often doubled.

Violet-crowned Hummingbird
Ramosomyia violiceps | VCHU | L 4¼" (11 cm)

■ APPEARANCE Medium-large hummer with unusual color scheme and long bill; sexes alike. Adult completely white below with crown violet and rest of upperparts bronze-olive; bill mostly red. Young similar to adult, but plumage duller above and scalier; bill not as red.

■ VOCALIZATIONS Issues sharply down-slurred *tsup* notes like other hummers, but also gives thin, squeaky notes: *tyeee, nyeee*, etc.

■ BEHAVIORS Nests in sunny foothills and canyon mouths where Arizona sycamore, *Platanus wrightii*, abounds.

■ POPULATIONS U.S. birds mostly migratory, with spring migration relatively late; winter records increasing.

Berylline Hummingbird
Saucerottia beryllina | BEHU | L 4" (10 cm)

Like somewhat smaller White-eared, mostly resident in Mid. Amer.; even rarer in U.S. than White-eared. Sexes and ages fairly similar. Head, back, and underparts mostly green, darker on male than female; bill, slightly decurved, shows red tones in good light. Coppery wings prominent in flight; tail dark overall, a mosaic of rufous and violet. Barely annual here in summer at lower elevations in mountain forests, with most records Southeast Ariz.; has bred. Song of male a short, shrill buzz, steadily repeated.

RAILS, GALLINULES, AND COOTS | RALLIDAE

With their plump bodies, ambling gaits, sturdy legs, bobbed tails, and general disinclination to take flight, these birds bear superficial resemblance to chickens. But they are only distantly related to the upland fowl (pp. 52-67), and all are mostly or wholly aquatic. Although rarely seen in flight, many are migratory, and some are prone to impressive vagrancy.

Ridgway's Rail
Rallus obsoletus | RIRA | L 15½" (39 cm)

■ **APPEARANCE** Orangey-gray with long, drooping bill; largest rail in West. Like an oversize Virginia Rail, but paler and grayer. Three subspecies: nominate *obsoletus* (San Francisco Bay); darker *levipes* (southern coastal Calif.); paler *yumanensis* (Ariz., Nev., interior Calif.).

■ **VOCALIZATIONS** Short, harsh, scratchy notes, repeated slowly and steadily: CH ... CH ... CH ... CH .... In excitement, a faster delivery.

■ **BEHAVIORS** Found only in wetlands: *obsoletus* and *levipes* in cordgrass, *yumanensis* in cattails. A generalist carnivore, but with a particular taste for crustaceans.

■ **POPULATIONS** Nonmigratory; all three subspecies imperiled by loss of marshland habitats. Recently split from Clapper Rail (p. 467) of Atlantic coast and Gulf Coast.

Virginia Rail
Rallus limicola | VIRA | L 9½" (24 cm)

■ **APPEARANCE** In most of West, the only long-billed rail. Adult dark rufous overall, with face gray (orangey on Ridgway's) and flanks barred black and white; Ridgway's larger, paler, and grayer than Virginia. Downy young small and black; compare with Black Rail (p. 100).

■ **VOCALIZATIONS** Song an irregular series of *k'dick* notes, sometimes with terminal *kjurrr*; can suggest song of Black Rail and even one of the Red-winged Blackbird's (p. 418) many calls. Call a series of piglike grunts, each note descending; has taunting, derisive quality.

■ **BEHAVIORS** Common but overlooked in marshes large and small, mostly freshwater but can occur in brackish conditions alongside Ridgway's. Sometimes wanders into the open, creeping slowly or even standing still, then quickly darting back into cover.

■ **POPULATIONS** Migratory across much of West, but a few hang on well north in interior where seeps remain unfrozen in winter.

Accidental to Hawaii

Sora
Porzana carolina | SORA | L 8¾" (22 cm)

■ **APPEARANCE** Bill short and thick-based. Adult has black on face and breast, contrasting with otherwise gray plumage. Juvenile buffier overall, similar to elusive Yellow Rail (p. 100); juvenile Sora plainer above, lacking bold, gold lengthwise stripes and fine white lateral barring of Yellow Rail.

■ **VOCALIZATIONS** Song a low, rising whistle, *oooh-whee! (so-ra!)*. Call a long series of short, sharp whistles, descending in pitch after the first few notes; slows down and drags out toward end.

■ **BEHAVIORS** Often found in company of Virginia Rail, but accepting of even smaller wetlands; makes ready use of agricultural habitats like rice fields and wet hay meadows.

■ **POPULATIONS** More migratory than Virginia, with all withdrawing in winter from northern part of breeding range. Most abundant rail on the continent; accidental to Hawaii.

RAILS, GALLINULES, AND COOTS
RALLIDAE

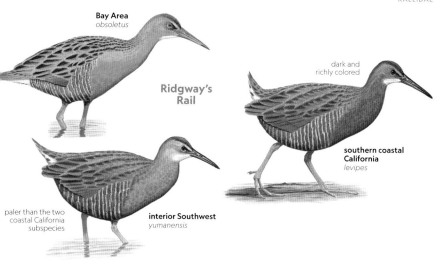

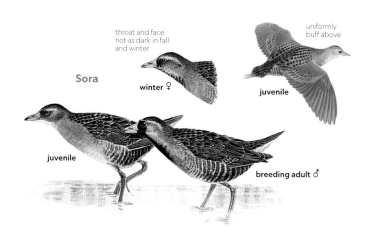

Endemic Hawaiian subspecies occurs on Kauai and Oahu

Common Gallinule
Gallinula galeata | COGA | L 14" (36 cm)

■ **APPEARANCE** More boldly patterned and a bit smaller than coots. All plumages dark slate and brown in good light, appearing black in most views. Breeding adult has red frontal shield, red bill with yellow tip, and straw-yellow legs with red "thighs"; bare parts duller in winter. Tarsi of Hawaiian subspecies *sandvicensis* more reddish. Immature grayer; all plumages show extensive white under tail and thin white striping on flanks.

■ **VOCALIZATIONS** Varied monosyllabic utterances, squeaky and nasal; given singly, *penk*, or in decelerating series, *p' p' p' peh peh peh penk penk penk* ....

■ **BEHAVIORS** Compared to other rallids (pp. 96–101) with the name rail, gallinules and coots are much more likely to be out in the open; this species has a penchant for freshwater marshes with pond lilies, duckweed, and other floating plants.

■ **POPULATIONS** Has undergone multiple name changes and taxonomic revisions in recent years; *G. galeata* refers to a species found only in the New World. Hawaiian subspecies, formerly more widespread, currently restricted to Kauai and Oahu.

Endemic; breeds on all main islands

Hawaiian Coot
Fulica alai | HACO | L 15" (38 cm)

■ **APPEARANCE** Coots are slate black and rotund with huge feet; their lobed toes are oversize and implausibly gray-green. Hawaiian Coot differs from slightly larger American Coot most noticeably in shape and color of frontal shield: usually all-white on Hawaiian and more extensive, extending farther rearward up forehead. Some Hawaiians have red knob like American, and a few Americans lack red knob like Hawaiian.

■ **VOCALIZATIONS** Little documented; the handful of published recordings reveal no differences from calls of American. More study needed.

■ **BEHAVIORS** Generalist denizen of diverse wetlands: Occurs singly or in small, loose flocks in natural settings like coastal lagoons and inland swamps, as well as at sewage treatment ponds and resort ponds. Swimming, both coots pump their heads jerkily.

■ **POPULATIONS** Classified as near threatened by the IUCN, but faring better than most other listed species in Hawaii; breeds on all the large islands, and disperses infrequently to Northwestern Hawaiian Is. Wanders among islands in response to drought and rainfall.

Accidental to Hawaii

American Coot
Fulica americana | AMCO | L 15½" (39 cm)

■ **APPEARANCE** Like Hawaiian Coot, a blackish ball of a bird. Adults are ivory-billed with a red knob on the white frontal shield, reduced in winter. Downy young are black with fiery red-orange highlights; they transition slowly to more adultlike proportions and plumage by fall, but are paler overall.

■ **VOCALIZATIONS** Outbursts, *frup* and *frap*, with undeniable flatulent quality. Also higher squeals, rising, often in series.

■ **BEHAVIORS** Adaptable; aggressive. Adept at both dabbling and diving; also grazes on berms, ball fields, and golf course fairways. Nests mostly in freshwater settings; in winter, perfectly accepting of brackish situations coastally.

■ **POPULATIONS** Breeding range expanding westward in western Canada. Accidental to Hawaii, where similar Hawaiian Coot occurs.

RAILS, GALLINULES, AND COOTS
RALLIDAE

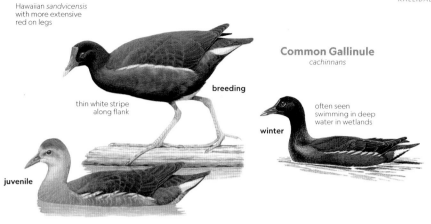

Common Gallinule
cachinnans

Hawaiian Coot

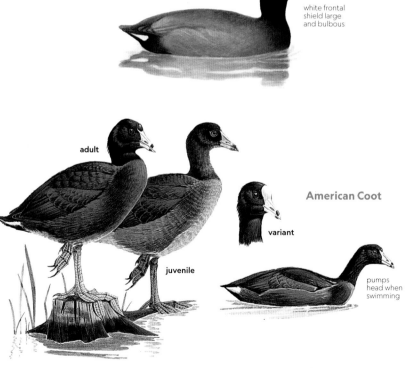

American Coot

Yellow Rail
Coturnicops noveboracensis | YERA | 7¼" (18 cm)
- **APPEARANCE** Rarely glimpsed; flushing, flashes white in secondaries. Sora (p. 96) is larger with plainer back; Yellow Rail has yellow lengthwise stripes and thin white transverse barring on back.
- **VOCALIZATIONS** At night and on cloudy days, an arrhythmic clicking, *tik ... tik ... tik-tik ... tik ... tik-tik ... tik ... tik ... tik ...*, like two stones clicked together.
- **BEHAVIORS** Notoriously elusive. Although capable of swimming, this rail nests and feeds in damp meadows with sedges, eschewing deeper cattail marshes.
- **POPULATIONS** Breeding range extends across Canada west to B.C.; elsewhere in West, occurs in small numbers in summer in such far-flung places as northeastern Calif., southwestern Mont., and south-central Colo.

Black Rail (EN)
Laterallus jamaicensis | BLRA | L 6" (15 cm)
- **APPEARANCE** Our smallest rail, the size of a large sparrow. Dark and slight, with a small bill; in good view, difficult to attain, note white speckles above and rich chestnut on nape. Recently fledged young of all rails are small and black; beware confusion with chicks of more common rail species, like Sora and Virginia Rail (both on p. 96).
- **VOCALIZATIONS** Two or three piping notes in rapid succession, followed by a lower, harsher whistle: *ki'ki'JJrrrr* or *ki'ki'ki'JJrrrr*.
- **BEHAVIORS** Nocturnal; loath to emerge from dense vegetation, although high tides sometimes push displaced birds into view.
- **POPULATIONS** Range in West disjunct; breeders in Ariz., Nev., and Colo., discovered only in later decades of 20th century, may be refugees escaping rising sea levels.

CRANES | GRUIDAE
Although they are popularly confused with herons (pp. 206–213), cranes are related to rails (pp. 96–101). They number only 15 species worldwide, but they are widespread and figure prominently in human culture. Cranes are tall like herons, yet share similarities with their smaller rallid relatives in gait, calls, overall body plan, and general habitat requirements.

Accidental to Hawaii

Sandhill Crane
Antigone canadensis | SACR | L 41–48" (104–122 cm) WS 73–84" (185–213 cm)
- **APPEARANCE** Nearly as tall as Great Blue Heron (p. 206) and similarly gray overall, often stained with rust. Cranes fly with necks outstretched and often soar; herons do neither. Adult Sandhill has red crown; immature plain-crowned with extensive rust in plumage. All have "bustle" of long tertials. Size variable: "Greater Sandhill" longer-legged, longer-necked, and larger overall than "Lesser"; wing tips of latter darker and proportionately longer.
- **VOCALIZATIONS** Rich, rolling *g'r'r'roo*, given on the wing or by pairs prancing and capering in agricultural fields; urgent and wild-sounding, carries tremendous distances. A shrill, reedy whistle is given in flight, apparently only by young birds.
- **BEHAVIORS** Nests on tundra and in wet meadows; in winter and during migratory stopovers, gathers on farmland in large, dense flocks. Birds on passage fly at dizzying heights, calling powerfully as they go over.
- **POPULATIONS** "Lessers" breed mostly northern Canada and Alas.; some even reach Russia. "Greaters" breed mostly U.S. and southern Canada. Accidental to Hawaii.

RAILS, GALLINULES, AND COOTS | CRANES
RALLIDAE | GRUIDAE

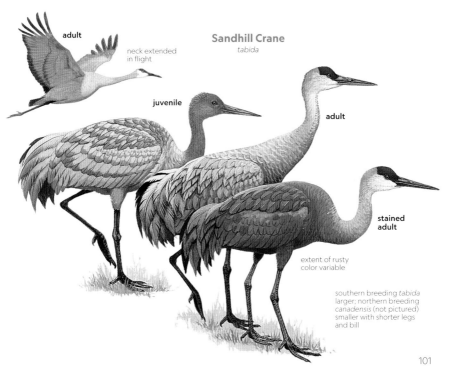

STILTS AND AVOCETS | RECURVIROSTRIDAE

Along with the oystercatchers (this page), plovers (pp. 104–109), and sandpipers (pp. 110–133), stilts and avocets are shorebirds: generally long-billed probers of mud and shallow water. Recurvirostrids, numbering less than a dozen species globally, are thin-billed and are among the most spectacular of all shorebirds.

Hawaiian subspecies widespread on main islands

Black-necked Stilt
Himantopus mexicanus | BNST | L 14" (36 cm)
- **APPEARANCE** Bright pink legs very long; straight bill exceedingly thin. Adult male jet-black above, snow-white below; female a bit browner above. Face of Hawaiian *knudseni* more extensively dark. All juveniles have buff feather edges to dark brown back.
- **VOCALIZATIONS** An explosive *kek,* often repeated in long series.
- **BEHAVIORS** Opportunistic; haunts sewage lagoons, evaporation ponds, and naturally hypersaline marshes; also coastal lagoons in winter.
- **POPULATIONS** Hawaiian *knudseni* endangered but recovering. Mainland birds have established new wintering grounds and breeding colonies in recent decades.

Accidental to Hawaii

American Avocet
Recurvirostra americana | AMAV | L 18" (46 cm)
- **APPEARANCE** Head, neck, and breast of breeding adult pastel orange; wings and back black and white. Legs long and ghostly gray-blue. Orange fades by winter; some are strikingly black and white. Female's bill more sharply upturned than male's.
- **VOCALIZATIONS** Sharp *peet,* less nasal than Black-necked Stilt's note, sometimes in frenzied series.
- **BEHAVIORS** Feeds on mudflats, but also in deeper water; often swims. Avocets, like stilts, are aggressive around nests.
- **POPULATIONS** Accidental in winter to Hawaii; breeding and wintering ranges have expanded in recent decades on mainland.

OYSTERCATCHERS | HAEMATOPODIDAE

These heavyset shorebirds have brilliantly colored bills that are laterally compressed for prying open bivalve shells; the stout bills can also be used to hammer right into the shells.

American Oystercatcher
Haematopus palliatus | AMOY | L 18½" (47 cm)

Unlike all-dark Black Oystercatcher, shows extensive white below; in flight, wings and tail strikingly pied. Barely reaches coastal Southern Calif. from Mexico. Most birds in our area—typically long-staying adults at jetties—show introgression with Black Oystercatcher: blotchy black on belly and black flecking on vent.

Black Oystercatcher
Haematopus bachmani | BLOY | L 17½" (44 cm)
- **APPEARANCE** Hulking like American Oystercatcher, but slightly smaller overall. Bill fiery orange; otherwise dull black and dark brown.
- **VOCALIZATIONS** Shrill peeps, usually in series; also a longer, wavering whistle.
- **BEHAVIORS** Part of the West Coast "rockpiper" guild, roosting and foraging on surf-sprayed jetties and sea rocks.
- **POPULATIONS** Resident throughout entire range, but many from B.C. northward concentrate at sheltered sites sometimes far from breeding areas.

PLOVERS | CHARADRIIDAE

These shorebirds have a distinctive run-stop-run gait: They trot quickly, then abruptly pause, then trot again. Most are patterned black and white below, at least in breeding plumage.

Regular to Hawaiian Is.

Black-bellied Plover
Pluvialis squatarola | BBPL | L 11½" (29 cm)

■ **APPEARANCE** Our largest plover; heavyset and block-headed, with stout bill. Breeding adult black below and extensively white above. Nonbreeding adult and juvenile plain gray overall, with blank face and large staring eye; nonbreeding American and Pacific Golden-Plovers show dark cap. In flight, all plumages of Black-bellied high-contrast; note white rump, bold white wing stripe, and especially black axillaries ("wingpits").
■ **VOCALIZATIONS** Flight call a rich, slurred whistle, *uhweeee-ee* or *weee-eee*.
■ **BEHAVIORS** Nests on Arctic tundra, both dry and wet; winters mostly coastally, right along seashore but also in protected lagoons inland a few miles. Roosting flocks bunch up tight, staying motionless.
■ **POPULATIONS** Generally scarce migrant inland in West, but flocks of hundreds put down in spring in western Great Basin. Rare but regular to Hawaiian Is.

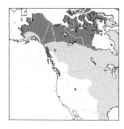

American Golden-Plover
Pluvialis dominica | AMGP | L 10¼" (26 cm)

■ **APPEARANCE** Breeding adults of both golden-plovers are spangled golden above; American more uniformly black below than Pacific. In drabber winter plumage, compare structure: American longer-winged, shorter-legged, and slighter-billed than Pacific. American also averages less bright in winter plumage than Pacific. Both golden-plovers are low-contrast in flight, lacking black-and-white patterning of larger-billed, bigger Black-bellied Plover.
■ **VOCALIZATIONS** Flight call an abrupt *queedy-quee* or *queedy*, shorter and choppier than Black-bellied.
■ **BEHAVIORS** Nests on rocky tundra with low-stature vegetation. Migrants in West most frequently detected around waterbodies, but typically away from standing water.
■ **POPULATIONS** Scarce on migration in much of West, mostly fall and almost entirely juveniles; bulk of passage through Midwest (spring) and over Atlantic Ocean (fall).

Regular in Hawaii

Pacific Golden-Plover
Pluvialis fulva | PAGP | L 9¾" (25 cm)

■ **APPEARANCE** Differs structurally from American Golden-Plover: Pacific, a bit smaller overall, has slightly shorter wings, slighter longer legs, and slightly larger bill. Breeding adult Pacific has splotchy white on flanks, vent, and undertail coverts; breeding American more uniformly black below. Compare also with nonbreeding Black-bellied, which is larger overall and bigger-billed, with high-contrast flight pattern but plain face lacking both pale eyebrow and dark cap.
■ **VOCALIZATIONS** Flight call a powerful, rising *weee* or *w'weee*.
■ **BEHAVIORS** Like American, nests on tundra, but Pacific favors lower, wetter sites with more vegetation. Habituates to humans on Hawaiian wintering grounds, often at parks and resorts, even yards and curbs.
■ **POPULATIONS** A few winter in coastal Calif.; extremely rare east of Cascade-Sierra axis. Common in Hawaii Oct. to Apr., especially at low elevations.

PLOVERS
CHARADRIIDAE

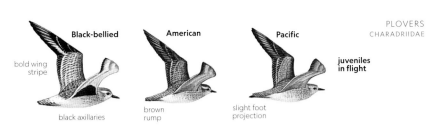

Black-bellied Plover

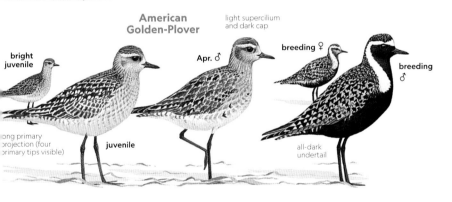

American Golden-Plover

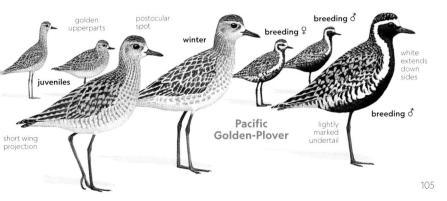

Pacific Golden-Plover

Casual to Hawaii

Killdeer

Charadrius vociferus | KILL | L 10½" (27 cm)

■ **APPEARANCE** A large, long-tailed plover. Has two broad, black breast-bands; our other *Charadrius* plovers have one, often broken, or none. Tail mostly rufous, conspicuous in flight. Just-fledged young, with single breastband, much smaller than adult; compare with Semipalmated Plover.

■ **VOCALIZATIONS** In agitation, a frequent state with this species, varied whistles: *deee* and *k'deee (kill-deer)* and *dee d'deee*. Especially when defending young, gives low, steady trill. Flight call, heard anytime day or night, a rising *deeeEEE*. Frequently imitated by European Starling (p. 340); reports of birds heard, but not seen, out of range in winter could be starlings.

■ **BEHAVIORS** Although perfectly at ease on mudflats and sandbars, is just as likely to be found in pastures and on golf courses, even around parking lots and construction sites. Near nest sites, engages in dramatic distraction displays, in which adult feigns a broken wing, luring dogs, dog-walkers, and others away from eggs.

■ **POPULATIONS** Hardy; lingers late into autumn, one of the first species to return in spring. Casual to Hawaii.

Common Ringed Plover

Charadrius hiaticula | CRPL | L 7½" (19 cm)

Old World counterpart of Semipalmated Plover; Common Ringed breeds widely at high latitudes, reaching eastern Canadian Arctic via Greenland. In the West, breeds in small numbers on St. Lawrence I., Alas.; very rare but annual migrant western Bering Sea region, casual south along coast to Southern Calif., and accidental to Hawaii. Differs from Semipalmated by broader black breastband, broader white supercilium, less colorful orbital ring, and slightly longer bill; flight call a flat-affect *tooee* or *tooip*, with little pitch change, lower-pitched and less strident than Semipalmated's. Away from Bering Sea region, records in the West of Common Ringed probably require photo or audio support.

Annual to Hawaii

Semipalmated Plover

Charadrius semipalmatus | SEPL | L 7¼" (18 cm)

■ **APPEARANCE** Like a subcompact, stripped-down Killdeer: shorter-tailed and notably smaller, with a single breastband. Beware recently fledged Killdeer, much smaller than adult, with single breastband; fledgling Killdeer notably long-legged and downy, and already long-tailed. Bill and feet of breeding adult Semipalmated orange; duskier in winter adult, even duskier in juvenile. Darker-backed than similarly proportioned Snowy and Piping Plovers (p. 108). Compare with Common Ringed Plover.

■ **VOCALIZATIONS** Flight call a strident *ch'wee!* or *chew-awee!* Spectrogram shows much more frequency sweep, typically falling then rising, than in Common Ringed.

■ **BEHAVIORS** Nests in the open, especially on tundra with loose gravel. On migration, fans out in loose, usually small flocks on mudflats. Away from breeding grounds, less given to drama than Killdeer.

■ **POPULATIONS** Widespread on migration inland as well as coastally; across much of the Interior West, it is the small plover most likely to be encountered. Annual in very small numbers to Hawaii.

PLOVERS
CHARADRIIDAE

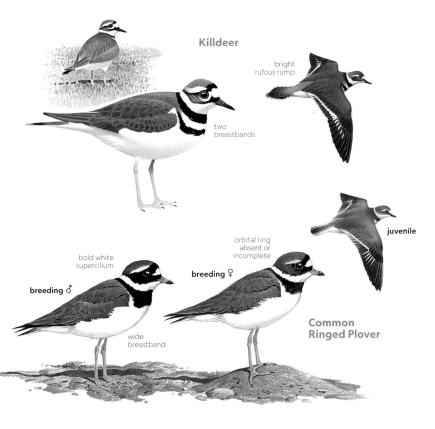

Piping Plover
Charadrius melodus | PIPL | L 7¼" (18 cm)

- **APPEARANCE** Thicker-billed, a bit larger overall, and paler above than Snowy Plover. Legs and bill yellow-orange breeding; nonbreeding adult and juvenile have black bill. Lacks dark ear patch of Snowy and has thinner breast patch.
- **VOCALIZATIONS** Flight call richer and lower than Semipalmated's (p. 106), a husky *julep* or *pieplow*, with second element lower than first.
- **BEHAVIORS** Lays eggs on exposed ground near lakeshores or large rivers; sensitive to disturbance when incubating. Migrants in West usually at mudflats where other shorebirds gather.
- **POPULATIONS** Declining range-wide, although strict protection in Colo. has permitted slow increase there. Breeders in our area pertain to interior *circumcinctus*; winters Gulf Coast, wanders less than annually to West Coast.

Lesser Sand-Plover
Charadrius mongolus | LSAP | L 7½" (19 cm)

East Asian plover that regularly reaches Bering Sea region and has bred mainland Alas. Casual in fall to U.S. West Coast, accidental to Hawaiian Is. About same length as Semipalmated Plover, but more bulked up. Bill thick and black. Breeding adult, with bright rusty color on breast and head, unique in West, but nonbreeders and juveniles trickier: gray above with fine scalloping; white below with gray bulge on sides of breast. Two flight calls, often interspersed: a short rattle and a rich, rising whistle.

Mountain Plover
Charadrius montanus | MOPL | L 9" (23 cm)

- **APPEARANCE** Among our *Charadrius* plovers, only Killdeer (p. 106) is larger. In all plumages, faded tan above, whitish below; essentially unbanded, although some adults show buffy smudge across breast. Black bill is long and thin; note small black patches on head.
- **VOCALIZATIONS** Calls harsh and descending, often short and clipped, *krr*, sometimes drawn-out, *kyeeurrr*; ternlike.
- **BEHAVIORS** Generally eschews aquatic habitats. Nests on disturbed shortgrass prairie far from water; in winter, occurs on barren flats near or far from coasts and lakeshores.
- **POPULATIONS** Numbers declining, range contracting. After breeding, stages in a few large flocks on western Great Plains, then disperses widely to wintering grounds.

Snowy Plover
Charadrius nivosus | SNPL | L 6¼" (16 cm)

- **APPEARANCE** A bit smaller than Piping Plover, almost as pale; thin-billed. In side view, Snowy has three dark patches: atop forehead, behind eye, and along sides of upper breast. Legs variably grayish; bill black. Piping has orange legs year-round and thicker bill, orange in breeding season but mostly black in winter.
- **VOCALIZATIONS** Flight call a rising, wavering whistle, *churwee* or *churawee*. Also gives a gruff *grit* or *grurt*, in agitation or in flight.
- **BEHAVIORS** Makes no pretense at all to nest-building; lays eggs right out in the open on alkaline flats and beaches with virtually no vegetation. Migrants on inland stopovers often find their way to particularly hypersaline microhabitats.
- **POPULATIONS** Declining across U.S. range. West Coast and interior breeders both nominate subspecies *nivosus*, listed as threatened.

SANDPIPERS | SCOLOPACIDAE

These are the quintessential shorebirds: flocky and sometimes frenetic, with long bills and gray-brown plumages. Most are powerful migrants, and many are threatened by habitat loss. With close to 100 species globally, the sandpipers are particularly well represented in the West, with a gratifying mix of vagrants (Appendix A) and regular breeders.

Upland Sandpiper
Bartramia longicauda | UPSA | L 12" (30 cm)
- **APPEARANCE** Small, short-billed, thin-necked curlew; potbellied, a pinhead. Tail and wings long; feet and bill yellow.
- **VOCALIZATIONS** Flight song a rough chatter followed by a rising then falling wail: *b'b'b'b'b'b'b'b'wheeEEE-wheEEEEEEeeeooo*. Flight call a rapid, far-carrying stutter, *quit-quit-quit* or *quiddy-quit-quit*.
- **BEHAVIORS** Least aquatic of all sandpipers; nests on open prairie and tundra. Flight direct and powerful.
- **POPULATIONS** Declining; formerly hunted in excess, but current threats are habitat destruction on Great Plains and climate change farther north.

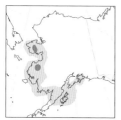

Bristle-thighed Curlew
Numenius tahitiensis | BTCU | L 18" (46 cm)
- **APPEARANCE** Distinguished from grayer, slightly smaller Whimbrel by bright buff tail; the eponymous bristles are hard to see.
- **VOCALIZATIONS** Flight call a two-note whistle, rich and low: *wlee-lyoo*.
- **BEHAVIORS** Nests in hilly meadows with low shrubs; winters along seashores. On wintering grounds, becomes flightless while molting.
- **POPULATIONS** Breeds only mainland western Alas. Fantastic migrant, wintering to islands in South Pacific. Regular to Hawaii fall through spring; casual to U.S. West Coast, especially spring.

Regular to Hawaii

Whimbrel
Numenius phaeopus | WHIM | L 17½" (44 cm)
- **APPEARANCE** A midsize curlew and quite large sandpiper; long bill droops sharply at tip. Cold gray-brown overall; legs blue-gray. Black crown has thin white median stripe; black eyeline separated from crown by pale eyebrow.
- **VOCALIZATIONS** Flight call a rich piping in rapid succession: *peep! peep! peep! peep! peep! peep!*
- **BEHAVIORS** Nests on open, vegetated tundra and on wetter taiga with stunted conifers. Winters coastally in salt marshes and on rocky beaches.
- **POPULATIONS** Breeding range in West expanding eastward; annual to Hawaiian Is. Asian subspecies *variegatus*, rare migrant to Bering Sea region and casual to Hawaiian Is., has whiter rump.

Annual to Hawaiian Is.

Long-billed Curlew
Numenius americanus | LBCU | L 23" (58 cm)
- **APPEARANCE** A colossal sandpiper; bill astonishingly long. Legs blue-gray. Warmer and paler overall than smaller Whimbrel, with plain face. Compare with Marbled Godwit (p. 112).
- **VOCALIZATIONS** Powerful flight call an urgent whistle that ends abruptly on a higher note: *eeeee-EE* or *ooeeee-EE (cur-lew!)*.
- **BEHAVIORS** Often nests far from water, in sandhills, shortgrass prairie, and even open juniper woodlands. In winter, more aquatic, with many coastal; also in agricultural fields.
- **POPULATIONS** Western breeding range contracting; loss of indigenous grasslands, accelerating with climate change, the gravest threat.

SANDPIPERS
SCOLOPACIDAE

Upland Sandpiper
- juvenile
- large, dark eye
- pin-headed appearance
- adult
- juvenile

Whimbrel
hudsonicus
- American *hudsonicus*
- light underwing
- Asian *variegatus*
- pale rump
- prominent stripes on head
- juvenile
- adult

Bristle-thighed Curlew
- adult
- warmer overall than Whimbrel
- bristles difficult to see
- adult

Long-billed Curlew
- juvenile ♂
- bill length variable; shortest in juvenile male
- adult ♀
- blue-gray legs
- adult
- cinnamon underwings

Annual to Hawaii

Bar-tailed Godwit
Limosa lapponica | BTGO | L 16" (41 cm)

■ **APPEARANCE** Large sandpiper with long, slightly upturned bill. Breeding male extensively rufous; suggests an oversize dowitcher (p. 126). Juvenile, nonbreeding male, and female year-round have little or no rufous. All plumages show low-contrast gray wings in flight; tail also gray, with eponymous barring.

■ **VOCALIZATIONS** Flight call a rich piping note, gull-like, singly or in rapid succession. In display on breeding grounds, a series of slurred whistles: *uh-REE-you, uh-REE-you, uh-REE-you* ....

■ **BEHAVIORS** Nests on open, damp tundra, where highly animated and aggressive. An astounding migrant, flying nonstop farther than any other bird; in 2022, a juvenile, only months old, flew from Alas. to Tasmania, a distance of 8,435 miles (13,575 km), without stopping once.

■ **POPULATIONS** Western breeders pertain to subspecies *baueri*, the most migratory of all. Rare but annual to Hawaii and West Coast, mostly fall; accidental in Interior West.

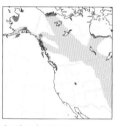

Accidental to Hawaii

Hudsonian Godwit
Limosa haemastica | HUGO | L 15½" (39 cm)

■ **APPEARANCE** The smallest godwit, but still a quite large sandpiper. Rufous of breeding male darker and less extensive than on Bar-tailed; other plumages mostly gray. All plumages distinctive in flight: upperwing flashes white, underwing coverts black; tail has broad black tip and bright white base. Broader-winged Willet (p. 130) also flashes black and white in flight, but lacks broad black tail tip and has straight, unicolored, shorter bill.

■ **VOCALIZATIONS** Flight call a squeaky *pip* or *pyep*, wimpier than Bar-tailed's. Display song like Bar-tailed's, but averages thinner, higher.

■ **BEHAVIORS** Builds nests in less open habitats than Bar-tailed; Hudsonian prefers taiga with medium-stature trees for perches.

■ **POPULATIONS** Breeding range in West strangely disjunct. Migration mostly to our east, although regular in spring in small numbers at eastern edge of our area. Casual elsewhere away from breeding grounds in West; accidental to Hawaii.

Accidental to Hawaii

Marbled Godwit
Limosa fedoa | MAGO | L 18" (46 cm)

■ **APPEARANCE** The largest godwit, about same length and heft as Whimbrel (p. 110). Buffy tones suggest Long-billed Curlew (p. 110), but bill shape different: slightly upturned on godwit, strongly decurved on curlew. Legs black; curlew's legs blue-gray. Breeding adult shows fine dark barring all over; juvenile and nonbreeding adult more washed-out, buffier. Compare with smaller, straighter-billed dowitchers.

■ **VOCALIZATIONS** Flight call nasal and rising, strongly slurred: *oh-wee!* or *ooh-week!* In display or in nest defense, gives short, slurred whistles.

■ **BEHAVIORS** The prairie godwit, nesting on grasslands near pothole marshes. On migration and in winter, forms dense flocks, sometimes with dowitchers (p. 126).

■ **POPULATIONS** Most migrate much shorter distances than other godwits. Adults leave breeding grounds early, by late June; huge numbers stage at Great Salt Lake. Widespread migrant across interior, with almost all in winter in West along coast. Despite short migration, several records for Hawaii.

SANDPIPERS
SCOLOPACIDAE

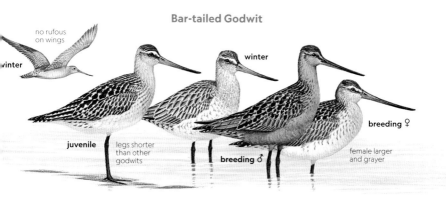

Bar-tailed Godwit

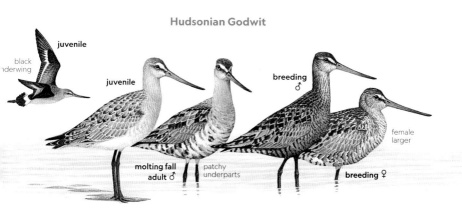

Hudsonian Godwit

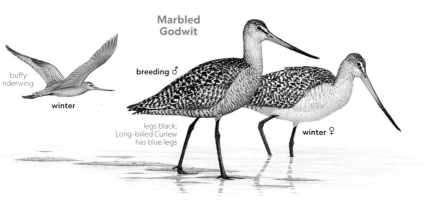

Marbled Godwit

Regular to Hawaiian Is.

Ruddy Turnstone
Arenaria interpres | RUTU | L 9½" (24 cm)

■ **APPEARANCE** Medium-large sandpiper with peculiar bill, stout and upturned; legs relatively short. Head and breast complexly patterned black and white, strikingly so in breeding plumage; nonbreeders retain general pattern. Rich rufous above in breeding plumage, otherwise darker above. Feet red-orange all year. Upperparts flashy in flight, a pattern shared with other West Coast "rockpipers."

■ **VOCALIZATIONS** Year-round, gives dull *chut* and sharper *pep* notes, often run into a short chatter. Defends nesting territory with short, rich whistles, rising or falling in pitch.

■ **BEHAVIORS** Year-round, searches for food by flipping debris (turning stones) with its odd bill. Nests on damp tundra, often with gravel. Away from breeding grounds, gathers at sea rocks and wrack lines, less frequently on open sandbars and mudflats.

■ **POPULATIONS** Most in continental West migrate along coast; rare migrant at best in interior. Winters coastally, including many on the Hawaiian Is.

Black Turnstone
Arenaria melanocephala | BLTU | L 9¼" (23 cm)

■ **APPEARANCE** Like Ruddy Turnstone, a medium-large sandpiper with upturned bill and fairly short legs. In breeding plumage, lacks extensive white on face and orange above of Ruddy. In winter plumage, appears largely black above in most settings; in good light, a dull green gloss is evident. In all plumages, "flash pattern" in flight similar to that of Ruddy. Year-round, feet are muddy red-brown, darker than Ruddy's.

■ **VOCALIZATIONS** Around nest, gives rising, gull-like squeals. Call notes, heard year-round, sharper and squeakier than those of Ruddy: *chrint* and *preep*, often run into short or long chatters.

■ **BEHAVIORS** Where breeding ranges of Black and Ruddy overlap, the former tends to be in wetter tundra. Haunts rocky coastlines when not breeding.

■ **POPULATIONS** Even more strictly coastal than Ruddy; accidental in interior.

Accidental to Hawaii

Surfbird
Calidris virgata | SURF | L 10" (25 cm)

■ **APPEARANCE** Larger overall and more heavyset than a turnstone. Orangey-based bill is short and thick; legs are short and dull yellow. Breeding adult densely spotted overall, with rufous on wings. Winter adult matches general plumage pattern of winter Black Turnstone, but Surfbird bulkier, thicker-billed, and grayer overall, with some spotting below; juvenile similar, but scalier. Flashes bold black and white on tail and wings in flight, but is wholly dark-backed; turnstones show much white on back in flight.

■ **VOCALIZATIONS** Fairly quiet away from breeding grounds; birds in winter chitter and squeak when flushing and interacting. Around nest, gives short, shrill whistles, often in series.

■ **BEHAVIORS** A bird of the surf only away from breeding grounds; nests on rocky tundra often well inland, to elevations more than a mile (1.6 km) above sea level. In winter, feeds singly or in very small numbers on surf-splashed sea rocks.

■ **POPULATIONS** Strictly coastal on migration and in winter. Migrants stage in large numbers at a few sites from Northern Calif. northward; accidental to Hawaii.

Casual to Hawaii

Red Knot
Calidris canutus | REKN | L 10½" (27 cm)

■ **APPEARANCE** Large *Calidris* sandpiper; as long as Surfbird (p. 114), but longer-billed and more slender-bodied. Straight, relatively short bill is black. Breeding adult smooth orangish below, rufous-and-gray above with weak scaling. Nonbreeding adult smooth gray above. Juvenile prominently scalloped above; scalloping of juveniles widespread among the many species of sandpipers, but is particularly pronounced, on the whole, in genus *Calidris*.

■ **VOCALIZATIONS** Rough *chup* and sharper *wheee* sometimes heard on migration and in winter; whoops and whistles on breeding grounds.

■ **BEHAVIORS** Most nest inland and upland a bit, on tundra exposed to the elements. Winters around coasts, feeding on sandy beaches in loose association with turnstones (p. 114); also in marine bays inland a few miles. Bunches up tight when roosting, sometimes with Dunlins (p. 118).

■ **POPULATIONS** Threats to eastern birds well publicized, but Red Knots in West seem to be in better shape. Migrates coastally; casual to Hawaii and interior.

Accidental to Hawaii

Stilt Sandpiper
Calidris himantopus | STSA | L 8½" (22 cm)

■ **APPEARANCE** Named for its legs, peculiarly long for a *Calidris* sandpiper. Bill, thin and decurved, also fairly long for the genus. Breeders, especially males, are dark and colorful, with orange ear patch and extensive dark barring. In nonbreeding plumage, mostly washed-out gray; pale eyebrow (supercilium) often prominent. Juvenile like nonbreeding adult, but scaly. Legs dull yellow in all plumages.

■ **VOCALIZATIONS** Fairly quiet on migration and in winter; flushing and socializing, sometimes gives harsh *chyurp*, descending. At nesting territory, issues anguished whistles.

■ **BEHAVIORS** Nests on wet tundra; on migration and in winter, occurs on mudflats. Year-round, feeds in standing water, often in the company of dowitchers (p. 126). With their long bills and rapid probing, nonbreeding Stilt Sandpiper and dowitchers are an unexpected ID challenge, but posture while feeding differs: On the shorter-billed and longer-legged Stilt Sandpiper, the tail sticks up higher than on the longer-billed and shorter-legged dowitchers.

■ **POPULATIONS** Bulk of migration mid-continent; can be fairly common west almost to Rockies. Rare west of Rockies, except Salton Sea, where flocks sizable even in winter; accidental to Hawaii.

Curlew Sandpiper
Calidris ferruginea | CUSA | L 8½" (22 cm)

Same size and heft as Dunlin (p. 118), but bill longer and more drooping, legs longer. Full-on breeding male unmistakable with deep-rust plumage, but sightings away from breeding grounds are usually of birds in transitional, nonbreeding, or juvenile plumage. Curlew Sandpiper has white uppertail coverts, conspicuous in flight, in all plumages; eyebrow of nonbreeding Curlew more prominent than on Dunlin. Varied calls include short trills, slurred whistles, and brief bouts of squeaky chatter; calls of Dunlin harsher overall. Has nested in Alas. Annual in recent years to West Coast; casual in interior and to Hawaii.

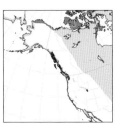

Annual to Hawaii

Sanderling
Calidris alba | SAND | L 8" (20 cm)

■ **APPEARANCE** Stocky and sturdy; bill straight, black, and fairly short. Most of the year, conspicuously pale; at rest, shows black wedge at leading edge of wing. Reddish breeding plumage not acquired until late in spring migration; compare with smaller congeners (pp. 120–123). White wing stripe always prominent in flight.

■ **VOCALIZATIONS** Industrious feeding flocks too busy to call, but flocks or individuals flushing give rough, rising *cheet?* or *chlit?*

■ **BEHAVIORS** Nests mostly on low-lying, high-latitude tundra in dry climes; haunts sandy beaches otherwise. Feeding behavior distinctive: Tightly bunched flocks run down the beach as a wave retreats, then just as quickly race back up as the next wave advances. Coordinated running alone identifies the species, but beware that individuals and even flocks often forage "normally" like other small sandpipers.

■ **POPULATIONS** Occurs coastally year-round, including nonbreeders in summer. Rare on migration in interior; annual in good numbers to Hawaii.

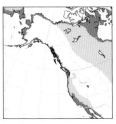

Regular to Hawaii

Dunlin
Calidris alpina | DUNL | L 8½" (22 cm)

■ **APPEARANCE** Same size as Sanderling; all plumages have long, drooping, black bill. Breeding adult reddish above with extensive black blotch on belly. Gray-brown above in winter; in flight, uniformly dark above with weak wing stripe. Juvenile plumage, warmer and browner than adult plumages, is worn only on and around breeding grounds. Western Sandpiper (p. 122) in winter smaller, paler, colder, and not as long-billed. Compare also with rare Curlew Sandpiper (p. 116).

■ **VOCALIZATIONS** Flight call a scratchy, slightly descending *kriih* or *prrriih*. Flocks on beach whisper softly while feeding.

■ **BEHAVIORS** Nests on damp tundra; migrates and winters widely wherever there are beaches and mudflats. Gregarious on migration and in winter; feeding flocks are disorganized and frenetic, but large flocks in flight twist and turn in perfect unison. Shares ocean beaches with Sanderling, but more likely in protected lagoons inland a bit.

■ **POPULATIONS** Rare on migration in most of interior, but thousands stop in western Nev. in spring. East of Great Basin, most records fall and winter. Regular to Hawaii.

Rock Sandpiper
Calidris ptilocnemis | ROSA | L 9" (23 cm)

■ **APPEARANCE** Slightly larger and stockier than Dunlin; weakly drooped bill not as long. In breeding plumage, splotchy black belly and reddish back can suggest Dunlin, especially in birds nesting on Pribilofs. Most in winter are dark gray-purple, with dull yellow legs and bill; Pribilof Is. breeders paler in winter. Juvenile, seen only on and near breeding grounds, splotchy buff and gray.

■ **VOCALIZATIONS** Fairly quiet in winter, but flushes and squabbles with rough scratchy notes, singly or in short series.

■ **BEHAVIORS** Nests on tundra; winters in wave-splashed intertidal habitats. Singly or in small flocks, feeds on rocks and debris, often in company of turnstones and Surfbirds.

■ **POPULATIONS** Distinctive Pribilof Is. breeders winter no farther south than Alas.; others range south regularly to Northern Calif.

SANDPIPERS
SCOLOPACIDAE

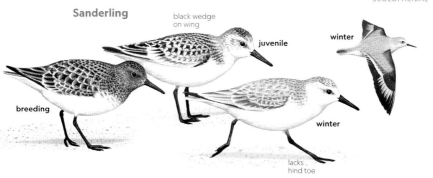

Sanderling

Siberian *sakhalina*

Dunlin *pacifica*

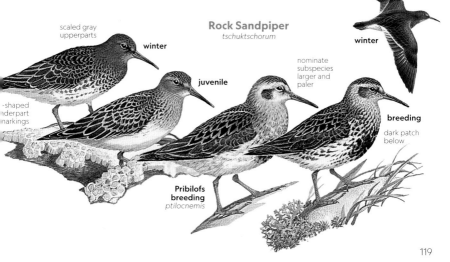

Rock Sandpiper *tschuktschorum*

Casual to Hawaii

Baird's Sandpiper
Calidris bairdii | BASA | L 7½" (19 cm)
- **APPEARANCE** One of the peeps (pp. 120–123), the smallest sandpipers. Baird's is long-winged and relatively large for a peep. At rest, the dark wing tips cross over like scissor blades; bill black and nearly straight. In all plumages, sandy gray-brown overall. White-rumped Sandpiper same size and shape, but grayer, with white rump; bill of White-rumped longer and droops slightly, shows pale at base. Smaller Least Sandpiper similar in color and pattern, but has thinner, slightly decurved bill and dull yellow legs.
- **VOCALIZATIONS** Flight call a short trill, rough and reedy, slightly falling in pitch: *brrrrrt* or *brriiip*. Song consists of scratchy, gasping notes, run together in longer trills.
- **BEHAVIORS** Nests on dry tundra, coastally and inland. On migration, often feeds on mudflats away from the water's edge a bit (like Least Sandpiper), but also wades into shallow water.
- **POPULATIONS** Migration, both spring and fall, mostly mid-continent. Flocks of hundreds stage at reservoirs at eastern edge of our area. Casual to Hawaii.

Annual to Hawaii

Least Sandpiper
Calidris minutilla | LESA | L 6" (15 cm)
- **APPEARANCE** The tiniest shorebird on Earth, a smidgen smaller than even Western and Semipalmated Sandpipers (p. 122). Bill thin, droops down, tapers to fine point. All plumages muddy and dark; adults gray-brown, juveniles warmer. Yellowish legs distinguish Least from other peeps, but lighting and mud can make yellow legs appear dark, and vice versa. Larger Baird's Sandpiper has longer wings, straighter bill, black legs.
- **VOCALIZATIONS** Flight call higher and sweeter than Baird's, a rising, rolling *prreee?* Song, given by both sexes, a rapid utterance of short, rich whistles.
- **BEHAVIORS** Nests in bogs. On migration and in winter, feeds at muddy edges of lakes and ponds; tends to occur a bit farther inshore than other peeps.
- **POPULATIONS** Lingers late, wintering inland across southern tier of U.S. states, north to Nev. and, sparingly, Utah and Colo. Rare but annual to Hawaii.

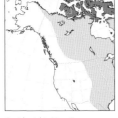

Accidental to Hawaii

White-rumped Sandpiper
Calidris fuscicollis | WRSA | L 7½" (19 cm)
- **APPEARANCE** Along with Baird's Sandpiper, this is one of the two large peeps, "large" being relative. Like Baird's, has long wings that cross over on the bird at rest; bill of White-rumped slightly decurved, pale at base. White rump distinctive, but beware that any fast-flying sandpiper can catch bright glare as it twists and turns. Gray in breeding plumage, with rufous highlights above and gray streaks on flanks; compare also with smaller Western Sandpiper.
- **VOCALIZATIONS** High-pitched flight call, squeaky and scratchy, drops in pitch: *squit* or *tsweek*. Rarely heard song likewise scratchy.
- **BEHAVIORS** Nests on wet coastal lowlands. On migration, typically in wetter microhabitats than Baird's, wading almost belly-deep into standing water.
- **POPULATIONS** Migration routes like that of American Golden-Plover (p. 104): north in spring up middle of continent, south in fall mostly over Atlantic. A quite late spring migrant, mainly June, at eastern periphery of our region. Accidental to Hawaii.

SANDPIPERS
SCOLOPACIDAE

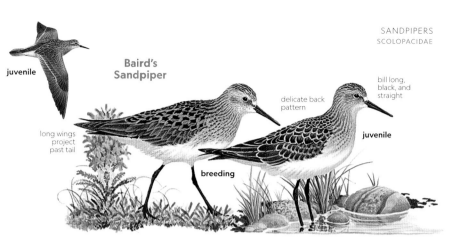

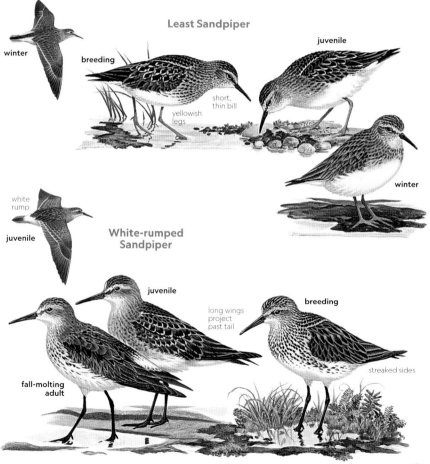

Casual to Hawaii

Red-necked Stint
Calidris ruficollis | RNST | L 6¼" (16 cm)

■ **APPEARANCE** Same length and heft as Semipalmated Sandpiper, but Red-necked Stint longer-winged and shorter-legged; feeding or roosting, looks rear-tapered, appears crouched. Rich rufous of breeding adult eye-catching, but breeding Sanderling (p. 118) can be unexpectedly similar. Breeding plumage begins to wear by July; vagrants away from Alas. often only vaguely rufescent. Drab juvenile, gray-brown above with dusky breast, resembles Semipalmated. Rarer Little Stint, *C. minuta* (p. 457), has darker-centered tertials; adult less colorful below than Red-necked, and juvenile has bolder white lengthwise stripes above. Records of stints away from Bering Sea region realistically require photos for credible ID.

■ **VOCALIZATIONS** Flight call a rough, wavering *fshlit*.

■ **BEHAVIORS** Nests on dry tundra with low vegetation. Vagrants find their way to gatherings of other peeps, conveniently staying with the flock for hours to days at a time.

■ **POPULATIONS** Breeds mostly eastern Arctic Russia, but also sparingly to Alas. mainland. Not quite annual to West Coast; casual to Hawaii.

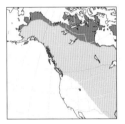
Casual to Hawaii

Semipalmated Sandpiper
Calidris pusilla | SESA | L 6¼" (16 cm)

■ **APPEARANCE** Midsize peep; bill relatively short, thick, and straight, with bulbous tip. Sandy gray in most plumages. Flanks largely unspotted in spring; Western has streaked flanks in spring. Eyebrow, slightly waved, a good mark on drab fall migrants. Semipalmated and Western are one of the most notorious pairs of similar species; compare calls, body structure, and migration timing, and keep in mind the effects of age, molt, and feather wear.

■ **VOCALIZATIONS** Flight call a simple, descending *churk* or *churp*, rougher than Western's.

■ **BEHAVIORS** Nests on low-lying wet tundra; puts down on migration at lake edges. Feeding, often wades into shallow standing water.

■ **POPULATIONS** Migration in West, both spring and fall, generally east of Rockies; quite scarce Great Basin, somewhat less rare coastally, and casual to Hawaii. Does not winter in West, except accidental at that season to Hawaii.

Annual to Hawaii

Western Sandpiper
Calidris mauri | WESA | L 6½" (17 cm)

■ **APPEARANCE** Slightly larger and lankier than Semipalmated. Bill longer, thinner, and more decurved than Semipalmated's, but bill structure variable. Adult in spring has rufous cap and "ears," heavy streaking down flanks; juveniles on southbound migration have rufous scapulars, showing as colorful band above, a mark not shown by Semipalmated.

■ **VOCALIZATIONS** Flight call scratchy, descending in pitch, *screet* or *jeet*, more like White-rumped's (p. 120) than Semipalmated's.

■ **BEHAVIORS** Nests on tundra, on sites a bit higher and drier than those favored by Semipalmated. Stages and winters, sometimes in huge flocks, where there is expansive shallow water.

■ **POPULATIONS** Common to abundant on migration west of Rockies; gathers coastally, but also 10,000+ western Great Basin and 100,000+ eastern Great Basin. Fall migration generally later than that of Semipalmated. Winters coastally; very rare inland in winter east of Cascade-Sierra axis and to Hawaii.

Regular to Hawaii

Pectoral Sandpiper
Calidris melanotos | PESA | L 8¾" (22 cm)

■ **APPEARANCE** Large *Calidris* with long wings and slightly decurved bill. Notably bibbed in all plumages; bib of breeding male like chain mail. Adults and especially juveniles on southbound migration more rufous than spring adults. Legs straw yellow; bill dull orange-brown. Compare with Baird's and Least (both on p. 120) and, in particular, Sharp-tailed Sandpipers.

■ **VOCALIZATIONS** Flight call a quick trill, *prrrrp*, on one pitch; not as growling as Baird's. In display on breeding grounds, low hoots and cackles.

■ **BEHAVIORS** Nests on marshy tundra in coastal lowlands; migrants put down at muddy lakeshores. The name "pectoral" refers to air sacs inflated by male in courtship.

■ **POPULATIONS** An astonishing migrant: Winters to Tierra del Fuego, with many crossing Bering Strait and some breeding as far as Yamal Peninsula, 70° E! Uncommon migrant in most of West, rather scarce in Great Basin; regular to Hawaii, sometimes in large flocks.

Sharp-tailed Sandpiper
Calidris acuminata | SPTS | L 8½" (22 cm)

Rare counterpart of Pectoral, nesting only in a sliver of coastal Arctic Russia. Like Pectoral, a large *Calidris* with slightly decurved bill. Most in West are juveniles in fall, told from Pectoral by bolder eyebrow, sharper cap, plainer breast, streaked rear flanks, and warmer tones overall. Flight call one to three short, rich, rising, whistled *wheet?* notes, distinct from Pectoral's. Annual in tiny numbers along West Coast; more common to Hawaiian Is., where it is not quite as common as Pectoral.

Buff-breasted Sandpiper
Calidris subruficollis | BBSA | L 8¼" (21 cm)

■ **APPEARANCE** An odd *Calidris*; erect posture, long neck, and beady black eye suggest Upland Sandpiper (p. 110). Bill slight, short, and black; legs relatively long and yellow. Plain pale buff below; darker above with buffy scaling.

■ **VOCALIZATIONS** Soft ticking and buzzing. Rarely heard, even at leks.

■ **BEHAVIORS** Nests on dry tundra, where males display communally at leks. On migration, part of the "grasspiper" guild, putting down around sod farms and other open habitats.

Accidental to Hawaii

■ **POPULATIONS** Migration is mainly up and down middle of continent. Rare but regular to western Great Plains; casual elsewhere on migration in continental West, accidental to Hawaii. Globally rare; fewer than 100,000 individuals.

Ruff
Calidris pugnax | RUFF | L 10–12" (25–30 cm)

Amazing sandpiper from Eurasia; has four genders, recently discovered, which are morphologically and behaviorally well-differentiated. Famously leks, like Buff-breasted Sandpiper, but not usually in West; individuals of one gender saunter around leks flaunting their insane "ruffs" of orange, black, and white, like something out of Elizabethan England. All ages and genders are potbellied, small-headed, and long-necked; most adults have orangey legs and at least some orange on bill. Bright buffy juvenile scalloped above and plain below, similar to Buff-breasted. Usually silent, even when lekking; flushing, occasionally gives soft, rough notes: *frrrt*, *brup*, etc. Very rare but annual to Alas., where breeding has occurred, and West Coast; accidental to Hawaii.

Accidental to Hawaii

Short-billed Dowitcher
Limnodromus griseus | SBDO | L 11" (28 cm)

■ **APPEARANCE** Despite its name, a very long-billed sandpiper. Differs from Long-billed Dowitcher in several aspects of structure: Slightly shorter bill of Short-billed is subtly kinked or curved (perfectly straight on Long-billed); eye is positioned high on face (lower on Long-billed); and wings on bird at rest extend just past tail tip (fall just short of tail tip on Long-billed). Feeds with flat-backed profile; humpbacked on Long-billed. Separating the two by plumage requires attention to age (juvenile vs. adult), plumage (breeding vs. winter if adult), and subspecies (Short-billed has two contenders in West, Long-billed only one). In all plumages, white bars on tail of Short-billed wider than on Long-billed, making Short-billed appear pale-tailed in flight. Breeding adult variably orangish and spotted below; winter adult dirty gray-brown, also spotted below. Fresh juvenile scaly above; tertials complexly marked with orange stripes.

■ **VOCALIZATIONS** Flight call a soft, descending *tyu*, typically in twos and threes, like that of Lesser Yellowlegs (p. 130).

■ **BEHAVIORS** Nests on ground near trees in boreal bogs. Migrants and winterers feed in shallow standing water. Where the two dowitchers co-occur, Short-billed favors more brackish conditions, but much overlap. Feeding flocks bunch tight and feed with rapid jabbing motion.

■ **POPULATIONS** Comprises three subspecies with disjunct breeding ranges. Two occur in West: Pacific *caurinus*, breeding Alas., migrating Pacific coast region; mid-continent *hendersoni*, uncommon west to Rockies. Breeding *hendersoni* washed orangey across underparts; breeding *caurinus* not as brightly washed, has more dark markings below and mostly white lower belly. Accidental to Hawaii.

Regular to Hawaii

Long-billed Dowitcher
Limnodromus scolopaceus | LBDO | L 11½" (29 cm)

■ **APPEARANCE** Bill averages longer and straighter than Short-billed's. Long-billed also shorter-winged, with eye lower on face than Short-billed; feeding, appears more humpbacked (Short-billed more flat-backed). Breeding adult dull brick red with short dark bars below. Compare carefully with breeding *hendersoni* Short-billed, which is softer, paler orange below and more spotted than barred; and especially with breeding, white-bellied *caurinus* Short-billed. Winter adult Long-billed sports unmarked plain gray "breastplate"; winter Short-billed more spotted below. Tertials of fresh juvenile Long-billed solidly dark-centered, plainer than in Short-billed.

■ **VOCALIZATIONS** Flight call a sharp, squeaky *pik* or *plik*. Feeding flocks twitter softly, producing notes like those of Short-billed.

■ **BEHAVIORS** Nests on open, swampy tundra. Migrates and winters widely where there is standing open water; in coastal landscapes, goes for less brackish microhabitats than Short-billed, but much overlap. Both dowitchers feed with rapid jabbing. Short-billed is the "confusion species" at all seasons, but Stilt Sandpiper (p. 116) and even Marbled Godwit (p. 112) can act like dowitchers.

■ **POPULATIONS** East of Cascade-Sierra axis, Long-billed is the default dowitcher on migration and winter; huge numbers stage in Lahontan Valley, Nev. Hardy, winters inland across southern tier of western states; regular in winter to Hawaii, often in small flocks.

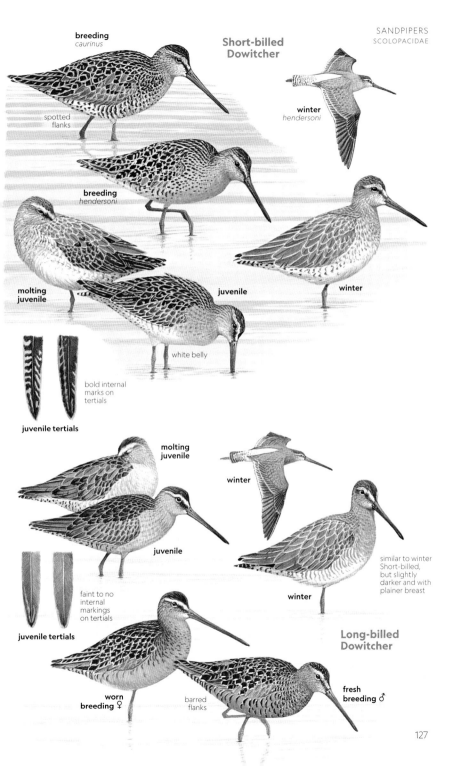

Casual to Hawaiian Is.

Wilson's Snipe
Gallinago delicata | WISN | L 10¼" (26 cm)
- **APPEARANCE** Similar in size and build to dowitchers (p. 126). Head striped lengthwise; back has long white streaks. Orange on short tail prominent in flight.
- **VOCALIZATIONS** Flushing, a startling *skrrrtt*. Two songs: spooky winnowing, *oohuhuhuhuhuhu*, given in flight and created by splayed-out tail feathers, and steady *wheep wheep wheep wheep* ....
- **BEHAVIORS** Nests in bogs and wet meadows, winters around seeps and springs. Dowitcher-like, feeds hunched over and jabs bill into deep water.
- **POPULATIONS** Winters well inland and fairly far north where open water persists; casual to Hawaiian Is.

Annual to Hawaiian Is.

Spotted Sandpiper
Actitis macularius | SPSA | L 7½" (19 cm)
- **APPEARANCE** Plump and short-legged. All plumages show thin white eyebrow and thin dark line through eye; bill variably dark to pink-orange, legs dull yellow to brighter yellow-orange. Boldly spotted below in breeding plumage. Juvenile and nonbreeding adult plain below, with white wedge in front of wing.
- **VOCALIZATIONS** A rising *wheee*, given in twos or threes as bird flushes, or run together in series in song.
- **BEHAVIORS** Nests in dense vegetation around ponds and rivers. Found anywhere water meets land, from tiny cattle ponds to ocean shores. Modes of locomotion distinctive: saunters along pond edge, teetering constantly; flushing, flies out over water on stiff, shallow, bowed wingbeats, then returns to shore.
- **POPULATIONS** Migrates across broad front, not flocking densely like most other migratory sandpipers. Rare but annual to Hawaiian Is.

Accidental to Hawaiian Is.

Solitary Sandpiper
Tringa solitaria | SOSA | L 8½" (22 cm)
- **APPEARANCE** Like a diminutive Lesser Yellowlegs (p. 130); Solitary has shorter, green-yellow legs and shorter, stouter bill. White eye ring often prominent. In flight, shows barred tail and dark underwings.
- **VOCALIZATIONS** Flight call, higher and squeaker than Spotted's, a strident *pee! wee!* or *pee! wee! wee!*
- **BEHAVIORS** Nests in trees in boreal forest; migrates and winters widely, with particular affinity for wooded ponds. True to its name, keeps some distance from other birds when feeding.
- **POPULATIONS** Alas. breeders, *cinnamomea*, have cinnamon-tinged spots above; widespread nominate subspecies white-spotted above. Uncommon in U.S. from Rockies westward; accidental to Hawaiian Is.

Common to Hawaiian Is.

Wandering Tattler
Tringa incana | WATA | L 11" (28 cm)
- **APPEARANCE** Structurally like a large Spotted Sandpiper; bill stout and straight, dull yellow legs relatively short. Dark gray in all plumages with weak white eyebrow. Breeding adult densely barred below; nonbreeding adult and juvenile smudgy gray all over.
- **VOCALIZATIONS** Flight call a series of four or five short whistles, descending a bit in pitch.
- **BEHAVIORS** Nests inland around streams and ponds in boreal bogs, winters coastally around sea rocks. Like Spotted, teeters while feeding.
- **POPULATIONS** Many migrate over Pacific Ocean; common to Hawaiian Is.

Annual to Hawaii

Lesser Yellowlegs
Tringa flavipes | LEYE | L 10½" (27 cm)

■ **APPEARANCE** Almost identical in plumage to Greater Yellowlegs, but differs structurally: Lesser is proportionately shorter-billed, longer-legged, and slightly longer-winged. In spring, not as boldly barred below as Greater. Compare also with smaller Solitary Sandpiper (p. 128).

■ **VOCALIZATIONS** Flight call a slightly descending *tyu*, often given singly; less urgent-sounding than Greater's. Similar to flight call of Short-billed Dowitcher (p. 126), but descends more sharply.

■ **BEHAVIORS** Nests on ground in boreal woods, dispersing widely after breeding to lakeshores. Wary and nervous, but less prone to drama than Greater.

■ **POPULATIONS** Uncommon on migration west of Rockies. Less hardy than Greater, with most wintering to our south; rare but annual to Hawaii.

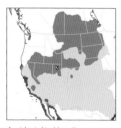
Accidental to Hawaii

Willet
Tringa semipalmata | WILL | L 15" (38 cm)

■ **APPEARANCE** Large and gray; mostly dark bill is long and stout. Legs grayish. Heavily barred in breeding plumage, plain gray in winter. High-contrast black-and-white wings striking in flight.

■ **VOCALIZATIONS** Flight call a whistled, anxious, far-carrying *eeewee eeewee ip (pill will willet)*.

■ **BEHAVIORS** Nests in grass near freshwater wetlands. Migrants often seen singly across broad front in West, but big flocks sometimes put down in spring snowstorms. Flushes noisily like a yellowlegs.

■ **POPULATIONS** All in West are subspecies *inornata*, distinct in many ways from nominate eastern *semipalmata* and perhaps a different species. Accidental to Hawaii.

Casual to Hawaii

Greater Yellowlegs
Tringa melanoleuca | GRYE | L 14" (36 cm)

■ **APPEARANCE** Bill proportionately longer than Lesser's, at least the length of its head, and typically upturned; Greater also has proportionately shorter legs and wings. Color and pattern nearly identical to smaller Lesser, but spring Greater averages more boldly marked.

■ **VOCALIZATIONS** Flight call a ringing *tyu!* like Lesser's, but Greater delivers it with greater oomph. Greater often belts out four or five such notes, the first couple especially loud: *TYU!! TYU! tyu tyu tyu*.

■ **BEHAVIORS** Habits like Lesser's: Nests on ground in wet boreal woods; migrates and winters widely where there is standing water. Active and alert like other *Tringa* sandpipers: Walks swiftly from danger, then flushes with strident flight calls.

■ **POPULATIONS** Migration extremely protracted; at midlatitudes inland in West, migrants may be seen from midwinter through late autumn, with just a brief window of absence in mid-June. Hardy; winters across southern tier of western states. Casual to Hawaii.

Wood Sandpiper
Tringa glareola | WOSA | L 8" (20 cm)

Widespread Old World *Tringa*. Fairly common to western Aleutians, where breeding has occurred; casual to West Coast and Hawaii. About the same build as Solitary and, like that species, a loner. Told from Lesser Yellowlegs and Solitary by heavy white spotting above, broad white eyebrow and contrasting dark cap, and short stout bill; in flight, wings pale smoky gray below. Flight call a short, clipped *plip* or *pleek*, descending, often in fast series.

Casual to Hawaii

Wilson's Phalarope
Phalaropus tricolor | WIPH | L 9¼" (23 cm)

- **APPEARANCE** Largest phalarope. A potbellied sandpiper with long neck and small head; long bill is very thin and straight. Bright breeding female has tricolored neck, with long swaths of burgundy, amber, and white; breeding male duller. Juvenile notably pale; richly scalloped above. Back of winter adult uniform pale gray.
- **VOCALIZATIONS** Muffled, nasal *uhrunk*, especially males around nests.
- **BEHAVIORS** All three phalaropes habitually swim; they spin rapidly to create vortices that pull small prey up to the surface, with multiple birds coordinating their actions to maximize feeding efficiency. Wilson's nests around marshes, freshwater or alkaline; puts down on migration at lakes. More terrestrial than the other phalaropes; in addition to swimming, also feeds on shorelines, darting and dashing as it goes.
- **POPULATIONS** Most widely encountered of the phalaropes inland. Females, not involved in care of young, leave breeding grounds early; can be on "fall" migration by mid-June. Immense flocks stage at Great Salt Lake midsummer. Not usually seen offshore, yet is casual to Hawaii.

Casual to Hawaii

Red-necked Phalarope
Phalaropus lobatus | RNPH | L 7¾" (20 cm)

- **APPEARANCE** The smallest phalarope; bill long and thin. Red-orange neck and upper breast of breeding female contrast with white cheeks; male duller. In breeding plumage, both sexes sport long golden bands, or "braces," on gray uppersides. Juvenile and winter adult show dark eye patch on mostly white head; note also dark back with white braces.
- **VOCALIZATIONS** Call a dull, clipped *pik* or *plick*, often run together in series as bird flushes.
- **BEHAVIORS** Nests on tundra, ranging farther inland and upslope than Red Phalarope. On migration inland, usually on deep water; Wilson's more likely on or close to shore.
- **POPULATIONS** Winters pelagically to our south, but migrates widely off West Coast; casual to Hawaii. Generally uncommon spring migrant inland, but huge numbers stage at a few large lakes in Great Basin late summer.

Annual to Hawaiian Is.

Red Phalarope
Phalaropus fulicarius | REPH | L 8½" (22 cm)

- **APPEARANCE** Intermediate in size between larger Wilson's and smaller Red-necked. In all plumages, bill of Red is much thicker than that of Red-necked and Wilson's. Breeding female deep rust all over with extensive white on face; breeding male paler, more orangey. Both sexes have straw-yellow bill in breeding plumage; bill retains some color in fall and winter. Winter adult, like Red-necked, shows dark eye patch; back of winter adult Red is plain gray. Juvenile Red dark above, with buff-fringed feathers; lacks dorsal stripes of juvenile Red-necked.
- **VOCALIZATIONS** Call a high, sharp *peek!* or *pleek!* strangely similar to Hairy Woodpecker's (p. 254).
- **BEHAVIORS** Nests on open tundra in coastal lowlands. On migration offshore, often seen in general company of Red-necked, but Red averages farther offshore.
- **POPULATIONS** The most pelagic phalarope, rare inland. Most inland records are from late fall, after the passage of Red-necked. Uncommon but annual to Hawaiian Is.

SANDPIPERS
SCOLOPACIDAE

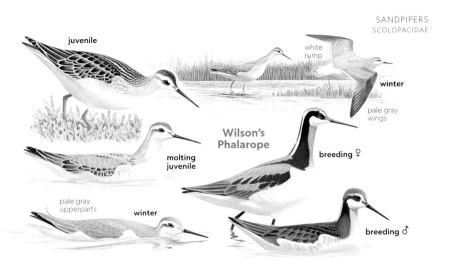

Wilson's Phalarope

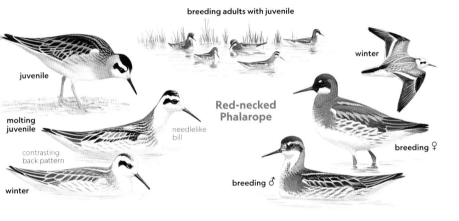

Red-necked Phalarope

Red Phalarope

SKUAS AND JAEGERS | STERCORARIIDAE

Encountered mostly at sea, these are the pirates of the bird world. Skuas and jaegers chase down gulls and other seabirds, forcing them to disgorge their prey. Clad in browns and grays, skuas and jaegers suggest large immature gulls (pp. 156–163), but they are more closely related to alcids (pp. 138–147). They number seven species globally, with all breeding at high latitudes. Identification is challenging, especially with the jaegers; gestalt is useful for birders with much experience, but nothing beats photos.

Nearly annual to Hawaii

South Polar Skua
Stercorarius maccormicki | SPSK | L 21" (53 cm) WS 52" (132 cm)

- **APPEARANCE** Big and bulky; bullnecked and barrel-chested. Most adults and all juveniles look quite dark, with bold white flashes at bases of primaries above and below; some adults are paler-bodied, but always flash white in dark wings. Jaegers are smaller and more svelte, often with contrasting pale underparts and conspicuous central tail streamers. Compare also with first-winter Herring Gull (p. 158), which is dark overall and bulky, with silver flash in inner primaries.
- **VOCALIZATIONS** Issues occasional barks and chatters in presence of other seabirds but is usually silent at sea.
- **BEHAVIORS** Seen only at sea in West; usually occurs singly. Flies fast and direct, but banks suddenly on spotting a prospective victim. Swimming, rides high on the water, like a cork.
- **POPULATIONS** An underappreciated "super migrant": breeds incredibly far south, only on Antarctic mainland; winters (our summer) north regularly to Gulf of Alaska, with records north all the way to Chukchi Sea! Recorded regularly well off West Coast late spring to fall; nearly annual to Hawaiian waters.

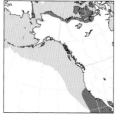

Regular in Hawaii

Pomarine Jaeger
Stercorarius pomarinus | POJA | L 21" (53 cm) WS 48" (122 cm)

- **APPEARANCE** Largest and heftiest jaeger; all "Poms" are large-billed, with bulbous expansion (gonys) of lower mandible prominent. The name of this bird derives from Greek words meaning "lid" and "nose"; the upper mandible, or maxilla, of all three jaegers is covered in a lid-like sheath. Light-morph adult Pomarine has extensive dark cap and smudgy breastband; uncommon dark morph mostly blackish. Central tail feathers of adult, often broken or missing, are rounded. Juvenile dusky all over like other jaegers; from below, note white at base of primaries and primary coverts, creating double wing flash.
- **VOCALIZATIONS** On breeding grounds, gives short, nasal wails; usually silent at sea, but sometimes protests other birds with disyllabic yaps.
- **BEHAVIORS** Nests on low-lying marshy tundra, especially where lemmings abound. At sea, often seen in purposeful flight, steady and direct; less inclined than Parasitic Jaeger (p. 136) to go at it with other seabirds. Migrates in small flocks.
- **POPULATIONS** Probably present year-round off West Coast, but status at sea in our region imperfectly known, clouded by bias toward summer and fall when many boat trips go out; regular in small numbers year-round off Hawaii. Booms and busts track regular interannual variation in lemming numbers, on three- or four-year cycles. Regular but rare in Interior West, mostly late fall at large waterbodies; averages later in season than the two smaller jaegers (p. 136).

SKUAS AND JAEGERS
STERCORARIIDAE

South Polar Skua

Pomarine Jaeger

Annual to Hawaiian Is.

Parasitic Jaeger

Stercorarius parasiticus | PAJA | L 19" (48 cm) WS 42" (107 cm)

■ **APPEARANCE** Averages intermediate in size and structure between Pomarine (p. 134) and Long-tailed Jaegers, but can overlap with either. Sharp central tail feathers of adult often broken or missing; light and dark morphs obviously different, but many adults intermediate. Parasitic lacks double wing flash below of Pomarine, flashes more white above than Long-tailed. Juveniles, a challenge to identify, are warmer and more golden overall than juveniles of other jaegers; also, the central tail feathers of juvenile Parasitic are pointed (rounded in juvenile Long-tailed and Pomarine).

■ **VOCALIZATIONS** Occasional yips and yaps in mixed-species assemblages at sea, but less inclined than Pomarine to vocalize. In defense of nesting territory, issues nasal squeals like Pomarine, but more yappy.

■ **BEHAVIORS** Nests widely on tundra, coastally and inland, even at considerable elevation; nest placed on ground, usually near water. Opposite of the other jaegers, Parasitic is a diet generalist on the breeding grounds, where it feeds ravenously on eggs, small birds, mammals, and even berries. On migration, tends to occur closer to shore than the other jaeger species; the most aggressive at sea of the jaegers, it is often seen stealing food from gulls and especially terns.

■ **POPULATIONS** Breeding season distribution of plumage morphs varies geographically, with light morphs farther north; in Aleutians (at southern limit of range), almost all dark morph. Despite geographic basis, along with well-supported genetic basis, for this polymorphism, Parasitic is treated as monotypic, without subspecies. Rare but regular migrant, almost entirely in fall, across Interior West; most are seen earlier than Pomarine and a bit later than Long-tailed. Very rare but annual to Hawaiian Is.

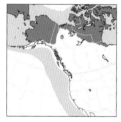
Annual to Hawaiian Is.

Long-tailed Jaeger

Stercorarius longicaudus | LTJA | L 22" (56 cm) WS 40" (102 cm)

■ **APPEARANCE** Discounting the adult's long central tail streamers, this is the smallest jaeger, petite and graceful overall, with slight bill, narrow wings, and long tail. In all plumages, flashes less white in wing than larger jaegers, especially above; shows only two or three white primary shafts. Very long central tail streamers of adult diagnostic, but are often broken or not fully grown; adult's dark remiges contrast with paler upperwing coverts. Juvenile, ranging from dark brown to quite gray, is extensively barred.

■ **VOCALIZATIONS** Around nest, gives excited squeals, purer and less nasal than those of the larger jaegers. Usually silent at sea, but occasionally gives soft, rising, nasal honks in interactions with other seabirds.

■ **BEHAVIORS** Nests on dry Arctic tundra, often far from sea; ecologically more like Pomarine (p. 134) than Parasitic, eating mostly lemmings when nesting. At large lakes inland, goes after Ring-billed Gulls (p. 154); at sea, harasses Sabine's Gulls (p. 148) and Arctic Terns (p. 170).

■ **POPULATIONS** Like Pomarine, subject to strong interannual variation in numbers. Leaves Arctic breeding grounds by late summer; sightings at sea and inland correspondingly early, in late summer. Winters much farther south than the larger jaegers. Like Parasitic, very rare but likely annual to Hawaiian Is.

AUKS, MURRES, AND PUFFINS | ALCIDAE

With their black-and-white plumages, often colorful bills and facial plumes, and erect postures, the birds in this family suggest penguins. And like penguins, alcids flourish in cold but productive oceans at high latitudes where they "fly" underwater with their wings and gather in crowded, smelly, noisy, densely packed nesting colonies. They number about two dozen species globally, and their center of diversity is right here in the West.

Dovekie
Alle alle | DOVE | L 7¾" (20 cm)

- **APPEARANCE** Chubby overall and stub-billed; not even one-fifth the heft of a murre. Breeding adult appears mostly black when swimming; in flight, white underparts obvious. Nonbreeders blotchy black and white above like other alcids, but small size, compact shape, and short bill always distinctive.
- **VOCALIZATIONS** Giddy wailing and chippering around nest, but almost never heard at sea.
- **BEHAVIORS** Most in West are seen at or near nesting colonies, where adults assemble in small numbers on outcroppings and snowbanks.
- **POPULATIONS** A classically East Coast alcid with puzzling distribution in West. Our breeders may be recently established, and where they winter is as yet unknown.

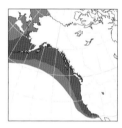

Common Murre
Uria aalge | COMU | L 17½" (44 cm)

- **APPEARANCE** All have long, thin, entirely dark bill; bill of first-winter birds, often seen at sea, not quite as long. Flanks mottled dusky. Breeding adult dull black or brown-black above; border between dark hood and white breast smoothly concave. In nonbreeding plumage, mostly white face marked by thin black line curving back from eye.
- **VOCALIZATIONS** In disputes at colonies, gives tremendous groans. Adults silent at sea, but recently fledged young solicits male parent with pure-tone *peedee* notes.
- **BEHAVIORS** Nests in dense colonies on rocky ledges by seashores, where adults stand upright like toy soldiers. On water, assumes very different profile, vaguely duck-like; often seen from fishing boats inshore, on jetties and headlands, and in harbors and marinas.
- **POPULATIONS** Numerous, but threatened by oil spills, entrapment in gill nets, and warming oceans; responds well to protection and restoration of nesting colonies, and numbers have increased recently in some places.

Thick-billed Murre
Uria lomvia | TBMU | L 18" (46 cm)

- **APPEARANCE** Bill, shorter and thicker than Common's, marked with white line. Flanks of Thick-billed pure white; Common's more smudgy. Adult Thick-billed, especially in breeding plumage, glistens jet-black; white breast pointedly juts up into black hood. Head and face of winter Thick-billed extensively dusky; Common more pale-faced with thin black line behind eye.
- **VOCALIZATIONS** Like Common, erupts in roars and groans at colonies; recently fledged young at sea whistles for father.
- **BEHAVIORS** Ranges farther offshore on average, feeding in deeper and colder waters, than Common, but the two species mix freely at sea and nest side by side.
- **POPULATIONS** Rare south of Alas., with most wintering Bering Sea; not even annual south of Canada.

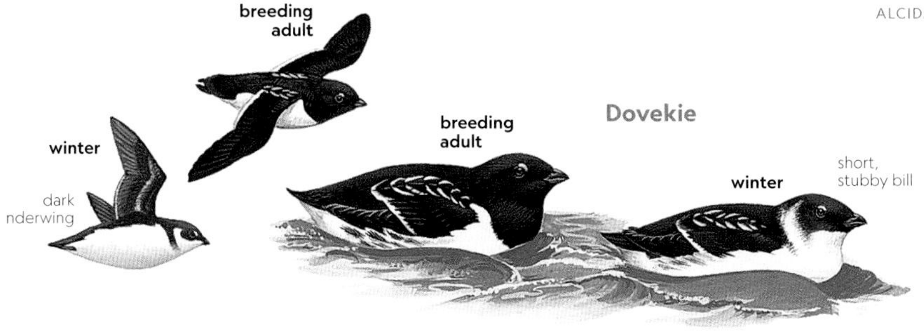

AUKS, MURRES, AND PUFFINS
ALCIDAE

Dovekie

breeding adult

winter
dark underwing

breeding adult

winter
short, stubby bill

Common Murre

breeding adult

matte gray upperparts

pale underwing with small dark center

breeding adult

winter

Thick-billed Murre

1st winter

pure white flanks

winter adult

breeding adult

white gape line

V-shaped mark on breast

white underwing

adult

Black Guillemot
Cepphus grylle | BLGU | L 13" (33 cm)

Chiefly North Atlantic species, but breeds sparingly along Arctic Ocean coast of Alas. Both guillemots are midsize alcids with scarlet feet and mouth linings; bill thin, tapered, and dark in both species. Breeding adult Black has entirely white wing oval; year-round, underwing of Black mostly white. Black Guillemots in West are subspecies *mandtii*, extensively white in first-winter and adult nonbreeding plumages. Calls like Pigeon's.

Pigeon Guillemot
Cepphus columba | PIGU | L 13½" (34 cm)

■ **APPEARANCE** A bit slimmer, longer-necked, and larger overall than Black Guillemot. Breeding adult like Black but white oval on wing broken by black bar; feet and mouth lining also scarlet like Black's. Nonbreeding adult gray-black and dirty white. Juvenile even duskier; compare with Marbled Murrelet, smaller and more compact, often found with it just offshore.

■ **VOCALIZATIONS** All calls thin and shrill; mixes long, high whistles with rapid piping notes.

■ **BEHAVIORS** A coast-lover, rarely seen out at sea; gets well into large estuaries. Guillemots, unlike many other alcids, nest as isolated pairs or in small, disorganized colonies in natural crevices, as well as wharfs, outbuildings, even nest boxes.

■ **POPULATIONS** Seasonal movements complex; many migrate north following breeding, with heaviest winter concentration from Puget Sound and Vancouver I. north through Inside Passage.

Marbled Murrelet 🇪🇳
Brachyramphus marmoratus | MAMU | L 10" (25 cm)

■ **APPEARANCE** Medium-small alcid with slight bill. "Marbled" plumage worn by breeding adult: finely scalloped brown and black above; mottled brown and white below. Winter adult black above and white below; note black cap and white collar. Juvenile duskier. Underwing linings gray-black in all plumages.

■ **VOCALIZATIONS** Gives loud, short, wavering whistles, *eee-eer* and *yeeay-ay*, especially around nest.

■ **BEHAVIORS** Most nest in dense forests. Astoundingly, many actually nest well up in tall redwoods (genus *Sequoia*); others nest on or near ground. Away from nest, seen singly or in pairs, just beyond shoreline.

■ **POPULATIONS** Declined in 20th century, due in large part to logging of old-growth forests where it nests; has recovered somewhat in 21st century.

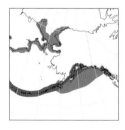

Kittlitz's Murrelet
Brachyramphus brevirostris | KIMU | L 9½" (24 cm)

■ **APPEARANCE** Similar build to Marbled Murrelet, but even shorter-billed. Belly, vent, and undertail coverts mostly white in otherwise mottled brown breeding plumage; Marbled darker below. Winter adult has mostly white face; juvenile duskier. Outer tail feathers white year-round; outer tail feathers of Marbled dark.

■ **VOCALIZATIONS** Gives low, nasal hoots; soft-sounding, but carries well.

■ **BEHAVIORS** Nests on ground in rocky habitats, sometimes far from shore and at considerable elevation. Like Marbled, typically seen near shore singly, in pairs, or, less commonly, in scattered small flocks.

■ **POPULATIONS** Accidental south of Tongass National Forest. Status in winter not fully understood; some or many may move into ice-free openings in Bering and even Chukchi Seas.

Scripps's Murrelet
Synthliboramphus scrippsi | SCMU | L 9¾" (25 cm)

- **APPEARANCE** Fairly small, slender-billed alcid. Guadalupe Murrelet has more white on face; on Scripps's, thin white eye arcs stand out against black cap. Underwing whitish in all plumages; underwing of Craveri's darker gray. Further distinguished from browner Craveri's, in close view, by white notch at base of cap in front of eye.
- **VOCALIZATIONS** Call a series of squeaky, chippering *peep* notes, run together.
- **BEHAVIORS** Nests on steep sea cliffs mostly on arid islands offshore; in breeding season pairs often seen at sea a few miles offshore. Heads farther out to sea after breeding.
- **POPULATIONS** The default *Synthliboramphus* murrelet off Southern Calif., especially winter to early summer.

Guadalupe Murrelet 🇪🇳
Synthliboramphus hypoleucus | GUMU | L 9¾" (25 cm)

- **APPEARANCE** Same size and shape as Scripps's Murrelet, but bill of Guadalupe slightly longer and thinner. Best mark is face pattern: White on face of Guadalupe extends to front of eye, wrapping back to meet upper eye arc. Both species have pale underwing.
- **VOCALIZATIONS** Call a rapid series of dry *twit* notes, sharply down-slurred, like Chimney Swift's (p. 84).
- **BEHAVIORS** Nesting and dispersal habits much like Scripps's. At San Benito I., Baja California, where both species breed, Guadalupe may nest slightly later than Scripps's.
- **POPULATIONS** Endangered; breeds regularly only on islands off Baja California. Rare even in Southern Calif. waters, yet has vagrated to waters as far north as B.C.

Craveri's Murrelet
Synthliboramphus craveri | CRMU | L 8½" (22 cm)

- **APPEARANCE** A bit smaller overall than Scripps's and Guadalupe, with slightly longer bill and plumage browner above than both. Underwing dusky gray; blackish sides bulge forward on upper breast. Blackish cap cuts cleanly across face, without small whitish notch of Scripps's or white in front of eye of Guadalupe.
- **VOCALIZATIONS** Call a short, rapid trill like Guadalupe's, but notes shriller.
- **BEHAVIORS** Like Guadalupe, breeds outside our area and normally occurs here only at sea, where typically seen in pairs.
- **POPULATIONS** Like Guadalupe, disperses to our waters late summer to fall; sightings increase in years with warmer ocean temperatures.

Accidental to Hawaii

Ancient Murrelet
Synthliboramphus antiquus | ANMU | L 10" (25 cm)

- **APPEARANCE** Largest *Synthliboramphus* murrelet; stubby bill is pale. In all plumages, gray back contrasts with darker head; white throat wraps broadly behind eye. The bird is "ancient" because of the wispy white plumes behind the eye, reduced on winter adult and lacking on juvenile.
- **VOCALIZATIONS** Variable, but most calls heard at sea are high and squeaky.
- **BEHAVIORS** Digs burrows at nesting colonies, often at edge of conifer forests by seashore. Many feed at sea, but many others enter bays and estuaries in scattered flocks.
- **POPULATIONS** Vagrants wander well inland every year in late fall; most are "one-day wonders," gone the next morning. Accidental to Hawaii.

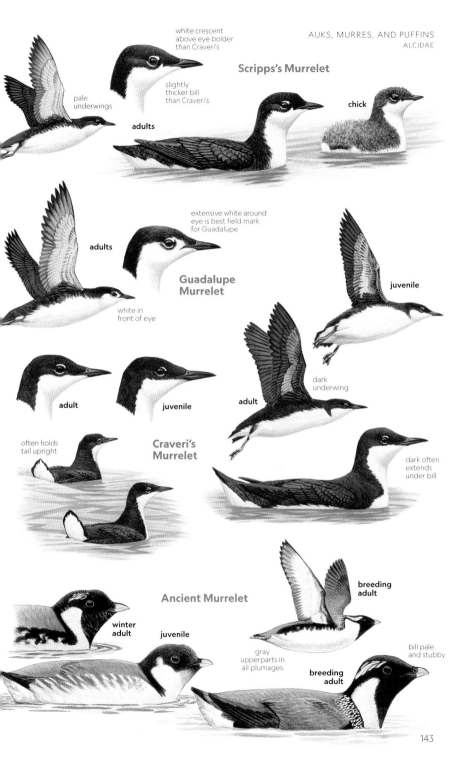

Least Auklet
Aethia pusilla | LEAU | L 6¼" (16 cm)

- **APPEARANCE** Tiny; smaller than Dovekie (p. 138) and, like that species, small-billed and compactly built. Breeding adult varies from mostly dark slaty overall to extensively white below; all breeders have white throat, red bill, and fine white plumes on face. Juvenile and winter adult mostly black above and white below with dark bill.
- **VOCALIZATIONS** Around nest, constant rough chatter; on arriving at or departing from colony, a short, descending screech, *RRrrreeee*.
- **BEHAVIORS** North Pacific ecological counterpart of Dovekie; both nest in huge colonies and feed at sea on copepods.
- **POPULATIONS** Abundant but concentrated; some colonies greater than one million birds. Winter range, well out at sea, largely unknown.

Parakeet Auklet
Aethia psittacula | PAAU | L 10" (25 cm)

- **APPEARANCE** Largest *Aethia* auklet, with odd, oversize bill. White eye and white plume behind eye contrast with otherwise dark upperparts. Bill of breeding adult red-orange; duller on winter adult and duller yet on juvenile.
- **VOCALIZATIONS** In addition to ill-defined chatter around nest, gives shrill trill, *r-r-r-r-r-r-r*, entering and leaving colony.
- **BEHAVIORS** Nests on rocky cliff faces with other auklets, but also in grassier, more protected sites. Less flocky than other *Aethia* auklets, and eats larger prey: fish, cnidarians, etc.
- **POPULATIONS** Prefers less cool ocean temperatures than other *Aethia* auklets; disperses south annually to latitude of Calif., well out at sea. Casual to Hawaii.

Casual to Hawaii

Whiskered Auklet
Aethia pygmaea | WHAU | L 7" (18 cm)

- **APPEARANCE** Small and compact, with small bill; built like Least Auklet, but larger overall. Adult dark overall with wispy facial plumes, ornate in high breeding plumage, more subdued otherwise; bill red-orange. Juvenile grayer, with plumes reduced or absent; bill dark gray.
- **VOCALIZATIONS** Around colony, descending squeals: *kyeeay, whyaaay,* etc.
- **BEHAVIORS** Nests on sea cliffs in company of other alcids. Feeds in well-mixed waters, sometimes close to shore, where copepods abound.
- **POPULATIONS** Unlike other auklets, stays close to breeding colonies year-round; can seem absent from a region except where a single huge feeding flock is concentrated.

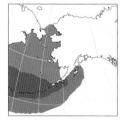

Crested Auklet
Aethia cristatella | CRAU | L 9¾" (25 cm)

- **APPEARANCE** Same heft as Parakeet Auklet, with thick, rounded bill. Dark overall, with bill orange on adult, nearly black on juvenile. Face marked with single white plume behind eye and curlicue of dark plumes jutting out of forehead, reduced on nonbreeding adult and juvenile.
- **VOCALIZATIONS** Short nasal yaps, usually given in flight as bird enters or leaves colony.
- **BEHAVIORS** Emits remarkable odor, a citrus-like "perfume," that both deters ectoparasites and attracts mates; closely related Whiskered Auklet does likewise. Nests in mixed-species colonies in company of other alcids; feeds on krill.
- **POPULATIONS** Disperses far from shore to cold, deep, well-mixed ocean waters, but does not migrate south. However, accidental south to Calif.

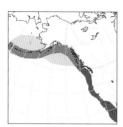

Cassin's Auklet
Ptychoramphus aleuticus | CAAU | L 9" (23 cm)
- **APPEARANCE** Round overall; dark bill is thick-based and fairly short. The plainest alcid; small white patch above white eye stands out against otherwise nondescript dark gray plumage.
- **VOCALIZATIONS** At colony, gives trills of rough chirps, *brrrrrttt*, repeated rapidly, often in chorus.
- **BEHAVIORS** Excavates burrows in varied microhabitats close to shoreline. At sea occurs both singly and in larger flocks; broad diet includes zooplankton and fish.
- **POPULATIONS** A Ring of Fire species, breeding from the Kurils all the way to Baja California Sur; breeders from both north and south converge in waters off B.C. fall to winter. Numbers vary with El Niño events.

Rhinoceros Auklet
Cerorhinca monocerata | RHAU | L 13¾" (35 cm)
- **APPEARANCE** Medium-large alcid, a bit smaller than our two puffins; large yellow bill of breeding adult has horn on upper mandible. Dark gray in all plumages. Breeding adult has long white plumes on face. Nonbreeding adult lacks plumes and horn. Bill of nondescript juvenile reduced and dusky gray-yellow; compare with Cassin's Auklet and juvenile Tufted Puffin.
- **VOCALIZATIONS** At colony, a sputtering rumbling, like a chain saw that won't start; also piercing *kyew* notes in rapid succession.

Accidental to Hawaiian Is.

- **BEHAVIORS** Nests in large colonies, visited at night. Eats fish in waters fairly close to shore; "Rhinos" and Cassin's Auklets, with similar feeding ecologies, often found in same waters.
- **POPULATIONS** Most breed off B.C. and Alas. Panhandle, with substantial movement southward in fall. Accidental to Hawaiian Is.

Horned Puffin
Fratercula corniculata | HOPU | L 15" (38 cm)
- **APPEARANCE** "Classic" puffin, with huge, parti-colored, laterally compressed bill; horn of breeding adult is a small protuberance above eye. Winter adult has dusky face and duller, but still impressive, bill. Bill of juvenile smaller and darker, like that of juvenile Rhinoceros Auklet, but juvenile Horned Puffin's plumage is white below.
- **VOCALIZATIONS** Low moans, heard only around nest; silent at sea and quiet even at colony.

Casual to Hawaiian Is.

- **BEHAVIORS** Nests in mixed-species colonies; unlike other puffins, does not dig its own burrows. At sea, typically seen sitting on water, away from other birds.
- **POPULATIONS** After nesting, goes far out to sea; tracks sea surface temperature, a proxy for fish availability. Casual to Hawaiian Is.

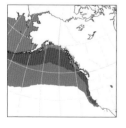

Tufted Puffin
Fratercula cirrhata | TUPU | L 15¾" (40 cm)
- **APPEARANCE** Largest puffin; the tuft refers to a swatch of blond behind eye on breeding adult. Nonbreeding adult lacks tuft; dusky juvenile, with bill smaller and duller, suggests Rhinoceros Auklet.
- **VOCALIZATIONS** Among even puffins, notably quiet; adults sometimes groan at nests.
- **BEHAVIORS** At mixed-species colonies, digs earthen burrows. Feeds singly in deep water on fish and squid.
- **POPULATIONS** Like Cassin's Auklet, enjoys wide breeding range. The only puffin likely to be seen on single-day boat trips out of ports south of Alas.

AUKS, MURRES, AND PUFFINS
ALCIDAE

GULLS, TERNS, AND SKIMMERS | LARIDAE

These are classic beach birds: gulls at seafood shacks, terns diving in waters just offshore, and skimmers roosting on spits. But they are strong fliers, and many get far inland; vagrancy is particularly pronounced in the gulls. "Shades of gray" is the color scheme with larids, except that many have brightly colored bare parts. Taxonomy is vexed; the family comprises about 100 species globally, with 50+ recorded in the West.

Casual to Hawaii

Black-legged Kittiwake
Rissa tridactyla | BLKI | L 17" (43 cm) WS 36" (91 cm)

- **APPEARANCE** Slightly smaller than Ring-billed Gull (p. 154); all plumages have black legs. Adult, pale gray above, has unmarked yellow bill; black primary tips contrast with rest of wing. Breeding adult has white head, smudgier on nonbreeding adult, with vertical gray splotch behind eye. First-winter has black collar and vertical black mark behind eye; in flight shows black M above and black-tipped tail.
- **VOCALIZATIONS** At colony, *eenie-eenk! (kittiwake)*, given repeatedly.
- **BEHAVIORS** Nests on sheer sea cliffs; feeds and winters widely offshore. Wingbeats of kittiwakes stiff and ternlike.
- **POPULATIONS** Very rare but regular inland, mostly late fall, in interior; casual to Hawaii.

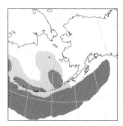

Red-legged Kittiwake
Rissa brevirostris | RLKI | L 15" (38 cm) WS 33" (84 cm)

- **APPEARANCE** Stockier than Black-legged Kittiwake, with shorter bill. Short legs are bright red on adult, duskier on juvenile. Adult darker-mantled than adult Black-legged. First-winter, with white triangle on wing in flight, suggests Sabine's Gull, but tail is all-white.
- **VOCALIZATIONS** Calls like Black-legged's, but *kittiwake* call ends on higher, squeakier note.
- **BEHAVIORS** Nests colonially on sea cliffs. Feeds at sea over deep water, often at night.
- **POPULATIONS** Breeds only at a few sites in Bering Sea. Winter range poorly known, thought to be mostly south of Aleutians.

Ivory Gull
Pagophila eburnea | IVGU | L 17" (43 cm) WS 37" (94 cm)

Supremely well-adapted to high latitudes, normally spends its entire life around Arctic pack ice, where it feeds on offal and keeps company with polar bears. Bulky for its size; short-billed, black-legged. Adult feathers pure white, bill yellow-gray; first-winter mottled dusky above. Uncommon wanderer in winter to Chukchi and northern Bering Seas; casual farther south.

Sabine's Gull
Xema sabini | SAGU | L 13½" (34 cm) WS 33" (84 cm)

- **APPEARANCE** Small, the size of Bonaparte's Gull (p. 150); tail notched. Striking in flight, flashing broad swaths of dark and white. Adult has dark bill with yellow tip; wears slaty hood in summer, smudgy half hood by fall. Juvenile brown above with fine white barring. Tail of adult white; tail of juvenile has black tip.
- **VOCALIZATIONS** Calls grating and ternlike, some notes rising, others falling.
- **BEHAVIORS** Nests on marshy tundra; precocial young can fly before fully feathered. Migrates mostly well offshore.
- **POPULATIONS** Regular inland passage in very small numbers in fall involves mostly juveniles.

Casual to Hawaii

Bonaparte's Gull
Chroicocephalus philadelphia | BOGU | L 13½" (34 cm) WS 33" (84 cm)

- **APPEARANCE** Smallest gull likely to be encountered in most birding situations in the West. Slight bill is black in all plumages; legs red on breeding adult, duskier on immature and winter adult. Breeding adult has black hood; mostly white-headed nonbreeding adult and immature have black blob behind eye. Wings distinctive in flight, but differ between adult and first-winter: Adult has flickering wedge of white on outer half of wing; first-winter has black M across upperwing like Black-legged Kittiwake (p. 148).
- **VOCALIZATIONS** Harsh, rising snarls, seemingly not befitting so graceful a gull.
- **BEHAVIORS** Wingbeats fluttery and tentative. Nests in conifer forests and, surprisingly, *in* the trees. In winter in West, mostly along and near coasts, foraging around breakwaters, marinas, and bays.
- **POPULATIONS** Widespread but uncommon inland, with more sightings fall than spring in most places; casual in winter to Hawaii. Occurrence inland reflects availability of open water with small fish.

Black-headed Gull
Chroicocephalus ridibundus | BHGU | L 16" (41 cm) WS 40" (102 cm)

Scarce visitor from Eurasia; similar to Bonaparte's Gull. Black-headed is longer-billed and larger overall. Bill of adult red, paler pink on immature; hood of adult dark brown. Adult in flight shows white wedge on outer wing, but note contrasting black below, which is lacking on Bonaparte's. First-winter bird has black smudge behind eye like Bonaparte's, but wing different: upperwing with broader black trailing edge and smudgy, ill-defined carpal bar; remiges dark below. Gives ternlike squawks, higher and less grating than Bonaparte's. Rare but annual in Bering Sea region, casual to U.S. West Coast; accidental inland in West and to Hawaii.

Little Gull
Hydrocoloeus minutus | LIGU | L 11" (28 cm) WS 24" (61 cm)

Like Black-headed Gull, a chiefly Old World species. The tiniest gull on Earth, smaller than Bonaparte's. Breeding adult has dark hood and unique wings: pale gray above with white tips; black below, also with white tips. Winter adult has smudge behind eye like Bonaparte's. First-winter in flight like first-winter Black-legged Kittiwake (p. 148), but not as "neat." Calls higher, more strident than Bonaparte's. Annual in West, mostly lower 48 states, with preponderance of records from Pacific coast states.

Accidental to Hawaii

Ross's Gull
Rhodostethia rosea | ROGU | L 13½" (34 cm) WS 33" (84 cm)

- **APPEARANCE** Very small gull, size of Bonaparte's, with wedge-shaped tail and pointed wings. Breeding adult has pink breast and unique black collar. Winter adult uniform soft gray above; underwing darker gray. First-winter, with black tail tip and dark M in flight, suggests Black-legged Kittiwake (p. 148).
- **VOCALIZATIONS** Call notes nasal and squeaky, calling to mind a coot or gallinule (p. 98).
- **BEHAVIORS** Flight buoyant, ternlike. Nests outside our region on marshy tundra, winters mostly at edge of pack ice.
- **POPULATIONS** Migrates east across Beaufort Sea, with impressive passage early to mid Oct.; single-day counts of thousands at Utqiagvik. Casual in West south of Alas.; accidental to Hawaii.

Laughing Gull
Leucophaeus atricilla | LAGU | L 16½" (42 cm) WS 40" (102 cm)
East Coast counterpart of Franklin's Gull, yet common in season at Salton Sea. Long-winged, with long, drooping bill. Legs black; bill dark red in breeding adult, black otherwise. White eye arcs of adult not as prominent as on Franklin's. On adult in flight, black wing tips blend smoothly with dark gray of rest of upperwing; primaries dark below. Juvenile and first-winter smudgy brown, including breast. Winter adult and immature dusky-headed, lacking half hood of Franklin's. Calls like Franklin's, but lower and richer. Visitants to Salton Sea, mostly summer to fall, number in hundreds. Annual in small numbers, mostly winter to spring, to Hawaii; casual elsewhere in West.

Casual to Hawaii

Franklin's Gull
Leucophaeus pipixcan | FRGU | L 14½" (37 cm) WS 36" (91 cm)
■ **APPEARANCE** Similar to congeneric Laughing Gull, but Franklin's has shorter bill, shorter wings, and smaller body. Black band on outer primaries of adult bordered by white tips and by white bar that separates it from gray mantle. Primaries from below pale in all plumages; dark in Laughing. White eye arcs of adult more prominent than on Laughing. Breeding adult has dark red bill, pink-tinged breast. Juvenile and first-winter mostly white on belly and breast; Laughing duskier below. All winter plumages show half hood, largely absent on winter Laughing. Compare also with juvenile Sabine's Gull (p. 148), especially standing or swimming.
■ **VOCALIZATIONS** Nasal squeals like Laughing's, but higher, more strident and urgent-sounding.
■ **BEHAVIORS** Nests around large cattail marshes; occurs in marshes on migration too, but also gets into agricultural districts. Like most gulls, feeds on almost anything, but this species and Laughing often "flycatch" for airborne arthropods.
■ **POPULATIONS** Migration generally east of Pacific coast states; arrives early spring, departs mostly late summer; casual to Hawaii. Winters along west coast of S. Amer.

Heermann's Gull
Larus heermanni | HEEG | L 19" (48 cm) WS 51" (130 cm)
■ **APPEARANCE** A bit larger than Ring-billed Gull (p. 154); long-winged. Darker throughout life cycle than our other *Larus* gulls. Breeding adult light gray below, darker gray above, and white-headed; bill mostly red. Nonbreeding adult darker, especially on dusky-streaked head. Immature dark brown; two-toned bill is horn-colored at base, dark at tip. Transitions from scaly juvenile with warm tones to solidly colored third-winter with cold tones and adultlike bill. Young birds at sea, long-winged and dark-bodied, call to mind dark-morph jaegers (pp. 134–137).
■ **VOCALIZATIONS** Calls short and nasal. Some are low-pitched, with humanlike quality: *aww, uh,* etc.
■ **BEHAVIORS** Seen mostly right along coast or offshore; roosts in large flocks on beaches, but also forages singly and in small flocks. Follows Brown Pelicans (p. 204) north from Baja California and steals fish from them.
■ **POPULATIONS** Disperses north along coast following winter nesting, mostly in Mexico; present north annually to B.C., with decided uptick starting in late spring and running through fall. Regular in very small numbers to Ariz., casual elsewhere in Interior West.

Accidental to Hawaii

Short-billed Gull
Larus brachyrhynchus | SBIG | L 16–18½" (41–47 cm) WS 43½–49½" (110–126 cm)

■ **APPEARANCE** Smallest *Larus* gull in the West. Petite build; Ring-billed, also small, looks bulkier. Small bill unmarked yellow on adult, two-toned on immature. Adult year-round has dark eye, medium-dark mantle, and, at rest, prominent white wing crescents; adult Ring-billed has pale eye, paler mantle, and smaller wing crescents. Larger California Gull (p. 158) also has medium-dark mantle and dark eye. Head of winter adult Short-billed streaked dusky, often densely so. First-winter duskier overall and less contrastingly patterned than first-winter Ring-billed. Second-winter similar to adult, but bill not yet plain yellow.

■ **VOCALIZATIONS** Squeals like other *Larus* gulls, but shorter and yappier than those of the larger species.

■ **BEHAVIORS** Nesting behavior the most protean of any gull in the West: Nests on the ground, but also in trees; in saltwater habitats coastally, but also inland in freshwater marshes and meadows; and in colonies ranging from a handful of pairs to hundreds. Winters at estuaries and beaches, but also on ball fields and around sewage outlets. Rides high on water, like a cork; Ring-billed and other *Larus* gulls ride lower.

■ **POPULATIONS** Taxonomy complex; formerly lumped with Common Gull, *L. canus*, of Old World, rare but annual along East Coast from Long I. northward, as Mew Gull. One of the subspecies of Common Gull is "Kamchatka Gull," *kamtschatschensis*, of eastern Russia, very rare but annual to Bering Sea islands; it is, confusingly, intermediate in size and plumage, especially in first winter, between Short-billed and Ring-billed. Short-billed is rare but regular, mostly in winter, to Interior West; both subspecies accidental to Hawaii.

Annual to Hawaii

Ring-billed Gull
Larus delawarensis | RBGU | L 17½" (44 cm) WS 48" (122 cm)

■ **APPEARANCE** Second smallest *Larus* gull in West. Adult has yellow legs, pale eyes, and yellow bill with eponymous black ring; mantle about same shade of gray as Herring Gull's (p. 158). Juvenile uniformly pale brown; by first winter, gray and splotchy overall, with broad black tail band and boldly patterned wings. By second summer (one year old), is still splotchy, but getting grayer above. By second winter, resembles adult, except black wing tips lack white spots. Like Short-billed, reaches adulthood after second winter; most other *Larus* gulls do so after third winter.

■ **VOCALIZATIONS** Slurred squeals, nasal and distressed-sounding: *hyeee, hyee-ay*, etc. Also a lower *hyow*, given singly or in series.

■ **BEHAVIORS** Colonial ground-nester. Winters widely at reservoirs, landfills, strip malls, etc. Eats anything: offal, small vertebrates, even airborne arthropods. Easily approached; careful study of a flock of Ring-billed Gulls in winter, noting first-winter, second-winter, and adult individuals, is a superb way to learn gulls.

■ **POPULATIONS** In much of West, especially in winter and especially east of Cascade-Sierra axis, the default gull. Following earlier persecution, has increased in West; range still expanding and numbers still rising here. Nonbreeders frequently seen in summer far from mapped breeding range. Along immediate coast, can be relatively scarce, even in winter. Rare but annual, mostly winter, to Hawaii.

GULLS, TERNS, AND SKIMMERS
LARIDAE

Accidental to Hawaiian Is.

Western Gull
Larus occidentalis | WEGU | L 25" (64 cm) WS 58" (147 cm)

■ **APPEARANCE** Big-billed and large overall, same size as Herring Gull (p. 158). Adult Western is the default dark-mantled gull on West Coast; like paler-mantled Herring, has pink feet and yellow bill with red spot. Southern adults, with darker mantles and lighter eyes, a bit smaller than northern adults; southern adults in winter white-headed, northern adults with faint smudging. Compare juvenile, dark and muddy, with juvenile California Gull (p. 158). Plumage becomes gradually adultlike through first three winters; adult plumage acquired after third winter. Hybridizes extensively with Glaucous-winged Gull (p. 162), especially Wash. and B.C.; intergrades, called "Olympic Gulls," have wing tips and mantle intermediate in darkness between parental species.

■ **VOCALIZATIONS** Down-slurred, anguished, nasal squeals and higher, thinner notes; in many settings, the classic "seagull" cry of West Coast. Also gruff croaks, often in triplets.

■ **BEHAVIORS** A constant sight and sound of coastal districts, even in dense urban and commercial stretches; sightings from land usually at or near salt water. Gets out to sea too; is a common sight on day trips by boat.

■ **POPULATIONS** Breakpoint between breeding ranges of southern *wymani* and northern nominate *occidentalis* around Monterey Co., Calif., with many *occidentalis* getting south into range of *wymani* in winter. Huge colony of *occidentalis* breeds at Southeast Farallon I. off San Francisco. Species rarely vagrates: not even annual in Interior West; accidental to Hawaiian Is. Regular in small numbers to Salton Sea, where Yellow-footed Gull also occurs.

Yellow-footed Gull
Larus livens | YFGU | L 27" (69 cm) WS 60" (152 cm)

■ **APPEARANCE** Even larger and bigger-billed than Western Gull. Adult and older immatures differ most obviously from Western by bright yellow legs and feet; Yellow-footed also slightly darker-mantled. Odd for a large gull, appears almost adultlike after only the second winter. Juvenile and younger immatures pink-legged like corresponding age classes of Western, but colder gray and paler overall, especially below; belly almost entirely white. Compare also with Lesser Black-backed Gull (p. 160), very rare but increasing at Salton Sea.

■ **VOCALIZATIONS** Repertoire of calls lower, gruffer, and less squeaky than Western; Yellow-footed sounds mellower, less urgent.

■ **BEHAVIORS** Most in our area are seen at south end of Salton Sea, at or near water's edge; forages around debris-strewn beaches, often perches on tufa. Typically occurs singly or in small, loose flocks.

■ **POPULATIONS** Breeding range entirely within Sea of Cortez. Following nesting season, some migrate north, reaching Salton Sea in May; in this way, unexpectedly shares ecology with "Large-billed Savannah Sparrow" (p. 388), which also disperses north to Salton Sea following nesting season around Sea of Cortez. Numbers build to hundreds by midsummer; most depart again for breeding grounds by early winter. Formerly treated as subspecies of Western Gull, but molecular data affirm closer relationship with dark-backed Kelp Gull (p. 468) of Southern Hemisphere. Casual at best to coast of Southern Calif.; accidental in Interior West.

GULLS, TERNS, AND SKIMMERS
LARIDAE

Western Gull

Yellow-footed Gull

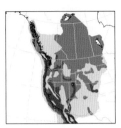

Accidental to Hawaii

California Gull
Larus californicus I CAGU I L 21" (53 cm) WS 54" (137 cm)

■ **APPEARANCE** Intermediate in heft between Short-billed (p. 154) and Herring Gulls. Adult has dark eye, dull yellow-green legs, and yellow bill with adjoining black and red spots; southern breeders about as dark-mantled as Short-billed, northern breeders not quite as dark above. First-year birds resemble subadult Herring, Lesser Black-backed (p. 160), and Ring-billed (p. 154) Gulls; but California has long, pink-based bill with dark tip. First-year California has pale legs and pale bill with dark tip, pale head and breast, and, in flight, double dark bars across inner wing. By second and third winters, begins to acquire adult characters: dark eye, yellow-green legs, medium-gray mantle.

■ **VOCALIZATIONS** Squeals midway in timbre between squeakier Ring-billed's and gruffer Herring's. Also a low *owl,* often doubled or trebled, a good indication of its presence amid Ring-billeds.

■ **BEHAVIORS** Nests colonially on islands in large waterbodies, often saline and sometimes at considerable elevation. Even among gulls, notably omnivorous: Famously rescued Mormons from a grasshopper plague in Utah in 19th century; "Cal Gulls" in our century still eat grasshoppers—along with anything else, animal or vegetable, alive or dead or processed, with a special penchant for cherries.

■ **POPULATIONS** Most find their way to Pacific coast region in winter; in Interior West, regular, but in most places scarce, in winter. Flocks on migration in interior, late winter to early spring and again late summer to early fall, number well into the hundreds. Accidental to Hawaii.

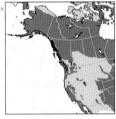

Casual to Hawaiian Is.

Herring Gull
Larus argentatus I HERG I L 25" (64 cm) WS 58" (147 cm)

■ **APPEARANCE** A bit smaller overall and slimmer than Western Gull (p. 156). Adult has pale mantle, dark wing tips, pink legs, yellow eye, and yellow bill with red spot; white-headed in breeding plumage, streaky-headed in winter. The "fun" starts with younger Herrings, easily confused with other large gulls. First-winter in flight shows contrastingly pale inner primaries, an excellent mark visible at some distance. Variation in first winter extreme, with some as dark as Lesser Black-backed Gull, others as pale as *kumlieni* Iceland Gull (p. 160). By second winter, most show pale eye and pale gray mantle of adult; third-winter even closer to adult appearance. ID at all ages complicated by hybridization, especially with Glaucous-winged and Glaucous Gulls (p. 162). Geographic variation a further complication: Old World "Vega" subspecies, *vegae,* is darker-mantled and darker-eyed as adult than widespread "American" subspecies, *smithsonianus;* immatures, with colder tones and paler rump, suggest Lesser Black-backed.

■ **VOCALIZATIONS** Anguished squeals, given singly or in series, average lower than those of Ring-billed.

■ **BEHAVIORS** Breeders in West nest mostly on the ground in colonies large or small. In Interior West on migration or in winter, often found amid Ring-billed Gulls. Birds making their way to coast in winter often get offshore 10+ miles (16+ km).

■ **POPULATIONS** Widespread south of breeding range late fall to early spring, but generally outnumbered inland by Ring-billed and coastally by Western. Casual to Hawaiian Is.; "Vega" regular in Bering Sea region; very rare but annual to lower 48 states.

GULLS, TERNS, AND SKIMMERS
LARIDAE

California Gull

- 1st winter — two dark bars across inner wing; primaries dark
- winter adult
- 2nd winter
- breeding adult
- breeding adult
- 1st winter — dark smudgy underparts
- pale juvenile
- dark juvenile

Herring Gull
smithsonianus

- winter adult
- juvenile
- 1st winter
- 1st winter — often pale-headed
- 2nd winter
- 3rd winter
- 1st winter
- breeding adult
- 1st winter *vegae*
- breeding adult *vegae*
- winter adult *vegae*
- 1st winter *vegae* — pale inner primaries

Accidental to Hawaiian Is.

Iceland Gull

Larus glaucoides | ICGU | L 22" (56 cm) WS 54" (137 cm)

■ **APPEARANCE** Built like a slimmed-down Herring Gull (p. 158); bill thinner. Most in West are subspecies *thayeri*, formerly treated as separate species, Thayer's Gull. Adult "Thayer's" has variably dark eye; dark on wing tips less extensive than on Herring, and often admixed with paler gray. First-winter, uniformly colored overall, ranges from pale fawn-gray to darker brown-gray; birds at dark end of spectrum resemble first-winter Herring or Western, whereas birds at pale end of spectrum resemble first-winter "Olympic Gull" (Western x Glaucous-winged; p. 156). Becomes progressively more adultlike through second and third winters. Eastern subspecies *kumlieni*, rare in West, is whiter than even pale *thayeri* and typically pale-eyed as adult, like Glaucous Gull (p. 162) at all ages; *kumlieni* smaller and slighter than *thayeri* and considerably smaller than Glaucous.

■ **VOCALIZATIONS** Squeals about the same pitch as Herring's, but not as nasal.

■ **BEHAVIORS** Nests mostly on sea cliffs. In winter, mainly coastal habitats; winterers in interior often find their way to large assemblages of other gulls, especially Ring-billed (p. 154) and Herring.

■ **POPULATIONS** Status of "Thayer's" as breeder in West not well known; gets at least as far west as Banks I., N.W.T., in Beaufort Sea. Greatest numbers of "Thayer's" in winter reported from Alas. Panhandle to Ore.; status in winter farther north, perhaps to edge of pack ice, unclear. "Thayer's" is rare but regular, widespread in very small numbers, in winter in interior; *kumlieni* very rare but probably annual west to base of Rockies. "Thayer's" accidental to Hawaiian Is. (Midway).

Accidental to Hawaii

Lesser Black-backed Gull

Larus fuscus | LBBG | L 21" (53 cm) WS 54" (137 cm)

■ **APPEARANCE** A bit smaller than Herring Gull (p. 158), with smaller bill and subtly odd build: long-winged, long-necked, and lanky overall. Mantle of adult about as dark as southern *(wymani)* Western Gull (p. 156); has yellow legs like adult Yellow-footed, but Lesser Black-backed is smaller and more slender overall, with a slighter bill. Mantle color on all gulls subject to viewing angle, tricks of light, and camera settings. First-winter distinguished from first-winter Herring by slighter build, longer wings, and mostly white head contrasting with cold, dark upperparts; in flight, first-winter Lesser Black-backed shows mostly dark wings, lacking pale panel of first-winter Herring. By second winter, starts to acquire yellow legs and dark mantle of adult; usually acquires adult plumage after third winter.

■ **VOCALIZATIONS** Like other large *Larus* gulls, issues slurred squeals and nasal squawks, averaging more nasal and lower-pitched than corresponding calls of Herring.

■ **BEHAVIORS** Wanderers to West find their way to gatherings of other *Larus* gulls at lake edges, ice shelves, and landfills.

■ **POPULATIONS** Sightings in West have increased greatly in recent decades, but still uncommon at best in most places, chiefly late fall to early spring. Until recently, bred only in Old World, but recently established on Greenland. In interior, the most expected dark-mantled *Larus* gull. Almost all in West known or presumed to be subspecies *graellsii*, but *heuglini* ("Heuglin's Gull") has reached Aleutians, and *taimyrensis* ("Taimyr Gull") has reached Northwestern Hawaiian Is.

GULLS, TERNS, AND SKIMMERS
LARIDAE

Iceland Gull
thayeri

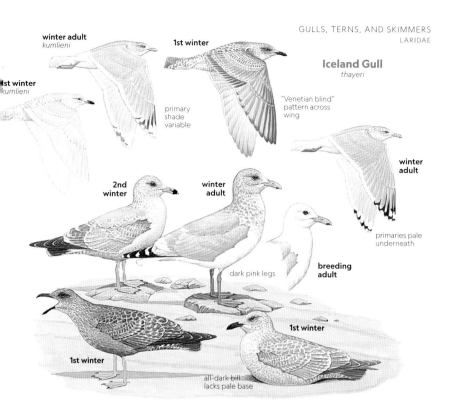

Lesser Black-backed Gull
graellsii

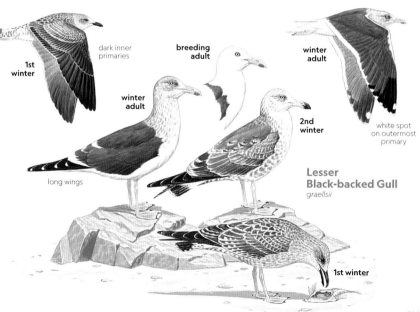

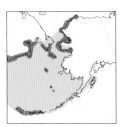

Casual to Hawaii

Slaty-backed Gull

Larus schistisagus | SBAG | L 25" (64 cm) WS 58" (147 cm)

■ **APPEARANCE** About same length as Herring (p. 158) and Western (p. 156) Gulls, but bulkier; bill large. Legs pink, brightly so on many adults. Adult in flight shows broad white trailing edge of most of dark wing, but outer primaries black with white spots. Winter adult dark-backed with dusky head and dark smudge across eye; breeding adult, seen in West mostly in Alas., has unmarked white head. First-winter combines dark tail and "milk chocolate" primaries with mostly white head. Transitions to adult plumage through second and third winters.

■ **VOCALIZATIONS** Call repertoire like that of other large *Larus* gulls, but notably low, rich, and nasal.

■ **BEHAVIORS** Associated with coastal lowlands throughout life cycle. Omnivorous like other large gulls, with penchant for crowberries, genus *Empetrum*.

■ **POPULATIONS** Breeds mostly northeastern Asia, with nesting records in recent years in Alas. Regular in winter in very small numbers, but increasing, along West Coast south to central Calif.; casual to Hawaii and interior.

Annual to Hawaii

Glaucous-winged Gull

Larus glaucescens | GWGU | L 26" (66 cm) WS 58" (147 cm)

■ **APPEARANCE** Size and shape similar to Western Gull (p. 156). At all ages, wing tips about same color as mantle. Adult has dark eye, pink legs, and yellow bill with red spot. Dusky mottling on head of winter adult relatively fine, not as splotchy as on winter adults of most other large gulls. First-winter variable in darkness; color of pale individuals can match color of dark first-winter Glaucous, but Glaucous larger, with two-toned bill. Transitions to adult plumage through second and third winters.

■ **VOCALIZATIONS** Repertoire like that of Western, but calls of Glaucous-winged less anxious, not as pure-tone.

■ **BEHAVIORS** Most nest colonially on sparsely vegetated islands, but the species has started using buildings and bridges. In winter, often right around coast, but also ranges farther from salt water for feeding than does Western.

■ **POPULATIONS** At southern edge of range hybridizes extensively with Western; also hybridizes with Herring and Glaucous Gulls where ranges overlap. A few wander far inland every winter; annual to Hawaii.

Casual to Hawaii

Glaucous Gull

Larus hyperboreus | GLGU | L 27" (69 cm) WS 60" (152 cm)

■ **APPEARANCE** Largest gull in West; bill huge. All have white wing tips; Iceland Gull (p. 160) in West usually has at least some dark in wing tips. Mantle of adult paler than that of Iceland; younger birds vary from pale fawn to nearly pure white. Adult has pink legs, pale eyes, and yellow bill with red spot. First-winter has dark eye and two-toned bill; by second winter, begins to acquire pale eye and mantle of adult.

■ **VOCALIZATIONS** The "seagull squeal" of Glaucous often slurs downward. Like other large gulls, gives low, rough growl, particularly nasal in this species.

■ **BEHAVIORS** Nests on open tundra as well as steep sea cliffs; even on far northern breeding grounds, finds its way to dumps, buildings, marinas, etc. In winter, typically in mixed-species flocks.

■ **POPULATIONS** The most Arctic of the *Larus* gulls, common across circumpolar breeding range. Widespread in interior in winter, but quite scarce; casual to Hawaii.

GULLS, TERNS, AND SKIMMERS
LARIDAE

Slaty-backed Gull

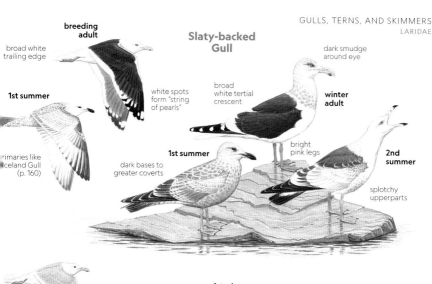

Glaucous-winged Gull

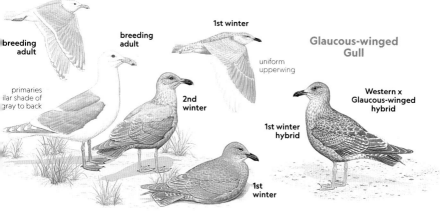

Glaucous Gull

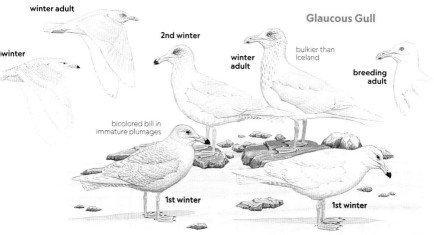

Brown Noddy

Anous stolidus | BRNO | L 15½" (39 cm) WS 32" (81 cm)

Colonies on Northwestern Hawaiian Is.; regular to main islands

- **APPEARANCE** Largest noddy; tail blunt-tipped and wedge-shaped. Brown Noddy longer-legged and stouter-billed than Black. Plumage mostly brownish; Black is blacker overall. Adult Brown has frosty gray crown, reduced on immature; pale cap of Black is more sharply demarcated. Compare also with juvenile Sooty (p. 166) and adult Black (p. 168) Terns.
- **VOCALIZATIONS** At nesting colony, a grinding, descending *grrraaaaw*, like a large alcid.
- **BEHAVIORS** Despite name, noddies are a kind of tern; their bowing, or "nodding," courtship display is arresting. Instead of plunge-diving like other terns, noddies glean fish and squid from surface, both in flight ("dipping") and while resting on water ("surface seizing"). Usually nests on ground.
- **POPULATIONS** Restricted in West to Hawaiian Is.; most colonies Northwestern Hawaiian Is., but regularly seen from main islands.

Black Noddy

Anous minutus | BLNO | L 13½" (34 cm) WS 30" (76 cm)

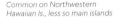

Common on Northwestern Hawaiian Is., less so main islands

- **APPEARANCE** Smaller than Brown Noddy, with thinner bill and shorter legs; tail of Black Noddy proportionately shorter. Black Noddy blacker overall than Brown, with more sharply demarcated pale cap, especially on immature. "Hawaiian Black Noddy," *melanogenys*, has orange legs; more widespread black-legged *marcusi* also occurs in Hawaiian Is.
- **VOCALIZATIONS** Repertoire like that of Brown, but Black rattles more and groans less; calls of Black higher and harsher.
- **BEHAVIORS** Habitually nests in shrubs and small trees; Brown does so less frequently. Like Brown, feeds by dipping and surface seizing.
- **POPULATIONS** Most on main Hawaiian islands and associated waters are *melanogenys*; dark-legged birds also on and near Northwestern Hawaiian Is.

Blue-gray Noddy

Anous ceruleus | BGNO | L 10–11" (25–28 cm) WS 18–24" (46–61 cm)

Widespread but infrequently encountered tropical Pacific tern. Tiny, with forked tail. Plumage soft blue-gray; bill thin and dark. Like larger noddies, feeds by dipping; flight fast, erratic. Call grating like larger noddies, rises slightly in pitch. Occurs in our area in Hawaiian Is., with almost all sightings restricted to Northwestern Hawaiian Is.; a few records, mostly from boat trips, off Kauai.

White Tern

Gygis alba | WHTT | L 11–13" (28–33 cm) WS 27–34" (69–86 cm)

Common on Northwestern Hawaiian Is.; locally common Oahu, casual other main islands

- **APPEARANCE** Slim, with shallowly forked tail. Bill, broad at base and sharply pointed, distinctive, like a large thorn; black eye, encircled by thin band of black feathers, appears huge; makes bird look like caricature of an extraterrestrial. Plumage snow-white; also called Fairy Tern.
- **VOCALIZATIONS** Calls, harsh and clacking, include rapping chatter and grinding buzz.
- **BEHAVIORS** Everything about this bird is bizarre and wonderful. Suddenly started nesting in downtown Honolulu in 1960s; precariously lays single egg on bare branch in improbably tall tree; no nest is built. After hatching, fledgling clings to tree trunk like a swift; paired adults cuddle in trees, indifferent to human admirers.
- **POPULATIONS** Characteristic sight in downtown Honolulu. Seen widely elsewhere, mostly at sea, in Hawaii region. Taxonomy unresolved; may actually comprise multiple species.

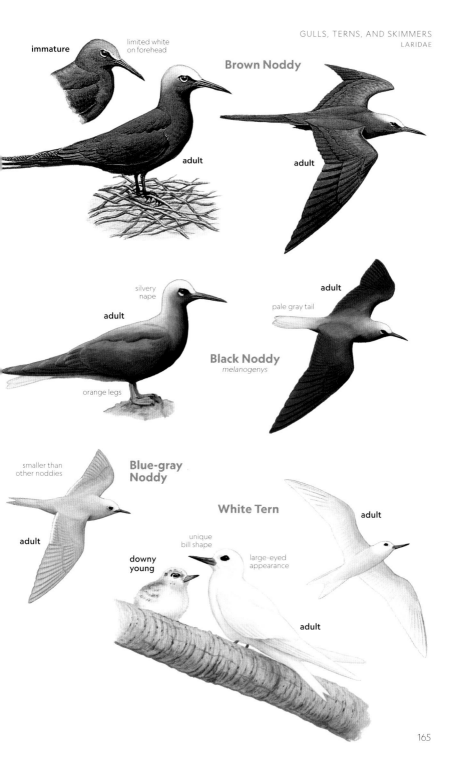

Sooty Tern
Onychoprion fuscatus | SOTE | L 16" (41 cm) WS 32" (81 cm)

Common but local breeder Hawaiian Is.

- **APPEARANCE** Bigger, bulkier than *Sterna* terns (p. 170). Adult nearly black above, white below. Black tail has bold white edges; nape and crown mostly black. White forehead patch ends at eye; extends behind eye on adult Gray-backed Tern. Juvenile, dark sooty all over, is heavily spotted pale above; darker-bodied overall than juvenile Gray-backed. Compare also with noddies (p. 164).
- **VOCALIZATIONS** Gruff squawks, rising in pitch, *riiit* and *riiih*. Also shriller squeaks, disyllabic or trisyllabic, *eeetie* or *ee-weetie;* an alternative name for the species, "Wide-awake," is an allusion to this call.
- **BEHAVIORS** Except at nesting colonies, almost always seen only in flight; sometimes alights on floating debris and even breached sea turtles, but not on water's surface. Catches food by "dipping" like noddies, but also snags airborne fish jumping from ocean.
- **POPULATIONS** Common within range, but numbers sometimes plummet following El Niño events. Adults and juveniles segregate to some extent at sea. Regular in West only in Hawaii region; casual to southern coastal Calif., accidental inland in West.

Gray-backed Tern
Onychoprion lunatus | GBAT | L 14–15" (36–38 cm) WS 29–30" (74–76 cm)

Breeds Hawaiian Is.; largely absent winter

- **APPEARANCE** Smaller than Sooty Tern. Back and wings of adult medium gray, markedly paler than black on nape and crown; white on forehead extends well behind eye, forming eyebrow. Juvenile paler than juvenile Sooty and not as prominently spotted.
- **VOCALIZATIONS** Repertoire matches that of Sooty, but calls average squeakier, less growling. *Wide-awake* call of Gray-backed even more clearly enunciated than that of Sooty.
- **BEHAVIORS** Shier and flightier than Sooty; at mixed-species colonies, harassed by more aggressive Sooty. For foraging, prefers waters with higher salinity than favored by Sooty, but much overlap. Perches on flotsam more than Sooty. Submerges on occasion; Sooty never does. Even forages on land, taking terrestrial arthropods.
- **POPULATIONS** Occurs only in tropical Pacific, with sizable numbers breeding on Hawaiian Is., mostly Northwestern Hawaiian Is. Sightings from main islands usually on summer pelagics; even then, rarer than Sooty.

Aleutian Tern
Onychoprion aleuticus | ALTE | L 13½" (34 cm) WS 29" (74 cm)

- **APPEARANCE** Smallest *Onychoprion* tern. More likely to be confused in West with Common and Arctic Terns, genus *Sterna*. Adult Aleutian mostly medium-gray above and below; white tail is relatively short, black bill is slight. Like Sooty and Gray-backed Terns, has black cap with white forehead. Distinctive juvenile warm cinnamon with sharp, pale scalloping above; feet orange.
- **VOCALIZATIONS** In aggressive encounters, a longspur-like rattle; around colony, descending whistles and relaxed twittering.
- **BEHAVIORS** Often nests in company of Common and Arctic Terns; during breeding season, feeds both offshore and inshore, getting 10+ miles (16+ km) inland to marshes and bogs.
- **POPULATIONS** Present May to Aug. at breeding grounds in Alas. Winter range not fully documented; mostly or entirely waters off Southeast Asia.

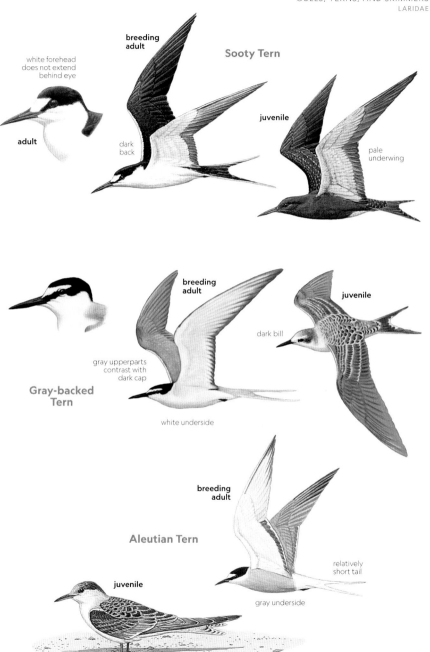

Annual to Hawaii; occasional breeder

Least Tern
Sternula antillarum | LETE | L 9" (23 cm) WS 20" (51 cm)
- **APPEARANCE** The smallest tern; tail short. Breeding adult has yellow bill and legs, darkening by summer's end; dark outer primaries create thin wedge on wing of bird in flight. Juvenile and first-summer show dark "shoulder" patch (carpal bar).
- **VOCALIZATIONS** Calls abrupt and irate; some are squeaky (*k'deet, kyeet*), others harsher, more nasal (*rreent, krrrp*).
- **BEHAVIORS** Nests in the open, where aggressive toward intruders, including humans. Wingbeats stiff, less fluid than larger terns'.
- **POPULATIONS** Two weakly differentiated subspecies, both endangered, breed in West: inland *athalassos*, uncommon on major river systems of western Great Plains; *browni* of coastal Calif. and Hawaii. Very similar Little Tern, *S. albifrons* (p. 459), occurs in tiny numbers on Midway Atoll.

Gull-billed Tern
Gelochelidon nilotica | GBTE | L 14" (36 cm) WS 34" (86 cm)

Enjoys wide range globally, but regular in West only at a few spots in Southern Calif. Morphology and behavior distinctive. Stocky, broadwinged, and short-tailed, with long, black legs; lower mandible of thick, black bill has gull-like expansion (gonys). Breeding adult gray above; juvenile and winter adult largely white. Forages on arthropods, which it captures in fields, in the air, and even from shrubs. Distinctive call a two-note *whit-week!* Breeds (subspecies *vanrossemi*) in small numbers at Salton Sea and in even smaller numbers in South San Diego Bay; accidental elsewhere in West, including Hawaii.

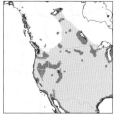
Casual to Hawaii

Caspian Tern
Hydroprogne caspia | CATE | L 21" (53 cm) WS 50" (127 cm)
- **APPEARANCE** Largest tern on Earth; size of California Gull (p. 158). Royal Tern (p. 172) nearly as long, but not as hefty. Bill of Caspian thick and dark red; crest small or absent. In flight, shows broad dark wedge across underwing. Cap solidly black in breeding adult, mottled black and white in nonbreeding adult and juvenile.
- **VOCALIZATIONS** Adult gives an impressive *kraawwr* or *krrock*, juvenile a thin wavering whistle. Flying in tandem, parent and offspring alternate those two utterly different calls.
- **BEHAVIORS** Nests colonially in both freshwater and marine settings. Eats fish, which it captures in impressive high dives; sometimes steals food from other terns and even gulls.
- **POPULATIONS** Increasing and expanding in West, although unevenly so, with particular expansion of breeding range coastward. Casual to Hawaii.

Casual to Hawaii

Black Tern
Chlidonias niger | BLTE | L 9¾" (25 cm) WS 24" (61 cm)
- **APPEARANCE** Quite small; the only dark tern found regularly in West away from Hawaii. Breeding adult mostly black, but with gray wings and white belly. Nonbreeding adult and juvenile paler above, with underparts largely white; head splotchy black and white.
- **VOCALIZATIONS** Clipped *pip* or *peep*, often heard from birds on the wing.
- **BEHAVIORS** Nests in freshwater marshes well inland, where it feeds mostly on insects captured in flight; also eats small fish. Migrates in large flocks.
- **POPULATIONS** Overland migrants widely noted in West, with "fall" passage underway by late July. Winters offshore, but well south of our region. Casual to Hawaii.

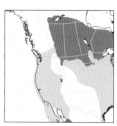

Casual to Hawaii

Common Tern
Sterna hirundo | COTE | L 14½" (37 cm) WS 30" (76 cm)

■ **APPEARANCE** Longer-winged and more slightly built than Forster's Tern; Arctic Tern, quite similar in build, has shorter feet and legs. Bill and feet of breeding adult Common dark red; wings plain gray above, underparts washed pale gray, and tail black-edged. Nonbreeding adult has extensive dark wedge on gray wings; note also dark "shoulder" patch on subadult and nonbreeding adult. Never shows dark eye patch of Forster's; instead, has black partial cap in nonbreeding plumages.

■ **VOCALIZATIONS** Calls, not as grinding as Forster's, are clipped and squeaky; also a longer, slightly descending *kyaaaaar*.

■ **BEHAVIORS** On migration overland, drops in at lakes and large rivers where Forster's also occurs; Common in flight has easier, more sweeping motions than Forster's.

■ **POPULATIONS** Very rare spring migrant inland in West; less uncommon in fall, with passage mostly Aug. to Sept. Uncommon, mostly summer to fall, off West Coast. East Asian *longipennis*, darker gray overall with blackish legs and feet, rare to Bering Sea region in spring. Casual year-round to Hawaii.

Casual to Hawaii

Arctic Tern
Sterna paradisaea | ARTE | L 15½" (39 cm) WS 30" (76 cm)

■ **APPEARANCE** Compared to other *Sterna* terns, short-billed and short-legged. Breeding adult has dark red bill and feet; underparts washed gray, wings plain gray above. Juvenile has weaker "shoulder" patch than juvenile Common. Arctic and Common ("Commic Terns") in flight give different impressions, but beware tricks of light: Upperwing of Arctic relatively uniform, pale and translucent, in all plumages; upperwing of Common splotchier, with dark wedge on primaries (adult) and dark leading edge (young, winter adult).

■ **VOCALIZATIONS** Mixes clipped, raspy *chit* notes with squeakier *kyeet* and *kyeep* calls.

■ **BEHAVIORS** Like many other terns, plunge-dives for fish, but also adds flycatching to behavioral repertoire; closely related Common rarely does so.

■ **POPULATIONS** Many migrate to Antarctic pack ice after breeding. Rarer inland on migration than Common, mostly in fall. Has nested as far from core breeding range as Wash. and Mont. Casual spring and fall to Hawaii.

Forster's Tern
Sterna forsteri | FOTE | L 14½" (37 cm) WS 31" (79 cm)

■ **APPEARANCE** Largest *Sterna* tern, longer-tailed and shorter-winged than Common. Bill and feet orange in breeding adult. Streamer-like tail is white-edged; wings are shimmering white in flight. Smooth black cap of breeding adult transitions to black face mask by late summer; bare parts darken after breeding. Juvenile sports extensive cinnamon after fledging, then shifts to plumage like that of nonbreeding adult. Juvenile and nonbreeder lack dark "shoulder" patch of Common.

■ **VOCALIZATIONS** Calls, grinding and growling, are down-slurred: *grrreeet* and *grrraaah*.

■ **BEHAVIORS** Intermediate in behavioral ecology between Black (p. 168) and Common Terns. Forster's is a lover of marshes, but also roosts in mixed-species flocks on spits and gravel bars.

■ **POPULATIONS** Only tern restricted to N. Amer., and only *Sterna* tern likely in West in winter. Some forgo reproduction and spend the breeding season on the wintering grounds. Generally does not get out to sea.

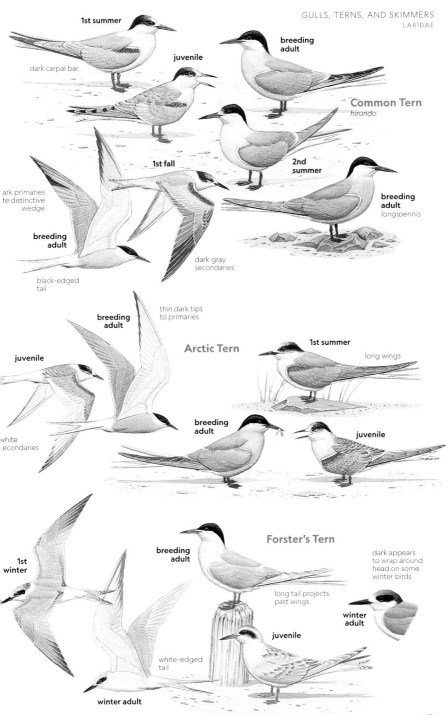

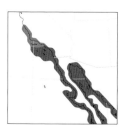

Royal Tern

Thalasseus maximus | ROYT | L 20" (51 cm) WS 41" (104 cm)

- **APPEARANCE** Along with Elegant, one of our crested terns; Royal is larger overall and thicker-billed. Crest of breeding Royal not as long as breeding Elegant's. Caspian Tern (p. 168), larger than even Royal, has huge red bill and smaller crest; tail of Royal more deeply forked than Caspian's, and wings paler below. Caspian has mostly or entirely dark cap in all plumages; nonbreeding adult and young Royal have white on forehead extending rearward.
- **VOCALIZATIONS** A rolling, wavering *keer-reet* or *keery-reet*, less harsh than *Sterna* terns' (p. 170) calls.
- **BEHAVIORS** A lover of salt water, especially where there are sandy beaches with warm waters. Nests in dense colonies, at which young are reared communally in crèches. Feeds chiefly by plunge-diving for fish.
- **POPULATIONS** Mostly southern coastal Calif. in West; some movement north along coast following breeding. Casual well inland, where some records may pertain to vagrants from Gulf of Mexico.

Accidental to Hawaii

Elegant Tern

Thalasseus elegans | ELTE | L 17" (43 cm) WS 34" (86 cm)

- **APPEARANCE** Smaller and slimmer than Royal Tern. Bill of adult Elegant notably thin and slightly decurved; bill of juvenile shorter, often tinged yellowish. Crest longer and wispier than Royal's, but beware the effects of wind, wear, and "bad hair." Breeding adult fully black-capped like Royal. Forehead and forecrown of nonbreeding adult and young whitish; Royal in corresponding plumages paler-capped.
- **VOCALIZATIONS** Calls include a clipped *k'd'rick*, falling in pitch; also more indeterminate clicks and squeals.
- **BEHAVIORS** Nests in mixed-species colonies, sometimes in heavily developed urban districts. Normally occurs only right along coast, roosting on beaches, or feeding inshore or out at sea a little ways. Like Royal, feeds by plunge-diving.
- **POPULATIONS** Many nest at single colony in Sea of Cortez, but new colonies have established recently in southern coastal Calif. Most along U.S. West Coast are present May to Oct.; numbers sometimes surge with El Niño events. Prone to vagrancy, with records well inland; accidental to Hawaii.

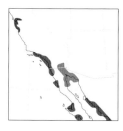

Black Skimmer

Rynchops niger | BLSK | L 18" (46 cm) WS 44" (112 cm)

- **APPEARANCE** About the size of a *Thalasseus* tern. Long-winged, with unique bill: laterally compressed, with lower mandible much longer than upper. Dark brown, almost black, above; white below. Bill reddish at base, black at tip; feet red.
- **VOCALIZATIONS** A plain, nasal *enk*, often given among flockmates when skimming.
- **BEHAVIORS** Feeding method alone identifies the species: Barely skims the water's surface with lower mandible; small flocks coordinate foraging with mesmerizing synchronicity. Feeding is tactile; many feed at dusk and even at night over inshore lagoons, canals, and inlets. Wingbeats graceful, strangely slow. Nests in mixed-species colonies.
- **POPULATIONS** A relatively recent arrival in the West. First detected in Calif. in 1960s, it began nesting almost immediately. Has expanded north to San Francisco Bay Area, where flocks in the hundreds now occur. Accidental inland to Ariz., even N. Mex. and Colo.

TROPICBIRDS | PHAETHONTIDAE

With their mostly white plumages, brightly colored bills, long tails, and habit of plunge-diving, these birds call to mind terns. But tropicbirds are in an entirely different order of birds, believed to be related to the Sunbittern, *Eurypyga helias*, of Neotropical swamps and the Kagu, *Rhynochetos jubatus*, of forests in New Caledonia.

Common on Hawaiian Is., especially Kauai

White-tailed Tropicbird
Phaethon lepturus | WTTR | L 30" (76 cm) WS 37" (94 cm)

■ APPEARANCE Smallest tropicbird. Central tail streamers of adult longer than tail of any tern. Bill of adult dull yellow (bright orange on most in East). In flight, adult has gleaming white upperparts with bold black patches across inner and outer wings. Juvenile lacks tail streamers; upperparts barred gray and white, and black on outer wing restricted to a few outer primaries.

■ VOCALIZATIONS At colony, drawn-out rasps, *r'r'r'r'a'a'a'a'*. Birds at sea sometimes utter an abrupt *rah* or *wah*.

■ BEHAVIORS Hawaiian breeders nest in small caves on steep cliffs; often seen from shore and at sea, but also over interior forests. Like other tropicbirds, feeds by plunge-diving. Flies well above surface; wingbeats faster, less languid than those of larger terns. Usually solitary at sea.

■ POPULATIONS Common in Hawaiian Is., especially Kauai; also breeds Midway. Accidental to coastal Calif. and even Ariz.

Casual to Hawaii

Red-billed Tropicbird
Phaethon aethereus | RBTR | L 40" (102 cm) WS 44" (112 cm)

■ APPEARANCE Larger than White-tailed Tropicbird. Extensive black on outer wing of streamer-tailed adult visible on flying bird at great distance; at closer range, note bright red bill, thick black line behind eye, and finely barred upperparts. Juvenile has yellow bill, black eyeline extending to nape, finely barred upperparts, and extensive black in outer wing.

■ VOCALIZATIONS Like other tropicbirds, rarely calls at sea; gives occasional yips and yaps *(reh, rah)*, sometimes in quick chatter.

■ BEHAVIORS Like all tropicbirds, feeds by plunge-diving and flies well above the ocean's surface; mixes gliding with rapid, shallow wingbeats.

■ POPULATIONS Regular off Southern Calif., usually far offshore, late summer to fall. Casual to Hawaii and, surprisingly, Ariz.

Breeds Hawaiian Is.

Red-tailed Tropicbird
Phaethon rubricauda | RTTR | L 37" (94 cm) WS 44" (112 cm)

■ APPEARANCE Heavyset tropicbird with relatively broad wings. Adult almost pure white, but with black primary shafts and splotchy black around eye. Red tail streamers of adult unique; but they are even thinner than those of other tropicbirds and thus can be hard to see. Juvenile combines black bill with barred upperparts and mostly white primaries.

■ VOCALIZATIONS At colony, clacks and squeaks, often run together in rapid series. Similar notes at sea; more vocal at sea than other tropicbirds.

■ BEHAVIORS Breeds colonially; nests often under shrubs. Wingbeats relatively slow for a tropicbird; mixes in short bouts of soaring. Like other tropicbirds, engages in plunge-diving and usually feeds solitarily.

■ POPULATIONS Widespread across Hawaiian Is.; common breeder Northwestern Hawaiian Is., local breeder on main islands. Casual far off U.S. West Coast, with preponderance of records off Southern Calif.

LOONS | GAVIIDAE

Known as "divers" outside the Americas, these birds submerge to considerable depths for minutes at a time. They catch fish with their daggerlike bills. Breeding adults are distinctive, but juveniles and winter adults can be a challenge to identify. All five of the world's loon species occur regularly in the West.

Red-throated Loon
Gavia stellata | RTLO | L 25" (64 cm)

- **APPEARANCE** Much smaller than Common Loon (p. 178), with thin bill, angled up. Breeding adult has gray head with maroon foreneck, but most adults seen away from breeding grounds are in winter plumage, strikingly white-faced. Juvenile like nonbreeding adult, but duskier. All plumages speckled white above in winter.
- **VOCALIZATIONS** Silent in winter when roosting or feeding, but in flight sometimes gives ducklike quacks. Yodel on breeding grounds shorter, harsher, and more nasal than famous wailing of Common.
- **BEHAVIORS** Nests in freshwater, low-elevation Arctic ponds and taiga peat bogs, avoiding Pacific Loons; switches to marine habitats on migration and in winter. Often feeds right along shoreline, even in surf.
- **POPULATIONS** Sensitive to annual variation in food availability; numbers fluctuate accordingly. Very rare migrant in West on large lakes.

Arctic Loon
Gavia arctica | ARLO | L 28" (71 cm)

- **APPEARANCE** More heavyset than closely related Pacific Loon, with larger bill. Resting on water, shows white tuft on flanks; Pacific at rest dark-flanked, but any bird can roll to the side, exaggerating or diminishing this mark. Compared to Pacific, breeding adult Arctic has darker gray nape, bolder streaking on neck, and greenish gloss on foreneck. Winter adult and juvenile apparently never show dark "chinstrap" of Pacific.
- **VOCALIZATIONS** Rising wail of breeders not as long as that of Common. In aggression, often gives low croaks and growls.
- **BEHAVIORS** Nests on brackish pools near coast. Winters coastally, sometimes finding shelter in bays and inlets.
- **POPULATIONS** Old World counterpart of Pacific; the two overlap as breeders in West in coastal western Alas. Annual in winter in recent years to Salish Sea, no doubt reflecting improved observer effort.

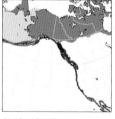

Accidental to Hawaii

Pacific Loon
Gavia pacifica | PALO | L 26" (66 cm)

- **APPEARANCE** A bit smaller than Arctic; much smaller than Common. Breeding Pacific and Common distinctive in respective breeding plumages, but winter adults and juveniles more similar: Pacific has smaller bill, smoothly rounded head, paler nape, and clean border between dark hindneck and white foreneck. Compare especially with Arctic: Breeding Pacific has paler nape, fainter streaking on neck, and purplish gloss on foreneck; winter adult and juvenile Pacific often have dark "chinstrap."
- **VOCALIZATIONS** Repertoire similar to Arctic's, but wailing usually longer.
- **BEHAVIORS** Nests on freshwater ponds coastally and in patchy boreal forest inland. Winters and migrates mostly coastally, with spectacular northbound concentrations of 2,500+ per hour.
- **POPULATIONS** Most winter off Mexico, but winter range shifting north. Widespread but rare overland migrant, especially in fall, inland. Accidental to Hawaii.

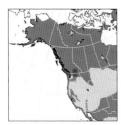

Common Loon
Gavia immer | COLO | L 32" (81 cm)

■ **APPEARANCE** Size of a large goose; appears hulking on the water, gangly in flight. Bill large; head blocky and angular. Breeding plumage crisp and high-contrast; winter adult and juvenile smudgy in winter, with uneven border between dark and white on neck. Bill black in summer, gray in winter with black tip and black upper ridge (culmen) of upper mandible. Pacific Loon (p. 176) in winter has sharp border on neck and grayer nape; note also smoothly rounded head, smaller bill, and smaller size overall of Pacific. Red-throated Loon, even smaller, has upturned bill; adults are white-faced in winter. Compare especially with larger Yellow-billed Loon.

■ **VOCALIZATIONS** Pure-tone wailing (yodel), heard on breeding grounds, is audible at well over a mile (1.6 km). A shorter, mirthful cackle (tremolo) is sometimes given away from the breeding grounds, especially by migrants in flight.

■ **BEHAVIORS** Nests on mounds of plant matter in freshwater bogs, on lake edges, and in marshes in the Northwoods, generally avoiding open tundra. Favors fish-filled lakes; water clarity is an important microhabitat feature for this species and probably other loons, all visual hunters. Most migrate and winter coastally, offshore as well as on bays and lagoons; migrants overland put down on reservoirs and large rivers. Away from breeding grounds, aloof and solitary, often all by itself on a waterbody or in loose association with a few other loons.

■ **POPULATIONS** Most widely encountered loon inland away from the breeding grounds, with more sightings fall to early winter than spring. Scarce inland on migration from eastern Great Basin eastward; larger stagings in western Great Basin threatened by drought and mercury contamination of major lakes. Summer sightings across midlatitudes raise possibility of breeding south of mapped range; formerly bred to Northern Calif.

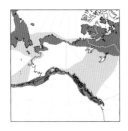

Yellow-billed Loon
Gavia adamsii | YBLO | L 34" (86 cm)

■ **APPEARANCE** Even larger than closely related Common Loon, but relative sizes of these two loons difficult to judge except in direct comparison. Bill structures different: Culmen of Yellow-billed straight (shallowly convex on Common) and bill upturned (held nearly horizontally on Common). Breeding Yellow-billed, rarely if ever seen by most birders, has pale yellow bill and much white on back; breeding Common has completely black bill and finer white checkering on back. Yellow-billed in winter, both adult and juvenile, usually appears quite pale, but beware effects of lighting and camera settings. Upturned bill of Yellow-billed duskier in winter than summer, but nevertheless pale; always pale-tipped and lacks dark culmen of Common.

■ **VOCALIZATIONS** Repertoire broadly matches Common's. Cackle (tremolo) is deeper and more nasal. Long yodel, also deeper, not as smoothly delivered; effect is oddly suggestive of chimpanzee, *Pan troglodytes*.

■ **BEHAVIORS** Like Common, nests on deep lakes with good water clarity, but almost always on treeless tundra, disdained by Common. Migrants overland put down at large reservoirs.

■ **POPULATIONS** Breeding range in West largely nonoverlapping with that of Common; Yellow-billed farther north. Winters coastally from Salish Sea northward. Very rare inland in Interior West, mostly fall to winter.

ALBATROSSES | DIOMEDEIDAE

These ocean wanderers are identified by their size alone: Even the small species have wingspans in excess of six feet. Albatrosses are classified as tubenoses, named for external coverings of the nostrils; two families of storm-petrels (pp. 182-187) and the one family of petrels and shearwaters (pp. 186-197) also are tubenoses. All are highly pelagic, usually feeding far from shore and coming to land only to nest.

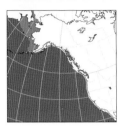

Common in Hawaiian Is.

Laysan Albatross
Phoebastria immutabilis | LAAL | L 28-31" (71-79 cm) WS 77-85" (196-216 cm)

- **APPEARANCE** Like all albatrosses, long-billed and very long-winged. Dark-mantled and mostly white otherwise; short blackish tail is separated from dark mantle by white uppertail coverts. Juvenile and adult similar.
- **VOCALIZATIONS** On breeding grounds, long whistles, short squeals, spooky braying, and nonvocal bill clicking; usually silent at sea.
- **BEHAVIORS** Nests alongside Black-footeds on islands and atolls. Feeds day or night.
- **POPULATIONS** Most breed on Hawaiian Is., especially Midway; range expanding. After breeding season, Nov. to June, disperses mostly north, occurring over cooler waters than Black-footed; fairly common off Aleutians late summer, rare but regular off West Coast fall to spring.

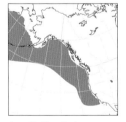

Widespread in Hawaiian Is.

Black-footed Albatross
Phoebastria nigripes | BFAL | L 28-33" (71-81 cm) WS 79-87" (201-221 cm)

- **APPEARANCE** Similar in size and structure to Laysan. Dark above and below in all plumages; most have white at base of bill, more extensive on breeding adults in summer. Older adults also acquire white on uppertail and undertail coverts. Compare with larger, rarer Short-tailed Albatross.
- **VOCALIZATIONS** Repertoire like that of Laysan, but calls average deeper. Gets rather vocal in disputes at sea, issuing groans and whistles.
- **BEHAVIORS** Nests colonially in loose association with Laysans; Black-footed is more aggressive at colonies and at sea. Off U.S. West Coast, seen singly or in very small groups from boats a few miles offshore; sometimes seen from land.
- **POPULATIONS** Default albatross off U.S. West Coast; present mostly late summer to fall. Situation reversed in Hawaii, where Black-footed seen less commonly than Laysan.

Annual to Hawaiian Is.; rare breeder

Short-tailed Albatross
Phoebastria albatrus | STAL | L 31-35" (79-89 cm) WS 87-94" (221-239 cm)

- **APPEARANCE** Larger than Laysan and Black-footed, with very large bill, light pink with pale blue tip. Juvenile dark brown above and below; smaller juvenile Black-footed has slighter, darker bill. Adult Short-tailed has white back and blond wash on head; mostly black upperwing broken by large white patch. Transition from juvenile to full adult plumage requires 10+ years; intermediate immatures told by huge pink bill and dark upperwing with white patch.
- **VOCALIZATIONS** Little documented, but apparently a bit deeper than those of Black-footed.
- **BEHAVIORS** Like Laysan and especially Black-footed, attracted to vessels at sea; favors cooler waters than even Laysan.
- **POPULATIONS** Almost driven to extinction in 20th century, but recovering; most breed off Japan, and a few have nested in recent years at Midway. Still very rare off West Coast; many or most concentrate off Aleutians after breeding.

ALBATROSSES
DIOMEDEIDAE

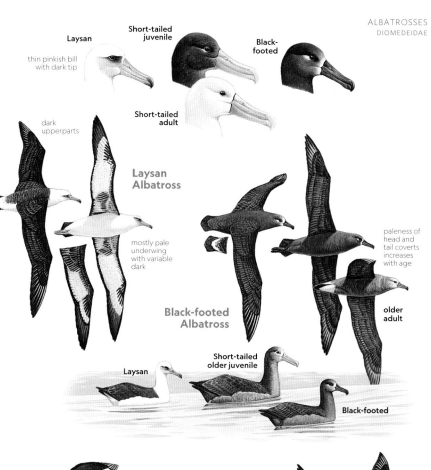

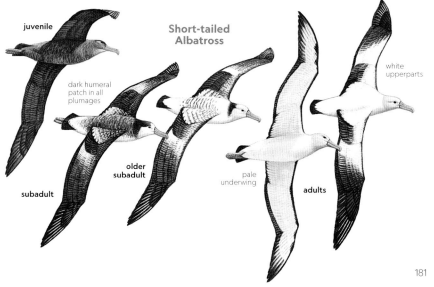

SOUTHERN STORM-PETRELS | OCEANITIDAE

Birds in the two families of storm-petrels are the smallest tubenoses (pp. 180-197). They are dark and relatively short-winged, with fluttery wingbeats; they fly just above the ocean's surface. The southern storm-petrels are, as their name implies, Southern Hemisphere breeders.

Wilson's Storm-Petrel

Oceanites oceanicus | WISP | L 7¼" (18 cm) WS 16" (41 cm)

Abundant breeder off Antarctica that migrates far north during boreal summer; formerly considered an East Coast–only species, but now known to be regular in small numbers off U.S. West Coast, especially central Calif. in fall. Wings short and rounded; legs extend beyond rounded tail. White rump, contrasting with otherwise blackish plumage, prominent at all angles. On foraging bouts, suggests a butterfly: wanders randomly, gently shifting course, then patters feet on water while plucking food from surface with bill. Accidental to Hawaii region.

NORTHERN STORM-PETRELS | HYDROBATIDAE

All our storm-petrels, save the regionally rare Wilson's, are in this family of Northern Hemisphere breeders. They come in shades of gray and black, especially the latter, and all have pale carpal bars, variably conspicuous in flight. Differences in flight style can be important in field ID.

Annual to Hawaiian Is.

Leach's Storm-Petrel

Hydrobates leucorhous | LESP | L 7-8" (18-20 cm) WS 17-19" (43-48 cm)

■ **APPEARANCE** Wings long and sharp-tipped, often angled; tail cleft or gently forked. Dark blackish brown overall; most in West show variable smudgy white on rump, split down the middle by smudgy central line, but southern *chapmani* tend to be dark-rumped.

■ **VOCALIZATIONS** Around nests, mellow purring and giddy chippering; usually silent at sea.

■ **BEHAVIORS** Excavates burrows for nesting; adults arrive and depart at colonies under cover of darkness, so most encounters are of birds at sea foraging in broad daylight. Flight powerful; zooms about in sudden bursts, suggesting a small shearwater (pp. 192-197) or even a nighthawk (p. 80). Tends not to enlist in big, mixed-species flocks and is not usually attracted to boats at sea.

■ **POPULATIONS** Common, but typically well offshore, sometimes out of range for short boat trips; annual in small numbers, fall to spring, to Hawaii region, where confusion with Band-rumped Storm-Petrel (p. 186) possible. Southern *chapmani* regular in very small numbers to Southern Calif., usually well offshore, summer to fall.

Townsend's Storm-Petrel 🇪🇳

Hydrobates socorroensis | TOSP | L 6-7" (15-18 cm) WS 16-18" (41-46 cm)

Until 2016 treated as conspecific with Leach's Storm-Petrel. Breeds only at Guadalupe I., off west coast of Mexico; ranges north into waters well off Southern Calif. summer to early fall. Townsend's is smaller, shorter-winged, shorter-tailed, and more compactly built overall than Leach's; as a result, flight style of Townsend's is steadier and less erratic than that of zigzagging Leach's. Many Townsend's have more extensive white on tail than Leach's, but some, confusingly, are dark-rumped and difficult to recognize. Declining; global population fewer than 10,000 birds.

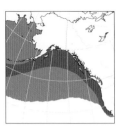

Accidental to Hawaii

Fork-tailed Storm-Petrel
Hydrobates furcatus | FTSP | L 8½" (22 cm) WS 18" (46 cm)

- **APPEARANCE** Fairly large storm-petrel; tail long and forked. Lovely blue-gray plumage unique among our storm-petrels, but beware effects of lighting at sea. Black mask stands out against otherwise pale head; sitting on water, suggests nonbreeding Red Phalarope (p. 132), but bill structure utterly different.
- **VOCALIZATIONS** Most calls harsh and grating, a few squeakier. Noisy at nest; usually silent at sea, but occasionally gives a descending chatter, rough and short.
- **BEHAVIORS** Breeds in colonies on offshore islands, tending nest at night. Feeds by hovering at ocean's surface, where it gleans oils and animal matter; olfaction critical for finding food, as for most tubenoses. Attracted to boats and artificial lighting.
- **POPULATIONS** Numerous but with higher numbers well north in range; northern breeders larger and paler than southern. Accidental to Hawaii.

Ashy Storm-Petrel EN
Hydrobates homochroa | ASSP | L 8" (20 cm) WS 17" (43 cm)

- **APPEARANCE** Smaller, shorter-winged, and longer-tailed than Black Storm-Petrel. Underwings and uppertail coverts grayer (more "ashy") than rest of body; Black Storm-Petrel more uniformly dark all over.
- **VOCALIZATIONS** Around nest, mellow chirps, higher squeaks, and slow trills. At sea, sometimes gives short cackles.
- **BEHAVIORS** Nests in natural cavities on rocky, fog-shrouded islands; does not excavate its own nest. Wingbeats shallow; glides only for short bouts, unlike larger Black and Fork-tailed Storm-Petrels. Attracted to lights; even seen at San Francisco Giants night games.
- **POPULATIONS** Although commonly spotted on boat trips off Calif., a globally rare bird, numbering fewer than 10,000. Feeds offshore, but not far out. Only weakly migratory, but with annual movements in response to variation in sea surface temperature.

Black Storm-Petrel
Hydrobates melania | BLSP | L 9" (23 cm) WS 19" (48 cm)

- **APPEARANCE** Larger and longer-winged than Ashy. Appears simply black, in flight or at rest, in most settings. Compare especially with dark-rumped Leach's and Townsend's Storm-Petrels (p. 182), both smaller.
- **VOCALIZATIONS** At colonies, toneless trilling, slow and steady. At sea, occasional squeaks and irritating chatter.
- **BEHAVIORS** Nests in cavities on rocky islands, often amid boulders. Feeds at ocean's surface like other storm-petrels, but dives a bit more. Wingbeats deep and fluid; Ashy has shallower wingbeats, and Leach's arcs and zigzags more in flight.
- **POPULATIONS** Breeds mostly to our south, then disperses north into waters off Calif. in late summer and fall.

Least Storm-Petrel
Hydrobates microsoma | LSTP | L 5¾" (15 cm) WS 15" (38 cm)

Breeds to our south, with most dispersing south following nesting season, but a few head north, reaching waters off Southern Calif. Smallest storm-petrel in the West, with short, wedge-shaped tail. Despite size, has deep wingbeats. In plumage and flight style, suggests a compact, quite small Black Storm-Petrel. Numbers of Least off Calif. vary annually: Almost unrecorded some years; widely noted, although never common, other years.

Band-rumped Storm-Petrel
Hydrobates castro | BSTP | L 9" (23 cm) WS 17" (43 cm)
Occurs regularly in West only in Hawaii region, where it is an uncommon breeder; apparently nests well inland. Most obvious mark is wholly white rump band; white rump of Leach's broken by dusky central line. Tail of Band-rumped shallowly forked, often appearing notched or rounded; tail of Leach's more deeply forked. Wings shorter and less pointed than Leach's too, and flight less bounding. Most sightings in Hawaiian waters in summer, when Leach's is mostly absent.

Tristram's Storm-Petrel
Hydrobates tristrami | TRSP | L 10" (25 cm) WS 22" (56 cm)
Like Band-rumped, occurs regularly in West only in Hawaii region. Large size, forked tail, and dark coloration suggest Black Storm-Petrel (p. 184), unrecorded from Hawaii. Rump of Tristram's dull gray-brown; Leach's in Hawaiian waters and Band-rumped have white rumps. Flight powerful and well-controlled for a storm-petrel. Breeds in Hawaii region in winter, with sightings at sea fall to spring; overlaps temporally with Leach's, not Band-rumped. Sightings from main islands almost entirely from Kauai.

PETRELS AND SHEARWATERS | PROCELLARIIDAE
With about 100 species worldwide, this family is the most speciose of the tubenoses (pp. 180-197). Procellariids range in size from the tiny diving-petrels to the enormous giant-petrels, but all species regular to the West are intermediate between those extremes. Procellariids are pelagic, and they are superb fliers.

Casual to Hawaii

Northern Fulmar
Fulmarus glacialis | NOFU | L 16-18" (41-46 cm) WS 37-45" (94-114 cm)
■ **APPEARANCE** Same size as Sooty Shearwater (p. 192), but more heavyset; bullnecked and barrel-chested. Also, wings more rounded than on narrow-winged Sooty Shearwater. Stocky bill is pale with prominent nostril tubes. Variable: Light-morph adult, gray-mantled and otherwise mostly white, suggests a *Larus* gull (pp. 152-163); dark-morph adult, soft dirty gray all over, calls to mind Sooty Shearwater. Both morphs flash pale inner primaries, reduced in dark morph.
■ **VOCALIZATIONS** Around trawlers, issues harsh cackles and nasal whines, audible above the drone of ship engines.
■ **BEHAVIORS** Nests on sea cliffs; routinely follows boats, day or night. Flight style like a shearwater (pp. 192-197), alternating short glides and quick flaps, typically just above ocean's surface.
■ **POPULATIONS** Light morphs breed farther north in West than dark morphs, but both morphs and their many intermediates mix freely following nesting season. Numbers increasing. Casual to Hawaii.

Murphy's Petrel
Pterodroma ultima | MUPE | L 14-15" (36-38 cm) WS 35-38" (89-97 cm)
Breeds on islands in south-central Pacific, with some dispersing to U.S. West Coast; most sightings Apr. to May, well offshore. Like other gadfly petrels, genus *Pterodroma* (pp. 186-191), arcs and bounds high above the ocean's surface, as if on a wild roller coaster ride. Murphy's is dark, dirty gray overall, with white chin, white underwing flash, and faint M pattern above. Compare with Sooty Shearwater (p. 192), dark-morph Northern Fulmar, and other gadfly petrels. Casual to Hawaii.

NORTHERN STORM-PETRELS : PETRELS AND SHEARWATERS
HYDROBATIDAE : PROCELLARIIDAE

Band-rumped Storm-Petrel

white rump band

short legs

Tristram's Storm-Petrel

grayish rump

dark gray above

Northern Fulmar

dark morph

intermediate morph

light morph

pale inner primaries

light morph

Murphy's Petrel

chin whitish

gray above

primaries pale underneath

gull-like when sitting on water

light morph

Regular to Hawaii

Mottled Petrel
Pterodroma inexpectata | MOPE | L 13-14" (33-36 cm) WS 33-36" (84-91 cm)
- **APPEARANCE** Stocky; broad-winged, short-tailed, and short-billed. Grayish above with faint M pattern across outstretched wings. From below, gray belly contrasts with white undertail coverts and gray-white breast; white underwing interrupted by thick black bar.
- **VOCALIZATIONS** Essentially silent at sea, but may give occasional brief bursts of soft chatter.
- **BEHAVIORS** Employs varied feeding methods, including surface-gleaning and shallow dives. Like other gadfly petrels (pp. 186–191), arcs high above the ocean, then zooms down to near the water's surface.
- **POPULATIONS** Breeds New Zealand. Migrants pass Hawaii both spring and fall, especially Apr. and Oct.; many gather in Gulf of Alaska and southern Bering Sea in summer. Sparse but regular well off West Coast, with concentration of sightings B.C. to Ore., fall to spring, particularly late fall.

Juan Fernandez Petrel
Pterodroma externa | JFPE | L 16½-17¾" (42-45 cm) WS 40½-45" (103-114 cm)
- **APPEARANCE** Long wings are narrow at base; long-tailed. Gray upperparts show faint M pattern; face and cap darker, forehead white. From below, almost entirely white; has just a small black smudge on underwing, not nearly as extensive as on Hawaiian and Mottled Petrels.
- **VOCALIZATIONS** Apparently not documented to vocalize at sea.
- **BEHAVIORS** A magnificent aerobat. Employs feeding methods similar to those of Mottled, but adds to its repertoire airborne capture of flying fish. More sociable than Mottled, sometimes enlisting in mixed-species flocks at boats.
- **POPULATIONS** Breeds only Juan Fernandez Is. off Chile; occurs year-round in Hawaiian waters, peaking Sept. to Oct. Accidental elsewhere in West (Ariz., Ore.).

Regular year-round in Hawaiian waters

Regular to Hawaii

Hawaiian Petrel 🅔
Pterodroma sandwichensis | HAPE | L 15-16" (38-41 cm) WS 38-41" (97-104 cm)
- **APPEARANCE** Long-tailed and long-winged like Juan Fernandez Petrel. Upperparts like Juan Fernandez's, but darker; contrast between cap and rest of upperparts relatively muted. Clean white below from chin to vent like Juan Fernandez, but with more black on underwing: Primaries and leading edge mostly black.
- **VOCALIZATIONS** At nest, a mix of long, low moans and higher chatter; silent at sea except for occasional cackling in interactions.
- **BEHAVIORS** Nests in forests in high mountains. Flight style and feeding methods like other gadfly petrels.
- **POPULATIONS** Endangered; breeds only in Hawaii, mostly Kauai and Maui. Occurs in Hawaiian waters year-round but mostly Apr. to Oct. Very rare but regular well off West Coast, especially central Calif., also Apr. to Oct.

White-necked Petrel
Pterodroma cervicalis | WNPE | L 17-18" (43-46 cm) WS 42-45" (107-114 cm)
Ranges regularly to Hawaiian waters from breeding grounds at Macauley I. off New Zealand. Most similar species in West is closely related Juan Fernandez Petrel, but White-necked is paler above with more prominent M pattern; black cap of White-necked completely separated from gray back by broad white hindcollar. Underparts clean white like Juan Fernandez's; tail of White-necked not as long. Outnumbered off main islands by Juan Fernandez about 10 to 1; records mostly fall to winter.

PETRELS AND SHEARWATERS
PROCELLARIIDAE

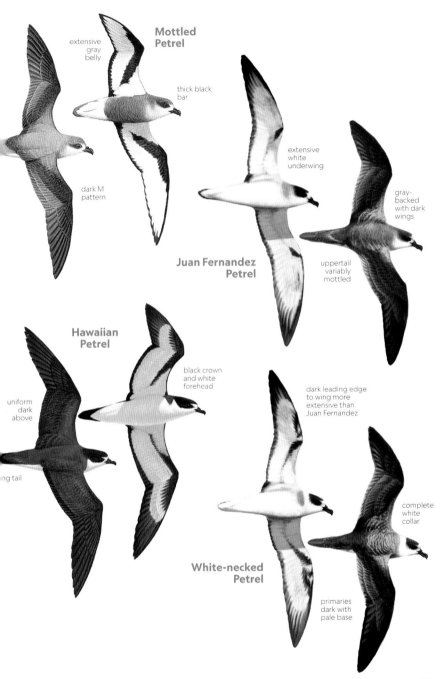

Bonin Petrel
Pterodroma hypoleuca | BOPE | L 12" (30 cm) WS 25–28" (64–71 cm)
Nests in winter on Northwestern Hawaiian Is.; disperses west following breeding. A small *Pterodroma,* or gadfly petrel (pp. 186–191); short-billed but long-tailed. Upperparts low-contrast gray with bluish hue, with darker cap; forehead pale. From below, patterned like larger, longer-winged Hawaiian Petrel (p. 188): Body pure white below from chin to undertail coverts; underwing also extensively white, but with gray primaries and broad black bar across "wrist." Pinkish feet distinctive, but hard to see. Not uncommon in Hawaiian region, but rarely seen at sea off main islands. Tends nests at night; except at remote nesting atolls, encounters are uncommon.

Black-winged Petrel
Pterodroma nigripennis | BWPE | L 12" (30 cm) WS 26" (66 cm)
- **APPEARANCE** Small-bodied and small-billed; short tail is wedge-shaped when flared, pointed otherwise. Mostly light gray above, darkening toward wing tips; gray extends to forehead on many individuals. White-bodied below, but with broad, pale-gray bulge extending onto breast; from below, white wing linings framed by black remiges, leading edge, and "wrist" bar.
- **VOCALIZATIONS** In aggression toward other seabirds, a short, shrill chatter; calls unrecorded in waters off Hawaii, however.
- **BEHAVIORS** Along with Bonin and Cook's Petrels, is placed within *Cookilaria* subgenus of gadfly petrels; compared even to their aerobatic congeners, these small petrels are notable for their steep dives and fast ascents.
- **POPULATIONS** Breeds mostly southwestern Pacific Ocean, dispersing to Hawaii region, peaking Sept. to Oct.; greatest concentration off west coast of Hawaii I.

Regular in Hawaiian waters

Cook's Petrel
Pterodroma cookii | COPE | L 12–13" (30–33 cm) WS 30–32" (76–81 cm)
- **APPEARANCE** Small-bodied and small-billed. Upperparts gray with dark M. From below, body pure white from chin to undertail coverts; underwing also mostly white. Combo of M across gray upperparts and extensively white underparts suggests larger Buller's Shearwater (p. 194).
- **VOCALIZATIONS** Like Black-winged, gives short chatter in aggression; but also like that species, not known to call in our waters.
- **BEHAVIORS** Like other *Cookilaria* petrels, especially energetic in flight. Often solitary, but also sometimes in flocks.
- **POPULATIONS** Breeds off New Zealand, dispersing widely after nesting. Only *Cookilaria* regular off West Coast, mostly Calif., spring to fall; in Hawaiian waters, scarce in spring, a bit more common early fall.

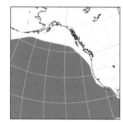
Uncommon migrant Hawaiian Is.

Bulwer's Petrel
Bulweria bulwerii | BUPE | L 11–12" (28–30 cm) WS 25–27" (64–69 cm)
- **APPEARANCE** Long-winged and long-tailed. Dark coloration with paler bar across upperwing suggests a large storm-petrel, but wings much longer.
- **VOCALIZATIONS** Around nest, gruff, doglike barks in slow sequence. Usually silent at sea.
- **BEHAVIORS** Despite name, more closely related to shearwaters than *Pterodroma* petrels. Flies close to surface like shearwater, with frequent but short-duration glides; zigzags about erratically like a storm-petrel.
- **POPULATIONS** Widespread but uncommon breeder Hawaii region in summer, with sightings at sea mostly spring to fall.

Regular in Hawaiian waters; breeds widely

PETRELS AND SHEARWATERS
PROCELLARIIDAE

Bonin Petrel

extensive dark markings on underwings

dark breast markings

gray back and uppertail separated by dark M pattern

dark "wrist" patch

Black-winged Petrel

pale head and faint breast markings

much shorter and thicker bill than Cook's

dark underwing markings

Cook's Petrel

dark tip to tail visible at close range

long, narrow tail

Bulwer's Petrel

long wings held forward

Common in Hawaiian waters

Wedge-tailed Shearwater
Ardenna pacifica | WTSH | L 18" (46 cm) WS 40" (102 cm)

- **APPEARANCE** Lightly built shearwater with distinctive tail: long and pointed in direct flight, flared when banking or soaring. Slender bill is dusky with darker tip. Polymorphic: dark morph brown overall with lighter flight feathers; light morph extensively white below. Some are intermediate. Compare with Buller's Shearwater (p. 194), similar in build.
- **VOCALIZATIONS** Calls at nest, heard dusk till dawn, are stirring: a mix of deep moaning and higher wailing.
- **BEHAVIORS** Flight style distinctive: graceful, seemingly effortless, with easy but powerful twists and turns. Nests in burrows mostly in remote locales, but some have adapted to humans; a few pairs nest at the heavily visited Kīlauea Point Light Station.
- **POPULATIONS** Warmwater species with a few records, summer to fall, off West Coast. The most common tubenose in Hawaii, but formerly was even more numerous; big flocks, sometimes visible from shore, stage in fall. Largely absent from Hawaii through the winter; seasonal movements vary among color morphs, with details not fully worked out.

Regular to Hawaii

Short-tailed Shearwater
Ardenna tenuirostris | STTS | L 17" (43 cm) WS 39" (99 cm)

- **APPEARANCE** A bit smaller than Sooty Shearwater. Despite the name, does not appear shorter-tailed than Sooty; better structural marks are the bill, thinner and shorter on Short-tailed, and the forehead, steeper on Short-tailed. Sooty flashes more white in underwing than Short-tailed; slightly paler throat of Short-tailed contrasts with darker crown and nape.
- **VOCALIZATIONS** Rarely calls at sea, but may issue tinny squeaks in interactions with other birds.
- **BEHAVIORS** Loose flocks of hundreds to thousands straggle past coastal headlands in Alas.; also occurs well out at sea. Feeds on surface, but also a superb diver, submerging to 200 feet beneath surface.
- **POPULATIONS** Breeds Australia, winters (our summer) mostly Bering Sea, with many reaching Chukchi and even Beaufort Seas. Migration back to Australia takes some along West Coast fall and winter; also occurs Hawaii fall to winter.

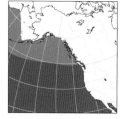
Regular to Hawaiian Is.

Sooty Shearwater
Ardenna grisea | SOSH | L 18" (46 cm) WS 40" (102 cm)

- **APPEARANCE** Very similar to slightly smaller Short-tailed Shearwater. Sooty Shearwater, with longer bill and less steeply angled forehead, flashes more white on underwing than Short-tailed; also, throat of Sooty averages less pale than Short-tailed's. All differences are slight; beware the effects of lighting, camera settings, and wishful thinking. Compare also with Flesh-footed (p. 194), Christmas (p. 196), and dark-morph Wedge-tailed Shearwaters.
- **VOCALIZATIONS** Usually silent at sea, but can get noisy in feeding frenzies, issuing nasal squeaks and squeals.
- **BEHAVIORS** Fairly social species: Singles and small flocks join mixed-species assemblages; also occurs in huge, dense flocks that rest on surface and dive to feed. Can be seen from land, but the large flocks are typically a mile (1.6 km) or more out.
- **POPULATIONS** Southern Hemisphere breeder that winters (our summer) well north. Most common off West Coast spring to summer; scarcer Short-tailed generally fall to winter. Off Hawaiian Is., mostly spring and fall.

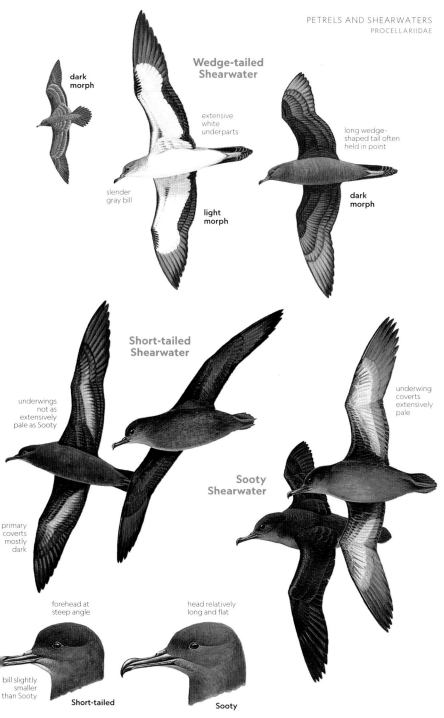

Accidental to Hawaii

Pink-footed Shearwater
Ardenna creatopus | PFSH | L 19" (48 cm) WS 43" (109 cm)

■ **APPEARANCE** Slightly larger than Sooty Shearwater (p. 192) and more heavyset; wings broader than narrow-winged Sooty's. Much larger than Black-vented Shearwater (p. 196), also gray-brown above and off-white below. Gray-pink feet of Pink-footed distinctive, but often tucked in; mostly pink bill, with dark tip, conspicuous even at a distance. Variable; a few are mostly dark below, inviting confusion with Flesh-footed Shearwater.

■ **VOCALIZATIONS** Like other *Ardenna* shearwaters, gives tinny squeaks, *eeen* and *iink*, in feeding disputes.

■ **BEHAVIORS** Classic shearwater in flight, mixing easy wingbeats with smooth glides; smaller Black-vented glides less, has faster, choppier wingbeats.

■ **POPULATIONS** Breeds off Chile. Closely related to Flesh-footed and treated by some as same species, but their breeding ranges are widely separated. After breeding season in austral summer (our winter), Pink-footeds move north, reaching waters off south-central Alas. Can seem quite common on boat trips off U.S. West Coast, but is declining and not common globally. Doesn't get far beyond continental shelf; accidental to Hawaii.

Flesh-footed Shearwater
Ardenna carneipes | FFSH | L 17" (43 cm) WS 41" (104 cm)

Rarer counterpart of Pink-footed Shearwater off West Coast, but Pink-footed almost always pale below; Flesh-footed dark below. Compare especially with Sooty: Flesh-footed is warmer overall, lacking silver-white flashes on underwing; note also pale bill of Flesh-footed and pinkish feet. Flies like Pink-footed; Sooty flaps faster on narrower wings, glides less. Range and timing off West Coast roughly similar to that of Pink-footed, but Flesh-footed peaks later in fall and occurs, on average, farther north. Relative statuses in Hawaiian waters reversed: More records of Flesh-footed than Pink-footed, no doubt reflecting birds intercepted on passage from Australian breeding grounds. Compare also with smaller Christmas Shearwater (p. 196).

Annual to Hawaiian Is.

Buller's Shearwater
Ardenna bulleri | BULS | L 16" (41 cm) WS 40" (102 cm)

■ **APPEARANCE** Slim overall and long-tailed like Wedge-tailed Shearwater (p. 192). Most distinctively marked shearwater in West: crisply patterned above in shades of gray, in a manner more like a gadfly petrel, genus *Pterodroma* (pp. 186–191), than a shearwater. Almost entirely white below, including underwings; bill cold gray.

■ **VOCALIZATIONS** Calls at sea higher, drier, and buzzier, *izzzz* and *iiinn*, than those of Sooty and Pink-footed.

■ **BEHAVIORS** Sociable, often gravitating to flocks of Pink-footeds; generally outnumbered in our waters by that species, but sometimes forms dense flocks of conspecifics. Flies like Wedge-tailed, with much soaring, sometimes well above water's surface; flight style, combined with high-contrast plumage, contributes to *Pterodroma*-like mien.

■ **POPULATIONS** Tremendous migrant; breeds New Zealand, circumnavigating much of Pacific Ocean in great clockwise circle. Present off West Coast, especially B.C. to central Calif., mostly fall; these are migrants on their return trip home. Buller's averages farther offshore than Pink-footed, but much overlap. Scarce but regular to Hawaiian Is., spring and fall.

Black-vented Shearwater
Puffinus opisthomelas | BVSH | L 14" (36 cm) WS 34" (86 cm)

■ **APPEARANCE** Smallest shearwater commonly seen off U.S. West Coast. Larger Pink-footed Shearwater (p. 194) similarly patterned; both are gray-brown above and paler, dirty, gray-white below with dark undertail coverts. Black-vented has dark bill, dusky gray-pink feet. Variable; many have whitish splotches or mottling, and a few are extensively, strikingly white. Compare especially with closely related Manx Shearwater, rare but increasing off Pacific coast.

■ **VOCALIZATIONS** In interactions at sea, occasional raspy notes, rougher than the squeaky calls of larger *Ardenna* shearwaters (pp. 192–195).

■ **BEHAVIORS** Alternates short glides with fast, choppy wingbeats, typically just above the water's surface; flight of Pink-footed more fluid and languid. Forms large flocks, often close to land; habitually dives for food.

■ **POPULATIONS** Breeds only on islands west of Baja California, then disperses at sea, with many heading north into our waters, mostly fall to winter. A warmwater species, with wanderings north to central Calif. and beyond corresponding to spikes in ocean temperature; almost never wanders beyond waters over continental shelf.

Manx Shearwater
Puffinus puffinus | MASH | L 13½" (34 cm) WS 33" (84 cm)

Closely related to Black-vented Shearwater; formerly treated as conspecific. Manx is a mostly Atlantic Ocean species that has been showing up with increasing frequency off West Coast. Plumage matches Black-vented, but is "cleaner" and more extensively white: Manx has white undertail coverts, white flanks often visible on bird at rest on sea, more and brighter white on underwing, and white on face that wraps behind ear coverts. In good view, can be picked out at sea, but photo documentation always a good idea with this still-rare species. Manx is the only *Puffinus* shearwater expected from Wash. north to Gulf of Alaska; Black-vented is irregular, absent some years, north of central Calif. to Ore.

Newell's Shearwater CR
Puffinus newelli | NESH | L 14–15" (36–38 cm) WS 30–33" (76–84 cm)

The default dark-and-light *Puffinus* shearwater in Hawaii region, although it is by no means common; faces multiple anthropogenic threats, including from introduced predators. Like an especially high-contrast Manx Shearwater: Newell's is gleamingly black above, with prominent white flanks and white on face behind ear coverts. Like other *Puffinus* shearwaters, flies close to surface with fast, choppy wingbeats. Breeds Apr. to Oct., nesting to 4,000 ft (1,220 m) above sea level on cliffs in dense forest; is noisy at night around colonies. Feeds well offshore. Global population about 20,000 pairs.

Christmas Shearwater
Puffinus nativitatis | CHSH | L 14–15" (36–38 cm) WS 28–32" (71–81 cm)

Along with Newell's, a *Puffinus* shearwater restricted in West to Hawaii region. All-dark like Sooty Shearwater (p. 192), but warmer; Sooty cooler gray-black. Flies with short, choppy wingbeats like other *Puffinus* shearwaters, but wingbeats of larger Sooty can be similar. Nests on low-lying islands and atolls, a bit earlier than Newell's; like that species, declining globally and threatened by anthropogenic impacts, especially introduced mammals. Feeds at sea where it rarely dives; Sooty frequently dives.

STORKS | CICONIIDAE
Huge and even intimidating, storks are recognized by their long, broad-based, often colorful bills; the head is partly unfeathered on many species, contributing to the sinister appearance. Storks feed along slow-moving waterways, roost in tall trees, and soar effortlessly. They number about 20 species globally, with only one in the West.

Wood Stork
Mycteria americana | WOST | L 40" (102 cm) WS 61" (155 cm)
Widespread in Neotropics to southeastern U.S.; formerly regular and locally common to southwestern U.S., but now rare. About same length as Great Egret (p. 208), but more hulking; often hunched over. Head and neck unfeathered; massive, horn-colored bill, very broad at base, droops downward. Perched, plumage appears mostly white; soaring, the black flight feathers of tail and wing are conspicuous. Recent sightings mostly Southern Calif., especially July to Aug.

FRIGATEBIRDS | FREGATIDAE
Piratical seabirds with hooked bills and incredibly long wings, the frigatebirds are a paradox: They feed exclusively over the ocean yet never get wet. They catch food in midair—flying fish and flying squid, as well as prey stolen from other birds. Frigatebirds also pick food from the surface, taking care to keep their feathers dry. Sometimes referred to as "man-o'-war birds," they number only five species globally.

Magnificent Frigatebird
Fregata magnificens | MAFR | L 40" (102 cm) WS 90" (229 cm)
Distributional story in West parallels that of Wood Stork: Widespread in Neotropics to southeastern U.S.; rare in West, with concentration of records Southern Calif., where less regular in recent decades. Identified as a frigatebird by long forked tail, long bill, and very long wings: male completely blackish except for red on chin (dispersers to West rarely poof out their throats); female blackish with extensive blotchy white on breast; juvenile white-headed. Wanders far inland, with scattered records to Great Basin and southern Rockies. Compare with Great Frigatebird, accidental to Calif.

Great Frigatebird
Fregata minor | GREF | L 37" (94 cm) WS 85" (216 cm)
■ **APPEARANCE** A bit smaller than Magnificent Frigatebird, a fact indicated by the specific epithet *minor*. Adult male shows broad buffy bar across upperwing, suggesting much smaller Bulwer's Petrel (p. 190). Female shows red orbital ring (blue on female Magnificent) and white on breast extending to throat (entire head black on female Magnificent). Juvenile pale-headed, but with rusty wash (head of juvenile Magnificent plain white).
■ **VOCALIZATIONS** In courtship, incessant squawks, nasal and descending; also longer purring and short whinnies.
■ **BEHAVIORS** Stays aloft indefinitely, soaring and wheeling most of the time; occasional wingbeats deep and powerful, but fluid and graceful. *Pteranodon*-like, mesmerizing to behold.
■ **POPULATIONS** Nests in West mostly Northwestern Hawaiian Is., but commonly sighted from all the main islands; impossible to miss at Kīlauea Point. A few records to coastal Calif.

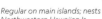
Regular on main islands; nests Northwestern Hawaiian Is.

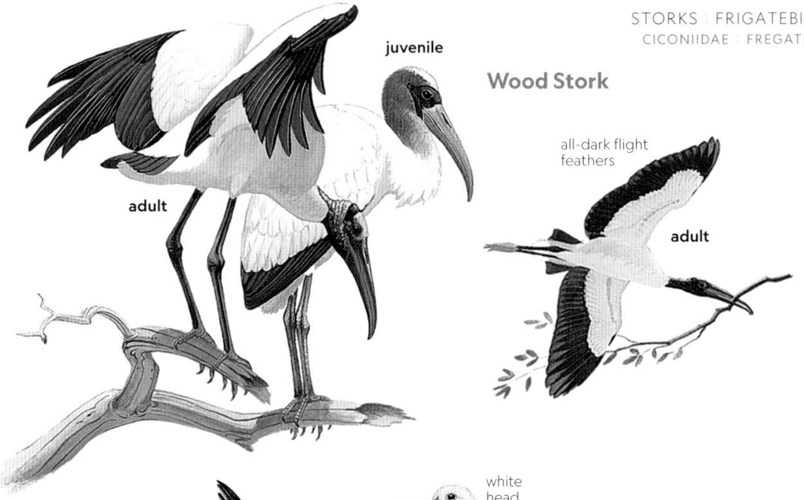

STORKS : FRIGATEBIRDS
CICONIIDAE : FREGATIDAE

Wood Stork

juvenile
adult
all-dark flight feathers
adult

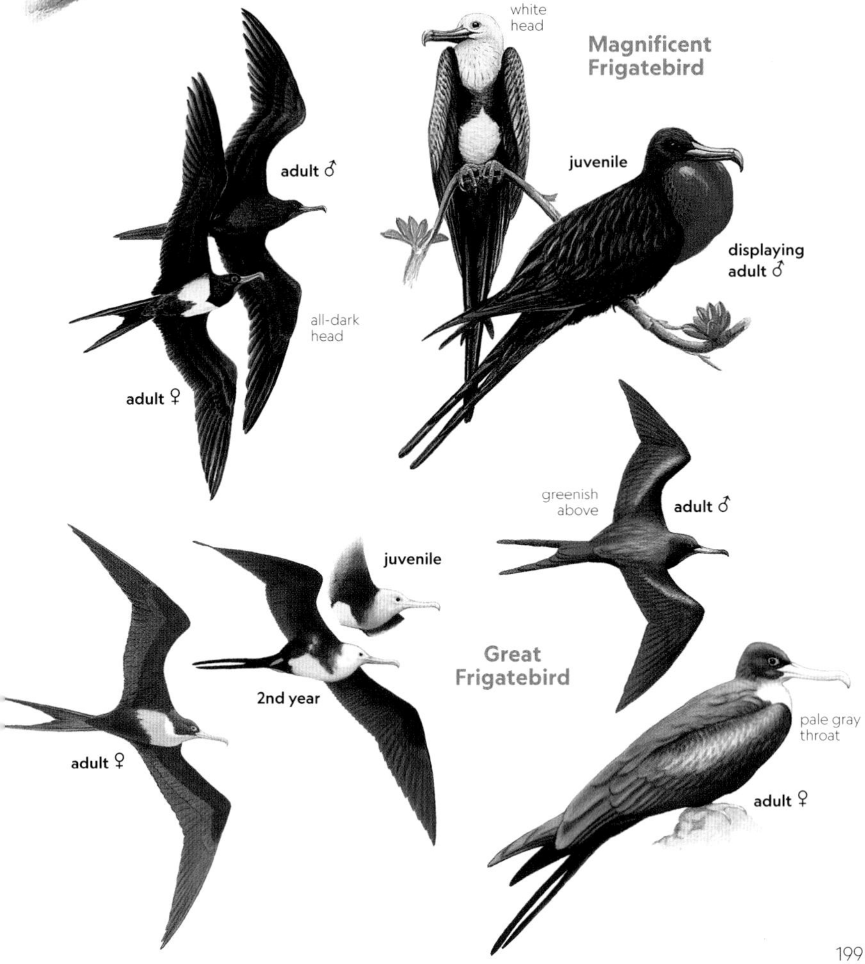

Magnificent Frigatebird

white head
adult ♂
all-dark head
adult ♀
juvenile
displaying adult ♂

Great Frigatebird

greenish above
adult ♂
juvenile
2nd year
adult ♀
pale gray throat
adult ♀

199

BOOBIES | SULIDAE

With their long wings, long tails, and long necks and bills, airborne boobies look like a flying cross or X. They fall to the sea like missiles in their pursuit of fish just below the water's surface. The sulids consist of the boobies and closely related gannets, and they number 10 species worldwide.

Common in Hawaiian Is.

Red-footed Booby
Sula sula | RFBO | L 28" (71 cm) WS 60" (152 cm)

- **APPEARANCE** Smallest booby; long-winged with slight bill. Variable adult may have brown or white head and body, and, independently, dark or white tail; all adults have red feet and bluish bill with red base. Juvenile dusky overall with smudgy breastband and dark underwing coverts; feet orangey and bill dusky pink with darker tip.
- **VOCALIZATIONS** At colony, rapid clacks, short growls, and whistles.
- **BEHAVIORS** Nests colonially in trees; feeds well offshore, but also easy to see around colonies. Attracted to boats, even landing thereon.
- **POPULATIONS** Variation complex, with multiple subspecies not well aligned with the different color morphs. Most in Hawaii are white-bodied, white-tailed morphs and subspecies *rubripres*. Increasing off Calif., though still rare; many there are brown-bodied morphs.

Regular in Hawaiian Is.

Brown Booby
Sula leucogaster | BRBO | L 30" (76 cm) WS 57" (145 cm)

- **APPEARANCE** Larger than Red-footed Booby, but shorter-winged. White belly of adult sharply demarcated from brown breast; bill and feet dull yellow. Head of adult, especially male, varies from solid brown to frosty white. Juvenile dark all over, including underparts and underwing.
- **VOCALIZATIONS** At colony, harsh honks and hoarse whistles.
- **BEHAVIORS** Nests on ground in colonies; feeds at sea, joining mixed-species assemblages and following boats.
- **POPULATIONS** Most in Hawaii are dark-headed *plotus*; most or all off Calif., where increasing rapidly, are slightly smaller *brewsteri*, in which males are frosty-headed.

Blue-footed Booby
Sula nebouxii | BFBO | L 32" (81 cm) WS 62" (157 cm)

Breeds disjunctly from Peru north to Sea of Cortez, reaching Southern Calif. annually, summer to fall. Large but short-tailed booby. Adult has bright blue feet; back brown, head finely streaked, underparts white. Juvenile has duller feet and is duskier overall; underwing coverts whitish like those of juvenile Masked, but Blue-footed has darker bill and lacks white collar. Formerly more frequent in West, with numerous records Salton Sea; casual farther inland.

Uncommon in Hawaii

Masked Booby
Sula dactylatra | MABO | L 32" (81 cm) WS 62" (157 cm)

- **APPEARANCE** Like Blue-footed Booby, large overall but short-tailed. "Mask" of adult comprises dark skin at base of yellow bill; tail black, entire trailing edge of upperwing black, and underwing linings white. Mostly brown juvenile has duller bill and white collar. Compare with Nazca Booby (p. 459), rare but rapidly increasing in region.
- **VOCALIZATIONS** Calls at nest squeakier, more nasal than other boobies.
- **BEHAVIORS** Nests on ground; even among boobies, an impressive diver.
- **POPULATIONS** Rarer in Hawaii than Brown or Red-footed; increasing, but still rare, off West Coast. Until 2000 treated as conspecific with Nazca.

CORMORANTS | PHALACROCORACIDAE

Dark overall with colorful eyes and facial skin, the cormorants have hooked bills and short legs set well back. They number about 40 species globally, and all catch fish and invertebrates by dives from the surface.

Brandt's Cormorant
Urile penicillatus | BRAC | L 35" (89 cm) WS 48" (122 cm)
- **APPEARANCE** Stockiest cormorant in the West. Breeding adult mostly dull black, but with thin white plumes on neck; cobalt throat pouch bordered below by buff. Nonbreeding adult retains buff on throat, but throat dull and white neck plumes absent. Juvenile like nonbreeding adult but browner.
- **VOCALIZATIONS** Juvenile at nest solicits adult with three-note whistle; adult responds with Wookiee-like snarl.
- **BEHAVIORS** Nests on sea cliffs, often offshore but sometimes in busy commercial districts along coast. Feeds mostly out at sea, only rarely in estuaries and inlets.
- **POPULATIONS** Annual movements complex and irregular, reflecting food availability; some move well north after nesting. Also engages in large-scale daily movements between roosting sites and feeding grounds.

Red-faced Cormorant
Urile urile | RFCO | L 31" (79 cm) WS 46" (117 cm)
Range is subsumed within that of similar Pelagic Cormorant. Red-faced a bit larger than Pelagic, with thicker neck and thicker, paler bill. Red facial skin of adult prominent when breeding, reduced otherwise. Typically seen right along coast or out at sea a few miles. Permanent resident from Valdez to Cordova area all the way out along Aleutians, with outposts to Pribilofs.

Pelagic Cormorant
Urile pelagicus | PECO | L 26" (66 cm) WS 39" (99 cm)
- **APPEARANCE** Our smallest cormorant; thin-necked, slender-billed. Adult glossy dark green, juvenile dull brown. Closely related Red-faced Cormorant a bit larger and thicker-necked, with more extensive red facial skin in breeding season, when both species also show thin white stripes on neck and white patch at rear flank.
- **VOCALIZATIONS** Quiet, even at colony. Calls of adult at nest nasal and honking; some notes short, others longer and wavering.
- **BEHAVIORS** Despite name, not especially pelagic; usually seen singly or in small numbers in flight or on sea rocks near shore. Nests on sea cliffs.
- **POPULATIONS** Uncommon but widespread coastally south of Ore.; numbers fluctuate with El Niño, especially in south of range. Accidental to Hawaii.

Accidental to Hawaii

Double-crested Cormorant
Nannopterum auritum | DCCO | L 32" (81 cm) WS 52" (132 cm)
- **APPEARANCE** Slimmer and longer-tailed than Brandt's Cormorant; neck kinked. Adult Double-crested, black overall, has yellow-orange throat patch, smoothly rounded and lacking white border. Juvenile has dark belly and variably pale breast. Compare with Neotropic Cormorant (p. 204) inland and with Brandt's coastally.
- **VOCALIZATIONS** At nest, low croaks and grunts, including one call that sounds unmistakably like human belching.
- **BEHAVIORS** Found singly and in flocks wherever there are fish-filled waters: bays, ponds, rivers, reservoirs. Nests colonially in treetops.
- **POPULATIONS** Following sharp losses in 20th century, has rebounded remarkably, especially inland. In much of interior, the only cormorant.

Neotropic Cormorant
Nannopterum brasilianum | NECO | L 26" (66 cm) WS 40" (102 cm)

■ **APPEARANCE** Smaller, slimmer than Double-crested, with proportionately longer tail and thinner bill. Yellow-orange throat patch, pinching back to sharp point, is thinly bordered by white feathering; throat patch of Double-crested more rounded, extending above eye. Juvenile brown-breasted; juvenile Double-crested paler gray on breast.

■ **VOCALIZATIONS** Soft croaking and oinking, like wood frogs or distant pigs.

■ **BEHAVIORS** A cormorant of inland waterbodies large or small. Neotropics in coastal Calif. have a knack for avoiding the actual seacoast, and sightings offshore are unheard of.

■ **POPULATIONS** Range expanding north; now rare but regular all the way to Great Salt Lake.

PELICANS | PELECANIDAE
These well-known piscivores have gigantic bills with distensible throat pouches used as dip nets for snarfing up fish. Pelicans are hulking and ungainly swimming or on land, but they are master aerialists. There are only eight species worldwide, but they are widespread.

American White Pelican
Pelecanus erythrorhynchos | AWPE | L 62" (157 cm) WS 108" (274 cm)

■ **APPEARANCE** Size alone is usually sufficient for field ID; wingspan matches that of California Condor (p. 214). Bill orange, adorned in breeding season with laterally compressed plate. Otherwise appears all-white on land or water, but in flight, the black remiges are prominent; soaring Wood Stork (p. 198), rare in West, shows black tail and long legs.

■ **VOCALIZATIONS** Even at colony, calls only rarely; however, whooshing of wings on bird in flight quite audible at close range.

■ **BEHAVIORS** Feeds in a manner utterly different from Brown Pelican; groups of 5–50+ swim to encircle a school of fish, then all plunge their faces into the water at the same time, with each pelican getting a fish. Flocks soar in bewitching 3D spiral formations at high altitudes.

■ **POPULATIONS** Breeds far inland, unlike marine Brown Pelican; lingers late and returns early, with ice-free, fish-filled waters the limiting distributional parameter.

Brown Pelican
Pelecanus occidentalis | BRPE | L 48" (122 cm) WS 84" (213 cm)

■ **APPEARANCE** Smaller than American White Pelican, but nevertheless a very large bird. Body gray-brown at all ages. Adult has white neck and cap with blond wash, intensifying in breeding season; bill gets redder too in breeding season. Juvenile dusky all over; bill and plumage always darker than juvenile American White.

■ **VOCALIZATIONS** Like American White, usually silent. One study has shown that the timbre of wing-whooshing in flight may be related to molt.

■ **BEHAVIORS** Feeds over the ocean; flocks fly single file in undulating waves, then suddenly fall to the water in marvelous plunge-dives. When not feeding, roosts at docks and marinas. Gets out to sea, where a distant bird could be confused with Black-footed Albatross (p. 180).

■ **POPULATIONS** Recovering rapidly from sharp crash during DDT era of mid-20th century. Breeds coastally, chiefly Feb. through Apr., then disperses north; a few wander far inland. Vagrants to Colo. and N. Mex. may refer at least in part to eastern *carolinensis* subspecies; our breeders are *californicus*.

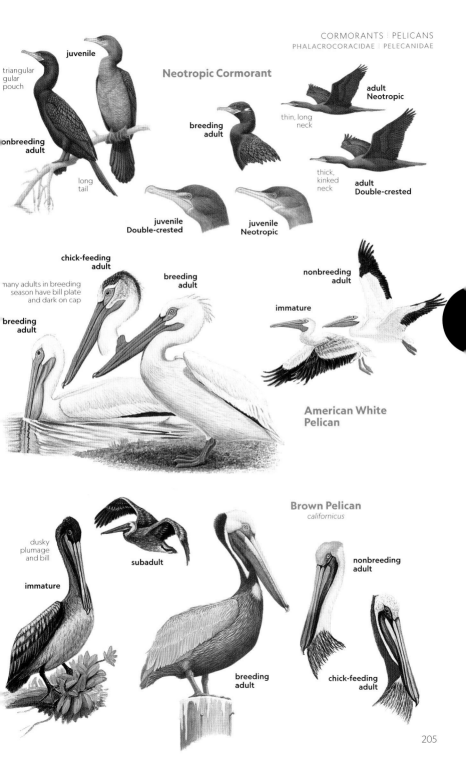

BITTERNS, HERONS, AND EGRETS | ARDEIDAE

The ardeids vary greatly in size and plumage, but all are adapted for life as hunters in shallow waters: They wade slowly or simply wait motionless, then strike suddenly with their daggerlike bills. Ardeids number about six dozen species worldwide, and their several names—bittern, heron, egret—are rather imprecisely applied within the family.

Accidental to Hawaii

American Bittern
Botaurus lentiginosus | AMBI | L 28" (71 cm) WS 42" (107 cm)

- **APPEARANCE** Medium-tall for a heron; heavyset. Muddy brown above, with long, thick, brown stripes below. Bill and feet dull yellow. Adult has long black stripe on face and neck, absent on juvenile. In flight, distinguished from immature night-herons (p. 212) by light brown wing coverts.
- **VOCALIZATIONS** Low-pitched song like someone swallowing air: *oonk-k'loonk*, repeated. Flight call an abrupt, monosyllabic *rrak*, similar to night-herons'.
- **BEHAVIORS** Solitary, secretive, and generally scarce; hides by day in bulrushes; more active dusk till dawn. Freezes in place when spotted, the bill pointing straight up. Breeds in freshwater wetlands. Accepts brackish wetlands in winter, when especially secretive.
- **POPULATIONS** Northern limit of nonbreeding range imperfectly known. Declining; accidental to Hawaii.

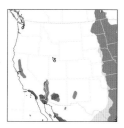

Least Bittern
Ixobrychus exilis | LEBI | L 13" (33 cm) WS 17" (43 cm)

- **APPEARANCE** Smallest heron in the world, barely the size of a robin (p. 348); long-necked, but typically seen hunched into a ball-like shape. Bright and buffy, although appears dark when glimpsed in dense cattails. All plumages have blurry streaks below and huge buffy oval on folded wing. Adult male black-backed; female and juvenile not as dark-backed.
- **VOCALIZATIONS** Song a series of three to seven clucking or cooing notes, fast but steady: *rook-rook-rook-rook*. Flight call a dull *ruck*.
- **BEHAVIORS** Like American Bittern, a recluse; sometimes comes out into the open, but more often a disembodied voice in the marsh.
- **POPULATIONS** Very local in West, especially in breeding season. Like American Bittern, in long-term decline; loss of marshlands the prime suspect in both species' ill fortunes.

Casual to Hawaii

Great Blue Heron
Ardea herodias | GBHE | L 46" (117 cm) WS 72" (183 cm)

- **APPEARANCE** Our largest heron; mostly gray-blue with impressive yellowish bill that brightens to orange-yellow in breeding season. Adult has thick black stripe on pale head; juvenile dark-capped. Dark legs of adult feathered at "thigh" with burgundy "trousers." Distinctive in flight, with neck tucked in, legs trailing, and dark remiges contrasting with paler wing coverts; compare with Sandhill Crane (p. 100).
- **VOCALIZATIONS** When flushing, a monstrous eruption of white noise: *rrRRRAAAHH!* Young at nest solicits food with weird, machinelike ticking.
- **BEHAVIORS** Eats almost any animal matter, not just fish and frogs, but also mammals and birds. Hunts solitarily; roosts and nests in loose groups. Assemblages of nests, composed of twigs and placed in treetops, are termed rookeries.
- **POPULATIONS** Across much of West, the most familiar and most frequently spotted heron—and indeed one of the most familiar of all waterbirds. Casual to Hawaii.

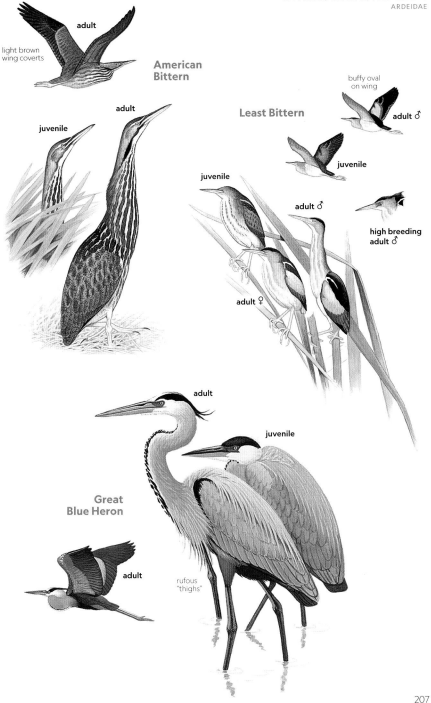

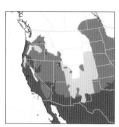
Casual to Hawaii

Great Egret
Ardea alba | GREG | L 39" (99 cm) WS 51" (130 cm)

- **APPEARANCE** Stately; proportioned like a slender Great Blue Heron, in the same genus. Legs black. Bill yellow like Cattle Egret's (p. 210), but Cattle Egret smaller and more compactly built. At beginning of breeding season, acquires bright green facial skin and displays filamentous plumes at rear.
- **VOCALIZATIONS** Especially when courting, emits loud rumbles like a mower or buzz saw starting up.
- **BEHAVIORS** Like Great Blue Heron, stands stock-still at edge of marsh or well out in shallow water. Occurs wherever there is shallow open water, fresh or brackish. Nests locally and often colonially in West.
- **POPULATIONS** Brought to precipice of extirpation from U.S. in early 20th century by plume hunters but has substantially recovered. After nesting, disperses extensively, often northward. Winter distribution delineated by availability of calm, continually ice-free waters for foraging. Casual to Hawaii.

Accidental to Hawaii

Snowy Egret
Egretta thula | SNEG | L 24" (61 cm) WS 41" (104 cm)

- **APPEARANCE** Slim and dainty. Adult readily separated from Great Egret by thin black bill and black legs with yellow feet ("golden slippers"). Like Great Egret, undergoes color intensification of bare parts at beginning of breeding season; lores and feet of Snowy become blood orange. Juvenile, with paler bill and greener legs, closely resembles juvenile Little Blue Heron, but the latter is not as elegantly built.
- **VOCALIZATIONS** Nasal groan, *raaa-aaah*, slightly wavering.
- **BEHAVIORS** Nests in mixed-species colonies in trees. Prances and capers while feeding; Great Egret does not.
- **POPULATIONS** Widespread but irregular in interior in warmer months. Like Great Egret, was nearly wiped out in early 20th century by plume hunters, but recovered quickly following legal protections and changing public attitudes. Accidental to Hawaii.

Accidental to Hawaii

Little Blue Heron
Egretta caerulea | LBHE | L 24" (61 cm) WS 40" (102 cm)

- **APPEARANCE** Similar in structure to but a bit stouter than Snowy Egret, a close relative. Adult mostly dark blue-gray with deep-plum neck and head; legs blue or greenish, bill blue-gray. Juvenile almost identical in plumage to juvenile Snowy Egret, but Little Blue Heron has gray tips to outer primaries, lacking in Snowy. As Little Blues mature to being all-dark, they pass through a splotchy white-and-gray intermediate plumage. Bare parts differ too: Little Blue has thicker bill, drooping slightly, pale blue-gray at base; bill of Snowy thinner and straighter, with yellow at base extending to lores. Legs of juvenile Little Blue dull yellow-green; legs of juvenile Snowy bicolored, black on "shins," dull yellow on "calves."
- **VOCALIZATIONS** Nasal groans like Snowy's, averaging higher-pitched.
- **BEHAVIORS** Not as inclined to flocking as Snowy, and does not usually engage in animated darting and dashing while feeding.
- **POPULATIONS** Scarce at best in much of West, but shows up widely in interior Southwest and western Great Plains every year in warmer months. Center of abundance in West is in coastal Southern Calif., especially San Diego area, where readily found throughout year. Accidental to Hawaii; one certain record, an individual present around Honolulu for 28 years!

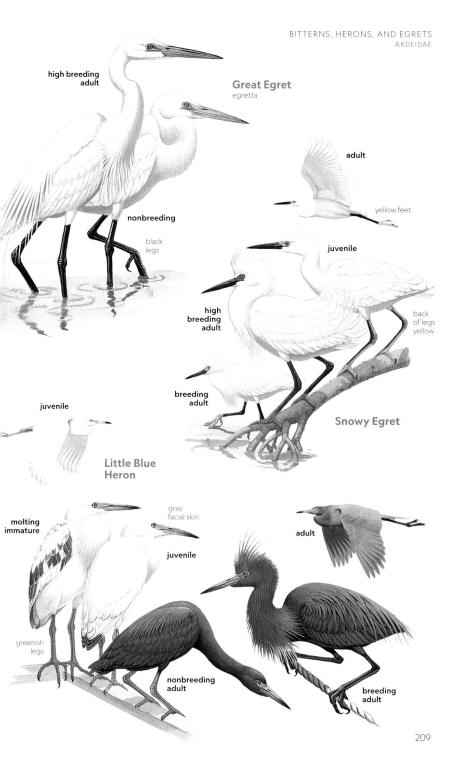

Tricolored Heron
Egretta tricolor | TRHE | L 26" (66 cm) WS 36" (91 cm)
Although fairly common well into northwestern Mexico, quite scarce north of international border in West. The "Trike" is long-necked and long-billed, even for a heron; despite the English name, it is a "true" egret. In all plumages, the white underparts and wing linings contrast with dark, colorful upperparts: Adult mostly slate blue above; juvenile extensively rusty above. Bill yellow most of year, becoming baby blue during courtship. Active by day and partial to saltwater habitats; relatively antisocial, feeding away from aggregations of waterbirds. Bulk of sightings in West along border with Mexico; barely annual north to Colo., casual farther north.

Reddish Egret
Egretta rufescens | REEG | L 30" (76 cm) WS 46" (117 cm)
Like Tricolored Heron, a rare but annual egret in West, with bulk of records along southern border of U.S. The largest and most frenetic egret; feeding, the bird lurches about, stamps its feet, and raises its wings to create an umbrella that lures fish to the surface and makes them easy to see. Head and neck of adult shaggy; long bill is bicolored in breeding season. Has two color morphs, but only dark morph documented in West: Adult of this morph slate gray overall with red-brown head and neck; juvenile dull maroon-brown all over. Even more than Tricolored Heron, a lover of salt water, with concentration of records coastal Southern Calif.; inland north of southern tier of U.S. states, even rarer than Tricolored.

Cattle Egret
Bubulcus ibis | CAEG | L 20" (51 cm) WS 36" (91 cm)

Well established in Hawaii

■ **APPEARANCE** Compact and relatively small; bill stout, legs short. Adult usually yellow-billed and often dark-legged like larger Great Egret (p. 208); neither Cattle nor Great is a "true" egret, genus *Egretta*. During courtship, adult acquires orange wash on feathers and bare parts brighten. Dusky-billed juvenile superficially resembles Snowy Egret (p. 208), but the latter always has "golden slippers" and dark "shins." Aberrant Cattle Egrets, with splotchy dark blue or dull green, although rare, occur with some frequency in this species; etiology not fully understood, but staining is the causal agent in many, perhaps most, instances.

■ **VOCALIZATIONS** Common call, typically given in flight, a nasal grunt, *unt* or *unk;* sometimes a squeakier *eent* or *aaayn*. Also quiet chatter at roost; usually feeds silently.

■ **BEHAVIORS** Most terrestrial of our herons, often in pastures with cattle and horses. Cattle Egrets feed right at the feet of livestock, pumping their heads as they go; often they feed while *on* livestock, "surfing" as the beasts lumber along. Nests in colonies in trees, often near water.

■ **POPULATIONS** Widespread in Americas, but believed to be a relatively recent arrival from Africa, in 19th century. Relative contributions of natural vagrancy and human assistance to establishment on mainland unknown. Species well established in Hawaii, where ancestors of today's birds were definitely introduced intentionally—a biocontrol initiative run amok. Opportunistic; still establishing in grazed landscapes that aren't too dry. Asian subspecies *coromandus*—perhaps a distinct species—more extensively rusty and with thicker bill, has been recorded from Aleutians and Midway.

BITTERS, HERONS, AND EGRETS
ARDEIDAE

Tricolored Heron

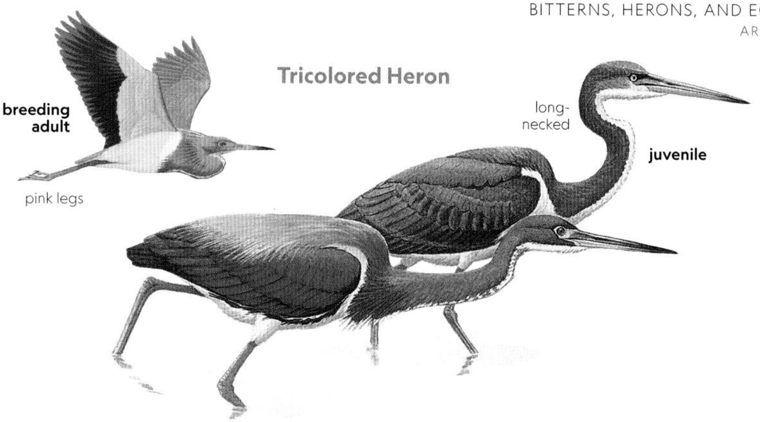

breeding adult
pink legs
long-necked
juvenile

Reddish Egret

bicolored bill
dark-morph breeding adult
dark-morph juvenile
only breeding adult shows bicolored bill
white-morph winter adult
dark-morph breeding adult

Cattle Egret

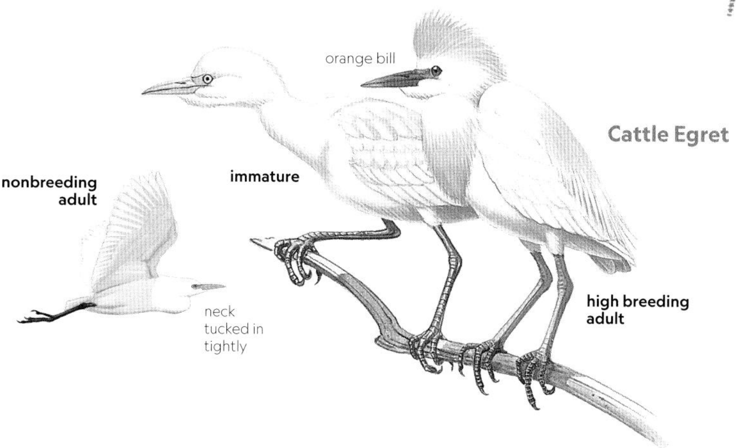

orange bill
nonbreeding adult
immature
neck tucked in tightly
high breeding adult

Accidental to Hawaii

Green Heron
Butorides virescens | GRHE | L 18" (46 cm) WS 26" (66 cm)
- **APPEARANCE** Smallish but sturdily built; bullnecked with short legs. Adult dark and glossy above, appearing more bluish than green in normal ambient light. Chestnut neck and face contrast with dark blue-green cap, occasionally erected as crest. Feet of adult yellow, brightening to orange-yellow in breeding season. Juvenile brownish, heavily streaked; Least Bittern (p. 206) smaller and brighter.
- **VOCALIZATIONS** On flushing, a piercing *kyark!*
- **BEHAVIORS** Solitary; occurs where woods and waterways meet. Waits quietly, often from overhanging branch, at bank of pond or slow-moving stream. Defecates explosively when flushed.
- **POPULATIONS** Widely distributed in West south of Canada, but quite scarce in much of arid interior; best represented here on Pacific Slope, where increasing and expanding north, and along southern border region north along major rivers. Accidental to Hawaii.

Common in Hawaii

Black-crowned Night-Heron
Nycticorax nycticorax | BCNH | L 25" (64 cm) WS 44" (112 cm)
- **APPEARANCE** Larger and stockier of the two night-herons; bullnecked and short-legged. Adult gray-bodied with black crown and black back. Juvenile brown with white speckling above, white streaking below; distinguished from juvenile Yellow-crowned by coarser speckling and streaking, spiky and mostly yellow bill, and short legs. Juvenile also recalls American Bittern (p. 206), which is more warmly colored overall and more strongly streaked below; wings of bittern in flight more obviously two-toned.
- **VOCALIZATIONS** In flight, usually at dusk or after sundown, a sudden *quok!*
- **BEHAVIORS** By day, roosts singly or in disorganized roosts in dense foliage near water; comes out at nightfall to hunt. Nests colonially, usually in trees; forages in both freshwater and saltwater habitats.
- **POPULATIONS** Stable, even increasing, across vast global range, but in decline in parts of West; widespread but local north of Great Basin and Central Rockies regions. Common breeding resident year-round in Hawaii.

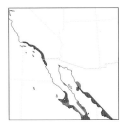

Yellow-crowned Night-Heron
Nyctanassa violacea | YCNH | L 24" (61 cm) WS 42" (107 cm)
- **APPEARANCE** Proportionately longer-necked and longer-legged, but a tad smaller overall, than Black-crowned Night-Heron. Adult has slaty body and bold black and white markings on head. Juvenile darker than juvenile Black-crowned, with finer white speckling and streaking; black bill of juvenile Yellow-crowned is thick and comparatively stout and blunt. As with Black-crowned, juveniles can be confused with American Bittern. Young of both night-herons slowly transition to adult plumage over a span of close to two years.
- **VOCALIZATIONS** Descending *kyaa* in flight, typically after dark.
- **BEHAVIORS** Active mostly at night; feeds on animal matter, especially crustaceans. Roosts by day, mostly in trees and shrubs, less commonly on docks and jetties. Establishing population in West not averse to urban habitats along coast.
- **POPULATIONS** Increasing steadily coastal Calif., north regularly to Santa Barbara Co. Casual at best west of Continental Divide in interior; annual in recent years along I-25 corridor, Colo. to N. Mex.

BITTERNS, HERONS, AND EGRETS
ARDEIDAE

Green Heron

adult / juvenile

adult — uniform wings above

Black-crowned Night-Heron

juvenile — dusky yellow bill

1st spring

2nd spring

adult

high breeding adult

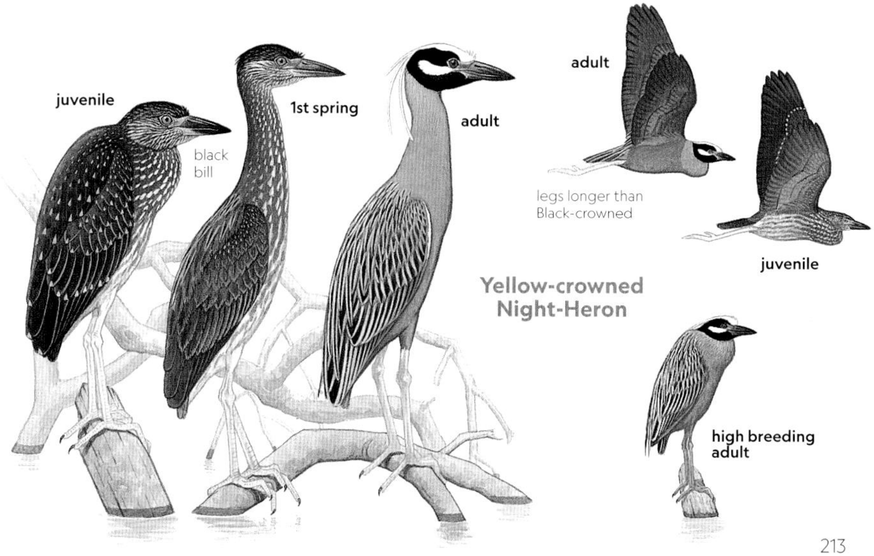

juvenile — bullnecked

adult

Yellow-crowned Night-Heron

juvenile — black bill

1st spring

adult — legs longer than Black-crowned

adult

juvenile

high breeding adult

IBISES | THRESKIORNITHIDAE

Long-legged waders with notably long bills, the ibises and related spoonbills are aquatic omnivores. Several species—perhaps most or all of them—in this family find food with motion detectors in nerve endings at the bill tip. The two species of regular occurrence in the West are hard to tell apart.

Glossy Ibis
Plegadis falcinellus | GLIB | L 23" (58 cm) WS 36" (91 cm)

Identical in build to White-faced Ibis, and vocalizations nearly the same. Reached U.S. East Coast in 19th century; has been expanding ever since. Is now rare but annual across lower 48, with strongest presence in West in Rocky Mountain states; may have bred in Colo. Breeding adult differs from White-faced by thin, pale, blue lines above and below dark eye; legs mostly dull with contrasting red "joints." Despite name, adult averages a bit less glossy than White-faced; both *Plegadis* ibises appear all-dark in low light.

Annual to Hawaii

White-faced Ibis
Plegadis chihi | WFIB | L 23" (58 cm) WS 36" (91 cm)

■ **APPEARANCE** Both *Plegadis* ibises are the size of a medium heron but proportioned more like a large rail; their long bills are decurved. Red eyes and red facial skin of breeding White-faced encircled by white feathering; feathering around eye of breeding Glossy dark, bordered by thin blue facial skin. Winter adult and older juvenile extremely similar to Glossy, but eye red; younger juveniles of the two species not safely separable in the field. Further complicating matters, hybrids are not rare.

■ **VOCALIZATIONS** Short grunts and groans, low-pitched and nasal; may average a bit higher than in Glossy, but much overlap.

■ **BEHAVIORS** Gregarious; compact flocks weave single file in flight, then put down at water's edge or in flooded fields where they feed hurriedly. Aided by tactile sensors in bills, are capable of feeding day or night.

■ **POPULATIONS** White-faced is the default ibis across the West, with most breeding in interior; rare but annual, sometimes in flocks, to Hawaii. Following range contraction and declining numbers mid-20th century, has rebounded and expanded in recent decades.

NEW WORLD VULTURES | CATHARTIDAE

Dark overall with unfeathered heads, these large to huge scavengers are seen in flight overhead or perched ominously on crags or snags. All are in the order Cathartiformes (from the same word that gave us "cathartic"), in distinction from raptors in the order Accipitriformes (pp. 216–231).

California Condor 🄲🅁
Gymnogyps californianus | CACO | L 47" (119 cm) WS 108" (274 cm)

Gargantuan; wingspan 30 percent greater than Golden Eagle (p. 218) with which it sometimes co-occurs. Short-tailed and bulbous-headed. Soars on huge, flat wings "like a flying piano." Adult, with pink-orange head, has white wing linings beneath that contrast with dark remiges, even at tremendous distance. Juvenile has grayer head; dusky wing linings contrast weakly with darker remiges. Transitions from juvenile to adult plumage in five to seven years. Key mark: Most have prominent numbered tags on both forewings. Entire population was captured in 1987; subsequently repatriated to Calif., Ariz., and Mexico. Is very slowly reestablishing, but the road to full recovery will be a long one.

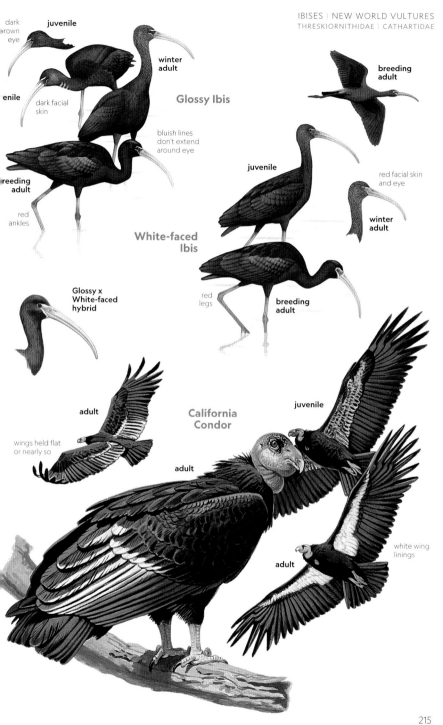

Black Vulture
Coragyps atratus | BLVU | L 25" (64 cm) WS 57" (145 cm)

- **APPEARANCE** A bit shorter in body length and quite a bit shorter in wingspan than Turkey Vulture. Tail short; legs long and whitish. Unfeathered head dark gray at all ages. On bird in flight, whitish outer primaries contrast with otherwise dark wings; long legs trail behind short tail.
- **VOCALIZATIONS** New World vultures lack syrinx, the sound-producing apparatus in most birds. So they are left with breathy grunts and hissing. At carrion, Black Vultures snap at each other with a gruff *woof*; defend young with drawn-out *HHHUUUUOOOOOHH*.
- **BEHAVIORS** More assertive and more adventurous of our two vultures; tends dumpsters and landfills. Lacks keen sense of smell of Turkey Vulture, and often associates with that olfactorily superior species in quest for carrion. Soars less than Turkey Vulture and has quicker wingbeats.
- **POPULATIONS** Expanding north range-wide. Expansion in West has been pronounced in Ariz., where now well established across southern tier of counties and north to Phoenix.

Turkey Vulture
Cathartes aura | TUVU | L 27" (69 cm) WS 69" (175 cm)

- **APPEARANCE** Longer-tailed and longer-winged than Black Vulture. Adult has unfeathered red head, gray in juvenile. Wings two-toned from below: wing linings dark, remiges paler.
- **VOCALIZATIONS** Like Black Vulture, lacks syrinx. Rarely makes sounds when feeding, unlike Black Vulture, but defends young with unpleasant groaning and hissing.
- **BEHAVIORS** Because it is large, common, and frequently airborne, this is one of the most frequently spotted birds across much of its range; a lone "hawk" or "buzzard" above deserts, forests, or roadways—or a kettle of several to many—often proves to be this species. In flight, tilts and rocks, wings locked in shallow V.
- **POPULATIONS** Like Black Vulture, slowly expanding north. Early spring migrant and fairly late fall migrant; wintertime sightings in summer parts of range quite rare.

OSPREY | PANDIONIDAE
This family comprises just one species, but it is nearly global in distribution. Ospreys eat only fresh-caught fish, which they catch in messy, feetfirst "cannonball" dives from considerable altitudes.

Annual to Hawaii

Osprey
Pandion haliaetus | OSPR | L 22–25" (56–64 cm) WS 58–72" (147–183 cm)

- **APPEARANCE** A large raptor, relatively long-winged; wings kinked in flight. Adult uniform chocolate above, mostly white below with variable "necklace" of dark feathering, especially on female; whitish head has messy crest and thick brown eyeline. Juvenile like adult, but upperparts spotted white.
- **VOCALIZATIONS** Powerful down-slurred whistle, often given in slow series: *kyew ... kyew ... kyew! KYEW!*
- **BEHAVIORS** Usually seen singly or in family groups in immediate vicinity of nests. Forages mainly over larger, deeper waterbodies, both marine and inland. Readily takes to human-made nest platforms, which are prominent along roadsides and lakeshores.
- **POPULATIONS** Suffered terribly during DDT era but has recovered impressively; large, fish-filled reservoirs inland have been important in species' recovery. Rare but annual, fall to spring, in Hawaii.

NEW WORLD VULTURES | OSPREY
CATHARTIDAE | PANDIONIDAE

HAWKS, EAGLES, AND KIN | ACCIPITRIDAE

Birds in this family include some of our best known raptors, such as the Bald Eagle (p. 222) and Red-tailed Hawk (p. 228). But the term "raptor" is taxonomically nearly useless: The only distantly related falcons (pp. 258–261) surely qualify, and even predatory songbirds like shrikes (p. 292) are fearsomely raptorial. Accipitrids are characterized by hooked bills and mostly muted colors; adult males and females are usually similar in plumage, but juveniles and adults are often distinct. The Accipitridae number 250+ species worldwide.

White-tailed Kite
Elanus leucurus | WTKI | L 16" (41 cm) WS 42" (107 cm)

- **APPEARANCE** Small hawk; long-tailed and long-winged. Gray and white overall, with black "shoulder" (upperwing coverts) and long, mostly white tail. Adult has gray back and white underparts; white face marked by black eye patch. Juvenile washed cinnamon but already shows black wing coverts and largely white tail of adult.
- **VOCALIZATIONS** Most calls short and clipped: *cleep* and *cleet*, whistled and descending. Also gives ternlike calls, grating and stuttering: *chree-ick* and *chreee-iiiih*.
- **BEHAVIORS** Hunts small mammals in open country; hovers on beating wings. In flight, can suggest a tern (pp. 164–173) more than another accipitrid.
- **POPULATIONS** Following range expansion in 20th century, especially Calif., has stabilized or withdrawn slightly.

Accidental to Hawaii

Golden Eagle
Aquila chrysaetos | GOEA | L 30–40" (76–102 cm) WS 80–88" (203–224 cm)

- **APPEARANCE** Same size as Bald Eagle, but proportioned differently: relatively small-headed and long-tailed; more likely to be confused at a distance with Turkey Vulture (p. 216). All plumages have golden wash on hindneck. Adult mostly dark; juvenile has striking white patches at base of tail and on center of wing, visible from above and below.
- **VOCALIZATIONS** Largely silent. Away from nest, gives down-slurred chirps, often repeated: *kyerp kyerp kyerp*.
- **BEHAVIORS** Solitary; often in lonely, arid country, where it preys on mammals, especially rabbits and hares.
- **POPULATIONS** Widespread but scarce across mainland in West; accidental to Hawaii. Enjoys broad distribution in Northern Hemisphere, where it is an emblem of royalty and majesty in many cultures.

Annual to Hawaiian Is.

Northern Harrier
Circus hudsonius | NOHA | L 16–20" (41–51 cm) WS 38–48" (97–122 cm)

- **APPEARANCE** Long-winged and long-tailed, with owl-like facial disc. All plumages have white patch near rump. Older males are gray above and white below with inky black wing tips; adult female dark brown with dense streaking below. Juvenile warm brown-orange below, mostly unstreaked.
- **VOCALIZATIONS** Mostly silent, but gets excited in nest defense, driving off intruders with scratchy yaps and harsh whistles.
- **BEHAVIORS** Usually seen flying low to the ground; holds wings in shallow V, tilts unsteadily as it searches for prey. Hunts over open terrain with dense cover; the facial disc is an adaptation for detecting prey by sound.
- **POPULATIONS** Distribution tightly coupled with annual variation in numbers of microtine rodents. Co-occurs with Short-eared Owl (p. 240), dependent on the same prey. Very rare but annual to Hawaiian Is., fall to winter.

Sharp-shinned Hawk
Accipiter striatus | SSHA | L 10–14" (25–36 cm) WS 20–28" (51–71 cm)
- **APPEARANCE** Smallest hawk in the West, with great sexual dimorphism in size: Female more than 50 percent larger than male. Sharp-shinned smaller-headed and rounder-headed than relatively square-headed Cooper's Hawk, with squared-off and proportionately shorter tail. Plumages similar to Cooper's: Adults of both are bluish above with reddish barring below; adult Cooper's often has more prominent cap, contrasting with paler nape. Juvenile Sharp-shinned dark brown above, with smudgy red-brown streaks below; juvenile Cooper's has sharper streaking below, often has a few white splotches on back.
- **VOCALIZATIONS** Shrill whistles, *pee* or *pyew*, in rapid succession, more similar in timbre to those of American Goshawk than Cooper's.
- **BEHAVIORS** An agile hunter; captures small birds in flight within woodlots and larger forests. Seen more often in open sky, flying across clearings or migrating, alternating short glides with several quick wingbeats.
- **POPULATIONS** Formerly persecuted for preying on small songbirds, but enjoys legal protection today. Breeding range probably expanding south.

Cooper's Hawk
Accipiter cooperii | COHA | L 14–20" (36–51 cm) WS 29–37" (74–94 cm)
- **APPEARANCE** Averages larger than Sharp-shinned, but male Cooper's and female Sharp-shinned can overlap in size. Tail of Cooper's quite long, with smoothly rounded corners; Cooper's is larger-headed than Sharp-shinned. Gray-black cap of adult Cooper's, especially male, more prominent than on Sharp-shinned. Juvenile Cooper's has crisp streaking below, often shows splotchy white on back; juvenile Sharp-shinned has blobby and reddish-brown streaking below, averages less white on back.
- **VOCALIZATIONS** Rich, nasal *kek*, given singly, in slow succession, or in rapid series.
- **BEHAVIORS** Like Sharp-shinned, hunts birds in woodlands. In normal flight in open sky, the long tail is flipped about, rudderlike, with ease.
- **POPULATIONS** Until late 20th century, a forest recluse, but has recently become routine in suburbs and even big cities. Urbanization has been aided by greenway initiatives and the species' acceptance of novel foodstuffs, including invading Eurasian Collared-Dove (p. 72).

American Goshawk
Accipiter atricapillus | AGOS | L 21–26" (53–66 cm) WS 40–46" (102–117 cm)
- **APPEARANCE** Largest hawk in genus *Accipiter* in West. Relatively short-winged and long-tailed like Cooper's and Sharp-shinned, but bulkier; size and build invite comparison with buteos (pp. 224–231). Adult blue-gray all over, finely barred below; white eyebrow contrasts with black patch behind eye. Juvenile separated from juvenile Cooper's Hawk by flared eyebrow, undertail coverts with brown splotches, and wavy tail bands.
- **VOCALIZATIONS** Sharp, slurred whistles, often in series; timbre like that of Sharp-shinned, but goshawk is louder and far-carrying.
- **BEHAVIORS** Scarce breeder in dense forests, often on north-facing slopes with conifers; defends nest ferociously. Alternates flapping and soaring flight, but with deeper flaps and longer bouts of soaring than Sharp-shinned and Cooper's.
- **POPULATIONS** Numbers boom and bust with 10- to 11-year cycles of prey. Possibly in long-term decline in Great Basin.

Bald Eagle

Haliaeetus leucocephalus | BAEA | L 31-37" (79-94 cm) WS 70-90" (178-229 cm)

- **APPEARANCE** Very large raptor with huge, hooked bill. Golden Eagle (p. 218) similar in overall heft, but Bald averages shorter-tailed and larger-headed. The white-headed, white-tailed, yellow-billed adult is known to everyone, but younger birds are trickier. Dark-billed juvenile is dark brown overall with variable white splotching on tail and wings, not as sharply delineated as on Golden. Immatures slowly transition to adult plumage over a four- to five-year period; many in third and fourth years have dark mask on smudgy white head, recalling Osprey.
- **VOCALIZATIONS** A mirthful cackle, descending in pitch, seemingly not befitting so regal a bird.
- **BEHAVIORS** Versatile and wide-ranging, but shows strong affinity for aquatic habitats; this species is a sea-eagle, whereas the Golden Eagle is a "true" eagle. Often stands on ice, hunched forward; also perches straight and tall on large branches. The nest, near water, is titanic, and usually placed out in the open, visible at great distances.
- **POPULATIONS** This national emblem of the U.S. is also the ultimate conservation success story. The Bald Eagle was endangered in the late 20th century but recovered swiftly thanks to muscular legal protections.

Mississippi Kite

Ictinia mississippiensis | MIKI | L 14½" (37 cm) WS 35" (89 cm)

- **APPEARANCE** About the same size and shape as White-tailed Kite (p. 218); the name "kite" is given to various small, often grayish accipitrids (pp. 218-231), but the term has essentially no meaning taxonomically. Adult Mississippi is gray-bodied and pale-headed, with tail dark (white in White-tailed); wings contrastingly patterned above, with whitish secondaries and, in good view, red-shafted primaries. Juvenile coarsely streaked rusty below, with tail barred black and white.
- **VOCALIZATIONS** Two-note whistle, the second note long and descending: *pee-peeeeeh*.
- **BEHAVIORS** Airborne much of the time, suggesting a falcon (pp. 258-261). Forages in loose aggregates of multiple birds; catches dragonflies on the wing, deftly plucks katydids and cicadas from treetops.
- **POPULATIONS** East of Continental Divide, steadily expanding north; seems to be tracking northward expansion of cicadas, genus *Megatibicen*. Also expanding west in southern border states. Highly migratory; occurs in West only spring to late summer, often nesting in towns and even cities.

Common Black Hawk

Buteogallus anthracinus | COBH | L 21" (53 cm) WS 50" (127 cm)

- **APPEARANCE** Long-legged, with wings broad and fairly short; tail also broad and quite short. Adult black with broad white band at base of tail; compare with adult Zone-tailed Hawk (p. 228), different in body structure. Juvenile, streaked below, has barred tail; note also pale supercilium (eyebrow), blotchy black malar (moustache), and splotchy dark flanks.
- **VOCALIZATIONS** Call a series of short whistles, sharp and piercing, dropping off in amplitude at end.
- **BEHAVIORS** Secretive; nests in woods near rivers and ponds, where it hunts frogs and crayfish.
- **POPULATIONS** Status in West not fully understood. Spring migration at Tubac, Santa Cruz Co., Ariz., impressive; many seem to wind up in Mogollon Rim region.

HAWKS, EAGLES, AND KIN
ACCIPITRIDAE

Bald Eagle

juvenile
2nd year
whitish underwing and "wingpits"
juvenile
larger bill than Golden Eagle
Osprey-like face pattern
adults
3rd year

Mississippi Kite

adult
white ...ndaries
adult
immature
juvenile
adult

Common Black Hawk

adult
short tail with broad white band
juvenile
short, broad wings
adult
juvenile

223

Harris's Hawk
Parabuteo unicinctus | HASH | L 21" (53 cm) WS 46" (117 cm)
- **APPEARANCE** Long-legged; tail, long and dark, has white band at tip and white patch at base. Dark overall with bright chestnut on wings; adult solid dark brown below with white undertail coverts, juvenile streaky below.
- **VOCALIZATIONS** Noisy in group settings; calls include a low roar, trailing off, *RRAAaaauh*.
- **BEHAVIORS** Occurs in shrublands, with or without scattered trees. Harris's Hawks hunt cooperatively, and they live in highly social assemblages with complex dominance hierarchies.
- **POPULATIONS** Mostly permanent resident within U.S. range, but a few wander north; casual to southern Nev.

Gray Hawk
Buteo plagiatus | GRHA | L 17" (43 cm) WS 35" (89 cm)
- **APPEARANCE** Smallish for genus *Buteo*; long tail and relatively short wings suggest an accipiter (p. 220). Adult gray overall, smoothly colored above, finely barred below; tail banded black and white. Juvenile dark brown above, streaked and spotted brown below; face boldly marked with white cheeks and dark moustache; tail finely barred with white at base.
- **VOCALIZATIONS** Call a simple whistle, usually descending in pitch.
- **BEHAVIORS** Occurs in riparian woods, a habitat affinity that contributes to its accipiter-like mien; eats small reptiles.
- **POPULATIONS** Increasing in West due to improving habitat quality, especially Southeast Ariz.

Red-shouldered Hawk
Buteo lineatus | RSHA | L 15–19" (38–48 cm) WS 37–42" (94–107 cm)
- **APPEARANCE** Long-tailed like an accipiter. All plumages show pale "window" near wing tips. Adult orange-rufous, especially on "shoulders," breast, and underwing linings; tail and remiges barred black and white. Juvenile darker and browner overall, but usually shows some rufous on shoulders and underwing like adult; also, tail bands narrower, less distinct on juvenile.
- **VOCALIZATIONS** Noisy. Call is a short, sharp, descending *kyeeah*, given both singly and in series.
- **BEHAVIORS** Usually near woods, including urban eucalyptus groves, but also perches in clearings, especially on utility wires.
- **POPULATIONS** Expanding and establishing in Nev.; annual to Ariz. Very rare but annual east of Continental Divide, where most if not all are slightly larger eastern subspecies, paler overall but with darker streaks below.

Broad-winged Hawk
Buteo platypterus | BWHA | L 16" (41 cm) WS 34" (86 cm)
- **APPEARANCE** A small buteo; wings broad but pointed. Most adults dark brown above and barred rufous-brown below, with white underwing bordered in black; tail banded black and white. Uncommon dark morph solid chocolate below. Juvenile dark brown above with weak tail bands; pale underwing edged in dark.
- **VOCALIZATIONS** Call a long, thin, high, monotone whistle preceded by short introductory note: *pit-eeeeeeee*.
- **BEHAVIORS** A woodland buteo. Nests within forests, often mid-successional; even on migration, will roost at night in woods if available.
- **POPULATIONS** Formerly considered very rare in West away from Canadian breeding grounds, but now known to migrate in modest numbers late Apr. to mid-May at and near base of east flank of Rockies.

HAWKS, EAGLES, AND KIN
ACCIPITRIDAE

Harris's Hawk

adult
white tail base and tip
adult
juvenile

Gray Hawk

distinct head pattern
whitish rump
juvenile
juvenile
adult
lightly marked underwings
juvenile
adult

Red-shouldered Hawk

juvenile
adult
juvenile
adult
barred secondaries
juvenile
pale crescent

Broad-winged Hawk

juvenile
adult
marked secondaries
dark-morph adult (uncommon)
adult
"crossbow" shape
juvenile
juveniles

225

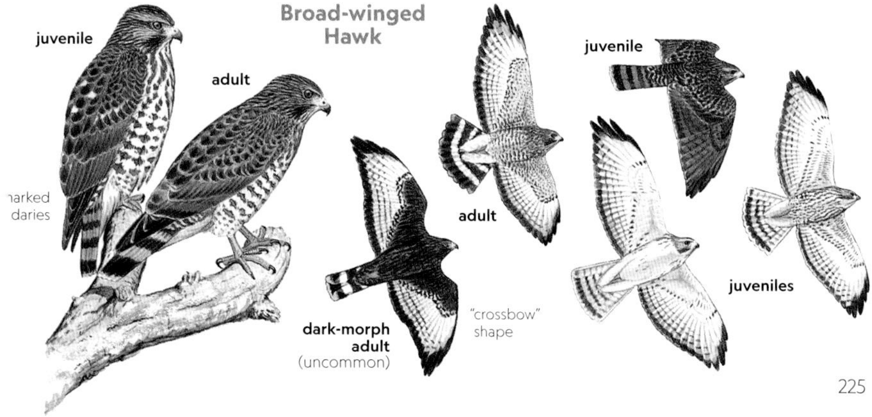

Endemic to Hawaii I.

Hawaiian Hawk
Buteo solitarius | HAWH | L 16–18" (41–46 cm) WS 34–40" (86–102 cm)
■ **APPEARANCE** A small buteo; wings broad and short, tail also broad and short. Sexes differ greatly in size, with female 30–40 percent heavier than male. Two color morphs: dark-morph adult cold brown above with gray highlights, warmer tan-brown below; light-morph adult also dark above, but mostly white from chin to undertail coverts. Juvenile light morph mostly white-headed; juvenile dark morph similar to dark-morph adult.
■ **VOCALIZATIONS** Call a short, nasal screech; Hawaiian name, 'Io, is onomatopoetic.
■ **BEHAVIORS** Nests in canopy in forest, where it hunts prey on ground or in trees. Often seen in soaring flight over clearings and forest edges.
■ **POPULATIONS** Endemic to Hawaii I.; not particularly numerous, but easily seen. Only other regularly occurring hawk on Hawaiian Is. is much rarer Northern Harrier (p. 218), present in tiny numbers fall to winter.

Short-tailed Hawk
Buteo brachyurus | STHA | L 15½" (39 cm) WS 35" (89 cm)
Began showing up in Ariz. in 1980s and is now rare but regular there in summer. A small buteo, occurring in two color morphs. Dark-morph adult in flight shows dark wing linings contrasting with paler remiges. Light-morph adult in flight patterned like Swainson's Hawk, but Swainson's is longer-winged, longer-tailed, and wears a prominent chestnut bib. Juveniles of both morphs resemble corresponding adults. Short-tailed is a forest buteo, has nested in pinewoods in Chiricahua Mts., Ariz., and probably elsewhere; difficult to spot perched in forest, but often soars well above clearings and edges. Along with records from Ariz., a few in recent years from N. Mex. and West Tex.

Swainson's Hawk
Buteo swainsoni | SWHA | L 21" (53 cm) WS 52" (132 cm)

■ **APPEARANCE** Slender buteo with long tail and long wings; on bird at rest, wing tips project beyond tail tip. Flight profile suggests Turkey Vulture (p. 216) or Northern Harrier, with long wings locked in shallow V. Most sightings in summer are of light-morph adults: uniformly gray-brown above, white-chinned and chestnut-bibbed below, with wing linings pale and remiges dark (opposite of Turkey Vulture); long tail with a bit of white at base above can suggest harrier. Uncommon intermediate-morph adult deep chestnut below, with splotchy chestnut underwing linings. Even rarer dark-morph adult dark brown above and below. Light-morph juvenile, commonly seen in late summer, quite different from adult: Head and breast suffused in warm caramel; lacks breastband, but shows broad dark bulges on sides of breast; back and upperwing coverts scalloped in caramel.
■ **VOCALIZATIONS** Call an anguished squeal, wavering but mostly on the same pitch: *eeeeeuh-eeeee*. Noisy around nest, notably in parent-offspring interactions.
■ **BEHAVIORS** Nests in isolated trees or small groves in open country, where it hunts diverse prey, especially snakes.
■ **POPULATIONS** Most winter south of Amazon Basin on the Pampas of Argentina, but increasing in winter in recent years in Calif., especially Central Valley. Otherwise absent from West except in warmer months. Breeds sparingly north of mapped range to Yukon and east-central Alas.

Zone-tailed Hawk
Buteo albonotatus | ZTHA | L 20" (51 cm) WS 51" (130 cm)

■ **APPEARANCE** In flight at any distance, suggests Turkey Vulture (p. 216): long-winged with a narrow tail; wings two-toned from below, with wing linings darker than remiges. However, all plumages have relatively large feathered head and yellow legs and base of bill, unlike Turkey Vulture; tail of adult has broad black and pale gray bands; tail of juvenile barred gray and black. Compare also with Common Black Hawk (p. 222), similar in color and pattern, but built differently.

■ **VOCALIZATIONS** Call a long, hoarse whistle, descending a bit. Also shorter, more pure-tone whistles, usually near nest or in interactions with other birds.

■ **BEHAVIORS** In addition to looking like a Turkey Vulture, also acts like one: Holds long wings in shallow V, rocking unsteadily as it soars in shallow circles. Zone-tailed Hawks often soar with vultures; it is believed that the predatory hawk "fakes out" its prey by appearing to be an innocuous vulture. Forages widely, but nests mostly in riparian settings, especially in canyon country.

■ **POPULATIONS** Uncommon in West, even within core range. Wanders a bit; annual in recent years to Colo. Does not have a light morph, perhaps because a light morph would gain no benefit from association with vultures.

Red-tailed Hawk
Buteo jamaicensis | RTHA | L 22" (56 cm) WS 50" (127 cm)

■ **APPEARANCE** Perched or flying, appears solid and stocky, with wings broad and rounded; tail fairly short, bill largish. Adults of most, but not all, populations have trademark red tail, striking above, much paler below. Most light-morph adults (several named subspecies) have brown back with white splotches and pale underparts with belly band, smudgy and variable; wings mostly pale from below, marked by darker patagial bar at leading edge of inner wing. Dark and intermediate morphs, widespread in West, can suggest other dark buteos, but note reddish tail on adults and, except on darkest individuals, patagial bar. Paler adults—for example, in subspecies *kriderii* ("Krider's Hawk")—can be confused with light-morph Ferruginous Hawk (p. 230). Dark-morph adult of subspecies *harlani* ("Harlan's Hawk") looks mostly cold black when perched, but with white splotches on breast; "Harlan's" has an uncommon light morph too, in which adults also are cold in tone. Even the darkest non-"Harlan's" have warmer tones than adult "Harlan's." Tail of adult "Harlan's" variably patterned with black, white, and reddish. All juvenile Red-tailed Hawks have tail finely barred and lacking red; note also belly band and, in flight, patagial bar.

■ **VOCALIZATIONS** Harsh, descending squeal, *RRReeeeeerr,* ubiquitous in movie soundtracks and television commercials, often attributed to the wrong species of raptor.

■ **BEHAVIORS** Bulky nest, constructed of twigs, placed near treetop; breeds early, incubating while it is still winter. A generalist predator, but mammals predominate in diet of many. Widespread and frequently seen, occurring in almost all terrestrial habitats: Watches from utility poles on the interstate, soars over deserts and forests, and roosts on bridges and buildings in big cities.

■ **POPULATIONS** Present year-round in much of West, but migratory, with a complex mosaic of breeders, winterers, passage transients, and permanent residents in many places. "Harlan's Hawk" breeds in Alas., winters mostly Great Plains.

Accidental to Hawaii

Rough-legged Hawk
Buteo lagopus | RLHA | L 21" (53 cm) WS 53" (135 cm)

■ **APPEARANCE** A lanky buteo with small feet and a small bill; the tarsi are feathered, an adaptation for cold temperatures. Plumage variation complex; all show pale tail below with broad black band (adult) or gray and blackish bars (juvenile), along with blackish blob on "wrist" (carpal patch) of underwing. All light morphs have considerable pale on head: Juvenile has blackish on belly; light-morph adult female similar, but has solid black tail band; light-morph adult male has belly flecked black and frosty white. Dark morphs nearly black when perched; in flight from below, black wing linings contrast with whitish remiges.

■ **VOCALIZATIONS** Usually quiet in winter, but occasionally gives a long squeal, fairly pure-tone and slowly dropping in pitch.

■ **BEHAVIORS** Nests in forested or unforested Arctic habitats; winters on desolate, nearly featureless prairie and other grasslands. Hovers habitually while feeding; when not hunting, perches on surprisingly flimsy shrubs or twigs, often at the tops of small trees.

■ **POPULATIONS** Our most Arctic buteo, breeding extensively across the Canadian Arctic Archipelago. Many don't arrive at midlatitudes of U.S. until Nov., and most are gone again by late Mar. Numbers fluctuate in response to winter prey availability and summer success on the breeding grounds. May be undergoing long-term shift on wintering grounds away from mid-continent and toward coasts. Accidental to Hawaii.

Ferruginous Hawk
Buteo regalis | FEHA | L 23" (58 cm) WS 56" (142 cm)

■ **APPEARANCE** Our largest buteo, likened by many to a small eagle. Long-winged and long-tailed, with a big bill and large feet; gape colorful and conspicuous (but note that juveniles of any buteo may show prominent gape). Like Rough-legged Hawk, probably its closest relative, has feathered tarsi. In flight, all plumages of Ferruginous show huge white panel on outer wings. Light-morph adult rusty above and on belly; feathered legs also rusty. Tail of adult varies from nearly white to extensively pale red, suggesting Red-tailed Hawk (p. 228). Light-morph adult shows splotchy rufous on underwing. Juvenile in flight mostly white below; note white in wings from above. Uncommon dark morph deep brown with a hint of chestnut; in flight from below, dark wing linings contrast with pale remiges. Adult dark morph pale-tailed above and below; juvenile darker-tailed above, but white wing panel stands out. Juveniles of both morphs have more dark marks on tail.

■ **VOCALIZATIONS** Descending whistle, richer than similar calls of other buteos. Heard mostly on breeding territory, but some in nonbreeding season defending winter feeding grounds do call occasionally.

■ **BEHAVIORS** Breeds around lonely steppes and badlands; solitary except at and near the nest. Preys on midsize mammals: jackrabbits, ground squirrels, and especially prairie dogs. When a prairie dog colony suddenly comes to life, that's often a good sign that a "Ferrug" is nearby.

■ **POPULATIONS** Most migrate short to medium distances, but others are nearly resident; local and annual variation in numbers reflects prey abundance. Range has contracted overall in the long term, but some regional expansions also reported. Rapid development of energy infrastructure (wind, oil field) is an emerging threat for this easily disturbed raptor.

HAWKS, EAGLES, AND KIN
ACCIPITRIDAE

Rough-legged Hawk

dark carpal patch

light-morph adult ♀

dark-morph adult ♂

small head and bill

adult ♂

hovers while hunting

juvenile

rusty leg feathering

pale primary "windows"

adults

Ferruginous Hawk

juvenile

dark-morph adult

pronounced gape

juvenile

adult

BARN OWLS | TYTONIDAE

These wraithlike inhabitants of outbuildings, cliff faces, and copses with standing deadwood have superb hearing. Their heart-shaped facial discs function as parabolic dishes, and their asymmetrically placed ears permit astonishingly precise orientation toward prey in total darkness. Most are long-legged and pale.

Established in Hawaii

Barn Owl
Tyto alba | BANO | L 16" (41 cm)

- **APPEARANCE** Legs long and dark; heart-shaped face outlined in a red-orange frame. Pale overall; washed butterscotch and leaden above, mostly white below; the eyes are dark and soulful. Unmistakable when spotted in broad daylight, but beware that any owl in car headlights or moonlight can appear white at night.
- **VOCALIZATIONS** A sudden shriek of white noise: *HHSSSSSSHHHH*. Call of juvenile Great Horned Owl (p. 234) is similar, but shorter, more nasal, less explosive.
- **BEHAVIORS** Voracious devourer of mice and rats. Roosts in barns and other outbuildings, but also caves and gnarled cottonwoods.
- **POPULATIONS** Widespread but patchily distributed within mapped range; established in Hawaii, where introduced in mid-20th century to control pests. Local population losses result from wide use of rodenticides, which sicken the owls.

TYPICAL OWLS | STRIGIDAE

Widely known, the typical (or "true") owls are cryptically brown, with large heads and bobbed tails. The eyes face forward, giving them a human visage. Like barn owls, strigids have facial discs and asymmetrically placed ears, adaptations for acoustic hunting. Some species have feather tufts that form visible "ears."

Flammulated Owl
Psiloscops flammeolus | FLOW | L 6¾" (17 cm)

- **APPEARANCE** Small, with indistinct "ears." Eyes large and dark. Ground color varies clinally from reddish in southern Rockies to grayish farther north and west.
- **VOCALIZATIONS** A simple hoot, *hoo*, often run out to two or three notes: *h'hoo* or *h'hoo hoo*. Sounds improbably low and powerful for so small an owl.
- **BEHAVIORS** Found mostly in conifer forests; favors dry ponderosa pine in southern Rockies, wetter spruce-fir in Great Basin. Nests in tree holes; hunts mostly moths.
- **POPULATIONS** Highly migratory; is in some respects the nocturnal counterpart of the insectivorous Neotropical migrant passerines.

Whiskered Screech-Owl
Megascops trichopsis | WHSO | L 7¼" (18 cm)

- **APPEARANCE** Smaller than similar Western Screech-Owl (p. 234); Whiskered is oddly small-footed, and its breast has more prominent transverse barring than Western's.
- **VOCALIZATIONS** Two songs: a steady series of 5–10 short whistles, not accelerating like Western; and a random-seeming stream of higher hoots, irregularly spaced like Morse code.
- **BEHAVIORS** Hunts mostly large nocturnal insects in mid-elevation pine-oak woods, especially around cabins, campgrounds, and lightly trafficked roads with supplemental lighting.
- **POPULATIONS** Generally at higher elevations than Western Screech-Owl. Sedentary, with little if any dispersal.

BARN OWLS | TYPICAL OWLS
TYTONIDAE | STRIGIDAE

Western Screech-Owl
Megascops kennicottii | WESO | L 8½" (22 cm)

- **APPEARANCE** Larger overall and with bigger feet than Whiskered Screech-Owl (p. 232); essentially identical in build to Eastern Screech-Owl. Note that the "ears" of any screech-owl vary from quite prominent to completely smooshed down. Dark-billed; vertical streaking below stronger than transverse barring.
- **VOCALIZATIONS** Two songs: "bouncing ball" series of mellow hoots, accelerating; steady trill, about two seconds in duration, with a break after the first few notes.
- **BEHAVIORS** Where Eastern and Western co-occur, Western goes for more arid landscapes; where Whiskered and Western overlap, Western usually at lower elevations.
- **POPULATIONS** Has expanded east in recent decades in eastern Colo. and parts of Tex., bringing it into contact with Eastern Screech-Owl.

Eastern Screech-Owl
Megascops asio | EASO | L 8½" (22 cm)

- **APPEARANCE** Like Western Screech-Owl, small and ovoid, big-headed and short-tailed. In daytime encounters, note pale gray-green bill of Eastern (darker on Western) and relatively weak vertical streaking below (stronger on Western).
- **VOCALIZATIONS** Two songs: a wavering, descending, pure-tone whinny; a long, low-pitched, monotone trill.
- **BEHAVIORS** In West, strongly associated with broadleaf groves, especially with willows, along creeks and rivers.
- **POPULATIONS** Mostly sedentary, but prone to some wandering fall to winter, and has probably expanded overall in past century-plus.

Great Horned Owl
Bubo virginianus | GHOW | L 22" (56 cm)

- **APPEARANCE** A massive predator; weighs much more than the fluff-and-feathers Great Gray Owl (p. 238); "ears" large. Variably gray-brown with heavy barring below; white bib prominent, especially when hooting.
- **VOCALIZATIONS** Male gives three to five slow, pure hoots (*hoo HOO... hoo ... hoo*), female four to eight faster, slightly higher, more nasal hoots (*hoo hu-hu HOO hu HOO hoo*). Female sometimes adds weird, nasal barks; rasping, wavering screech of juvenile can suggest Barn Owl (p. 232).
- **BEHAVIORS** Nocturnal, but readily spotted roosting by day in tall trees. Versatile; nests in rugged wilderness as well as in big cities.
- **POPULATIONS** Faring well, but many are hit by cars. Varied prey includes threatened species of raptors and other birds.

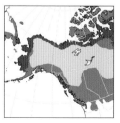

Accidental to Hawaii

Snowy Owl
Bubo scandiacus | SNOW | L 23" (58 cm)

- **APPEARANCE** Very large. Adult, especially male, mostly white. Lacks "ears"; yellow eyes stare from white face. First-winter female densely barred black, but nevertheless whitish overall.
- **VOCALIZATIONS** Male hoot a short, gruff lowing, normally heard only on breeding grounds. In interactions with other predators in winter, gives short yaps and a longer, gull-like squeal.
- **BEHAVIORS** Found in open country year-round. Breeds Arctic tundra, winters on prairie and beaches; roosts conspicuously on roofs and in open fields.
- **POPULATIONS** Southern limit of winter range varies annually; many at southern edge of winter range are first-winters. Accidental to Hawaii.

Northern Hawk Owl
Surnia ulula | NHOW | L 16" (41 cm)

■ APPEARANCE A midsize owl; the tail is tapered and quite long for an owl. Perched at a distance, suggests Cooper's Hawk (p. 220). Dark overall; blackish facial disc encircles close-set yellow eyes.

■ VOCALIZATIONS Song, spooky and far-carrying, is a slow trill of short whistles, the entire series 5–10+ seconds in duration.

■ BEHAVIORS Inhabits broken forest, especially spruce, where it hunts mammals in clearings and edges. Often out by day, where it watches from a prominent perch; nabs prey in short-distance, almost horizontal, sorties. Also hunts by night.

■ POPULATIONS Wanders after nesting, mostly within breeding range; a few drop south every winter, with larger irruptions in years with disruptions in food supply.

Northern Pygmy-Owl
Glaucidium gnoma | NOPO | L 6¾" (17 cm)

■ APPEARANCE Very small owl with a long tail; lacks "ears." Underparts coarsely streaked; dark tail has thin white bars. Eyes yellow; also has "false eyes" on nape, their purpose not agreed upon. Ground color varies clinally from gray-brown in southern Rockies to warmer red-brown in Salish Sea region. Compare with Northern Saw-whet Owl (p. 240).

■ VOCALIZATIONS Most in West give a simple hoot or whistle, repeated slowly by Pacific Slope birds, a bit faster by interior birds. In "Sky Islands" of Ariz. and southern N. Mex., the hoot is doubled. Call a rapid chippering, more like a squirrel than an owl. Wingbeats unmuffled, unlike most owls.

■ BEHAVIORS Inhabits mountain forests; most active dawn and dusk. Hunts small birds, which go berserk on discovering one roosting by day.

■ POPULATIONS Essentially resident throughout entire range, but a few move out onto lowlands fall to winter.

Ferruginous Pygmy-Owl
Glaucidium brasilianum | FEPO | L 6¾" (17 cm)

Widespread in Neotropics; reaches our area in saguaro forests of southern Ariz. Essentially identical in build to Northern Pygmy-Owl, but Ferruginous is a warmer brown, with the tail barred black (barred white on Northern); sings more rapidly than Northern, and even more prone than that species to be out in broad daylight. The subspecies in Ariz. is *cactorum*, the "Cactus Pygmy-Owl," declining and uncommon; current concentrations are in Santa Cruz R. drainage and Organ Pipe Cactus N.M.

Elf Owl
Micrathene whitneyi | ELOW | L 5¾" (15 cm)

■ APPEARANCE Tiny but well-built, with short tail and no "ears"; eyes yellow. Plumage intricately patterned in foxy colors: red-orange and leaden.

■ VOCALIZATIONS Song a yapping series of 8–12 notes, muffled at beginning and end, shrill in the middle: *yup-yup-yee-yee-YEE-YEE-yee-yee-yup*.

■ BEHAVIORS Nocturnal and easily overlooked; best seen at dusk, when it starts singing. Most famously associated with saguaro, but occurs widely in broadleaf groves with standing deadwood for nesting.

■ POPULATIONS Locally common. Migratory; gets back to U.S. breeding grounds by Mar., heads south in Sept. A few may winter.

TYPICAL OWLS
STRIGIDAE

Northern Hawk Owl

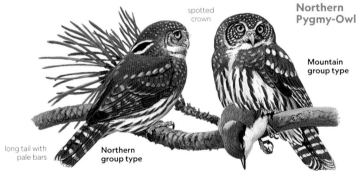

Northern Pygmy-Owl

Ferruginous Pygmy-Owl

Elf Owl

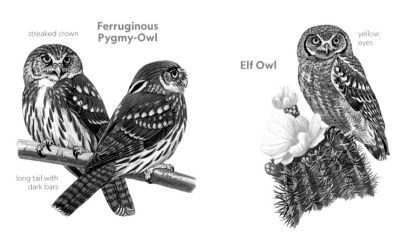

Burrowing Owl
Athene cunicularia | BUOW | L 9½" (24 cm)
- **APPEARANCE** Fairly small; long-legged. Lacks "ears"; eyes gloweringly yellow. Adult barred below, juvenile plain and buffy below.
- **VOCALIZATIONS** Two-note song, first note shorter: *hu hoooo*, repeated, not unlike Scaled Quail (p. 54). Also a rough chatter.
- **BEHAVIORS** A prairie dog mutualist. Nests in "dog" colonies, where visible by day; watches from short mounds, the lower half of the body often below ground level. Family groups stand shoulder to shoulder like meerkats.
- **POPULATIONS** Declining, due in part to historical and ongoing persecution of prairie dogs. Responds well to human help, however, including provisioning of artificial burrows for nesting and roosting.

Spotted Owl
Strix occidentalis | SPOW | L 18" (46 cm)
- **APPEARANCE** Midsize owl; big-headed, with dark eyes and no "ears." Dark overall, with neat white checkering above and below. "Mexican" subspecies paler and with larger white spots than "California" and "Northern" subspecies. In Pacific coast region, compare with Barred Owl.
- **VOCALIZATIONS** Song a series of hoots separated by exaggeratedly long pauses: *hoo . . . . hooo hu . . . . . hooo*.
- **BEHAVIORS** A forest owl; eats mostly mammals, including flying squirrels and bats. Hunts at night, but sometimes starts calling by late afternoon, especially in Southwest.
- **POPULATIONS** "Mexican" subspecies occurs in West in Four Corners states, "Pacific" from Northern Calif. northward, and "California" in Southern Calif. and the Sierra Nevada. "California" and especially "Pacific" subspecies imperiled by logging and invading Barred Owls.

Barred Owl
Strix varia | BADO | L 21" (53 cm)
- **APPEARANCE** Midway in size between smaller Spotted Owl and larger Great Gray Owl. Eyes dark; lacks "ears." Breast densely barred; belly broadly streaked.
- **VOCALIZATIONS** Noisy, with pairs duetting even during the day. Classic song a rhythmic, eight-and-a-half-note chant: *hu-HU-hu-HU hu-HU-hu-HUUaww*. Also various screams, squeals, and screeches.
- **BEHAVIORS** Aggressive and adaptable; hostile toward Spotted Owl.
- **POPULATIONS** Range has expanded dramatically in Pacific coast states, where it poses a double threat to the smaller Spotted Owl: directly by predation and indirectly through interbreeding and subsequent genomic dilution.

Great Gray Owl
Strix nebulosa | GGOW | L 27" (69 cm)
- **APPEARANCE** Appears huge, but weighs 25 percent less than Great Horned Owl (p. 234); puffy plumage of Great Gray an adaptation for cold climes. Mostly dark gray with yellow eyes, concentric rings on face, and white "bow tie."
- **VOCALIZATIONS** Deep, well-spaced hoots, up to 10 per series.
- **BEHAVIORS** Aided by excellent hearing, detects mammals *beneath* deep snow, then smashes into snowbank and captures prey. Even more amazingly, "jams" prey's hearing with acoustic deception.
- **POPULATIONS** Weakly irruptive; a few drift south from main Canadian range most winters. Note isolated outpost in central Sierra Nevada.

TYPICAL OWLS
STRIGIDAE

Long-eared Owl
Asio otus | LEOW | L 15" (38 cm)

- **APPEARANCE** Smaller than Great Horned Owl (p. 234). Roosting, assumes elongate posture; eponymous long "ears" prominent. Orange face is crossed vertically by black and white bands; orangey patch across outer wing prominent in flight.
- **VOCALIZATIONS** Male hoot a simple, rising *hooOO*, repeated slowly; also claps wings in display. Birds coming in to roost at first light make varied cackles, whines, and hoots.
- **BEHAVIORS** Secretive; may be seen at day roosts—sometimes a single owl pressed up against a conifer trunk, sometimes 10+ in a hedgerow. Flight erratic with jerky wingbeats. Hunts at night in open habitats next to woods.
- **POPULATIONS** Widespread but rarely detected. Declines in arid interior linked to degradation of riparian woods.

Short-eared Owl
Asio flammeus | SEOW | L 15" (38 cm)

- **APPEARANCE** A bit paler and stockier than Long-eared Owl; the "ears," set close, are tiny. Yellow-buff patch across upperwing and black bar on underwing prominent on bird in flight.
- **VOCALIZATIONS** Gives short barks when interacting: *chyurr* and *chyaap*. Rarely heard courtship song of male a fast series of simple hoots.
- **BEHAVIORS** An owl of open country; frequently abroad by day, especially when overcast. Seems to beat wings slowly, like a giant butterfly. Perches on haystacks, fences, and outbuildings; roosts in groups in winter.
- **POPULATIONS** Like Northern Harrier (p. 218), specializes on microtine rodents; numbers of the owl and harrier wax and wane in tandem. An endemic subspecies is resident on main islands of Hawaii; also, vagrants from Asia and N. Amer. have occurred throughout the Hawaiian Is.

Endemic subspecies resident in Hawaii; nominate subspecies vagrant to Hawaiian Is.

Boreal Owl
Aegolius funereus | BOOW | L 10" (25 cm)

- **APPEARANCE** Darker and larger than Northern Saw-whet, but female saw-whets approach the size of some male Boreals. Bill pale (dark on saw-whet); body of Boreal streaked and spotted dark chocolate and white. Whitish face bordered by black; forecrown speckled black and white. Rarely seen fledgling mostly uniform dark brown, but with white markings on face.
- **VOCALIZATIONS** Song a series of *hu* notes, increasing in loudness; similar to winnowing of Wilson's Snipe (p. 128), common in boreal bogs.
- **BEHAVIORS** Strictly nocturnal; favors boggy spruce-fir forests with aspen-birch admixtures.
- **POPULATIONS** Largely resident in West; minor irruptions in some years.

Northern Saw-whet Owl
Aegolius acadicus | NSWO | L 8" (20 cm)

- **APPEARANCE** A bit smaller than a screech-owl (pp. 232–235); lacks "ear tufts." Bill dark (pale on Boreal). Adult chestnut, coarsely streaked below. Fledgling rich buff-brown with white Y on dark face.
- **VOCALIZATIONS** Song a very long succession of short whistles, easily imitated by humans; supposedly resembles the whetting of a saw, an odd reference for the 21st century.
- **BEHAVIORS** Most in West nest in coniferous forests, especially pines; roosts unobtrusively in winter in diverse forest types.
- **POPULATIONS** Movements complex, with much interannual variation; winter distribution in West not fully worked out.

TROGONS | TROGONIDAE
Colorful birds of tropical forests, the approximately 50 species in this family are in their own order, Trogoniformes, of uncertain affinities with other bird groups. They are generally quiet and slow-moving, often sitting still in the shade for long periods of time. Only one species regularly reaches the U.S.

Elegant Trogon
Trogon elegans | ELTR | L 12½" (32 cm)

- **APPEARANCE** Large and ovoid, with long, broad tail; all have stout yellow bill and coppery tail. Red and green of adult male stunning in good light, but bird often hides in shadows. Female more brownish, but nevertheless with warm reddish wash below.
- **VOCALIZATIONS** Song a simple series of 5–10 gruff barks on same pitch: *rruhhrk ... rruhhrk ... rruhhrk ....*
- **BEHAVIORS** Typically seen sitting still, but sallies out for aerial insects or to glean fruit. When flushed, flies straight through forest. Found mostly in wooded canyons, often with Arizona sycamore, *Platanus wrightii.*
- **POPULATIONS** Most in Ariz. withdraw south in winter. Very rare to Bootheel of N. Mex., casual to Big Bend of Tex.

KINGFISHERS | ALCEDINIDAE
With 100+ species worldwide, the kingfishers are recognized by their blocky heads, large bills, mostly colorful plumages, and generally harsh calls. All kingfishers in the Americas are aquatic, but many in the Old World, including the legendary Laughing Kookaburra, *Dacelo novaeguineae,* are not.

Belted Kingfisher
Megaceryle alcyon | BEKI | L 13" (33 cm)

- **APPEARANCE** Only kingfisher in West away from a few waterways in southern Ariz., where Green is possible. Blocky head has shaggy crest; bill quite large. Blue and white all over; adult male has single blue breastband, adult female has orange and blue breastbands, and juvenile has blue breastband with orange flecks.
- **VOCALIZATIONS** Startling rattle, given in flight. Perched, gives a quieter, faster, more muffled rattle, starting with a quick burst and then slowing: *ch'CH'CH'CH'-chuh-chuh-chuh.*

Annual to Hawaii

- **BEHAVIORS** Usually occurs singly, watching from snag over open water or flushing noisily. Catches fish in almost vertical plunge-dives. Wary and difficult to approach, but accepting of built-up landscapes: parks, marinas, and fish hatcheries.
- **POPULATIONS** Widespread but uncommon. Many migrate considerable distances, but can also be found wintering as far north as there is open water. Annual fall to spring to Hawaii.

Green Kingfisher
Chloroceryle americana | GKIN | L 8¾" (22 cm)

Tropical kingfisher that regularly reaches the West only in southern Ariz. near the international border and only in very small numbers. Much smaller than Belted Kingfisher. Upperparts dark green; bill long and thin. Male has rufous breast; breast of female white with splotchy green. Call is a soft, two-note clicking, *tik ... tik ...,* repeated slowly by perched bird; easily overlooked. In flight, gives single notes, more widely spaced and erratic than Belted. Unlike Belted, is not given to hovering out in the open; flies swiftly along streams or pond edges, quite low to the water, flashing white in tail.

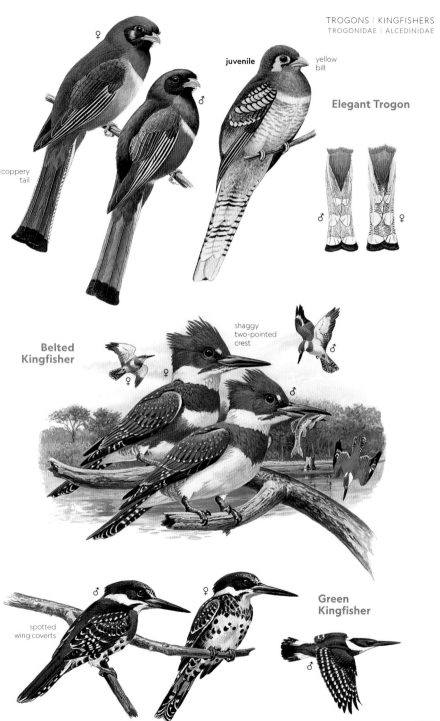

WOODPECKERS | PICIDAE

Famously industrious, the woodpeckers are equipped with stiff tails for bracing against tree trunks and chisel-shaped bills for digging into bark. They hitch their way up trees with jerky movements. Woodpeckers have simple calls, and all in the West sing nonvocally by drumming. The family numbers nearly 250 species worldwide.

Lewis's Woodpecker
Melanerpes lewis | LEWO | L 10¾" (27 cm)

- **APPEARANCE** Medium-large woodpecker. In good light, adult oily green above with gray collar, beet-red face, and rosy belly; but often appears simply dark, especially in flight. Juvenile dusky-headed.
- **VOCALIZATIONS** All calls squeaky and strained-sounding; has a *churr* call like other *Melanerpes* woodpeckers, but shriller and harsher. Drum soft and tentative. As woodpeckers go, a notably quiet one.
- **BEHAVIORS** Widespread but local in open woods, especially orchards and recent burns. A habitual flycatcher; all day long, sallies out in short forays for insects. Also caches nuts and berries. Flight smooth and graceful, with frequent glides; can suggest a crow.
- **POPULATIONS** Migratory, with movements complex: widespread some autumns, nearly absent others; a few winter well north of mapped range. Rare vagrant coastally north of Calif.

Red-headed Woodpecker
Melanerpes erythrocephalus | RHWO | L 9¼" (23 cm)

- **APPEARANCE** Midsize woodpecker with bold white flashes at all ages. Adults of both sexes instantly recognized by completely red head. Juvenile has streaky gray-brown hood and dusky white underparts; like adult, has large white patches on secondaries and rump, prominent in flight.
- **VOCALIZATIONS** Primary call an urgent *queeer-uh,* like nocturnal flight call of Gray-cheeked Thrush (p. 346). Also a soft, muffled, scolding rap, four to six notes, like a wren (pp. 330–335) or frog. Like Lewis's, drums weakly and tentatively.
- **BEHAVIORS** In very open woods in West; in limited zone of range overlap with Lewis's, occurs side by side with that species in burns, old orchards, and storm-damaged tracts. Like Lewis's, engages in flycatching and food caching.
- **POPULATIONS** Generally east of Rockies; locally common in suitable habitat on western Great Plains.

Acorn Woodpecker
Melanerpes formicivorus | ACWO | L 9" (23 cm)

- **APPEARANCE** A midsize, sturdily built woodpecker. Harlequin face pattern of adult accentuated by staring white eye; eye of juvenile darker. Rest of body appears mostly black on perched bird, but white in wings and on rump prominent in flight.
- **VOCALIZATIONS** Very noisy; main call an insistent, repetitious *wake-up wake-up WAKE-UP* or *Jacob Jacob JACOB.* Also a harsh taunting *churr* and short drumming.
- **BEHAVIORS** Highly social and incessantly vocal around communal granaries, where the birds store acorns; most granaries are in trees, but utility poles and even buildings may be used. Frequently engages in flycatching.
- **POPULATIONS** Widespread in oak groves and oak-pine woods, but disjunctly so: *bairdi* of Pacific Slope has longer bill and more extensive black on breast than nominate *formicivorus* of Ariz., N. Mex., and West Tex.

WOODPECKERS
PICIDAE

Lewis's Woodpecker

juvenile

gray collar

adults

crowlike wingbeats

Red-headed Woodpecker

adult

white secondaries

juvenile

barring on secondaries

distinct color pattern

adult

Acorn Woodpecker

♂

female has black bar on crown, lacking in male

♀

♂

245

Gila Woodpecker
Melanerpes uropygialis | GIWO | L 9¼" (23 cm)

■ **APPEARANCE** Along with Golden-fronted and Red-bellied, one of three "zebra-backed" *Melanerpes* woodpeckers in the West. Gila is the plainest of the trio: Male has only a small red skullcap atop gray-brown head, female entirely plain-headed. Compare with female Williamson's Sapsucker (p. 250), an uncommon but regular migrant through range of Gila; the sapsucker has a variably blackish breast and barred flanks (Gila plainer gray-brown below), but is otherwise superficially similar.

■ **VOCALIZATIONS** Noisy, with two main calls: a rolling, churring *prrreeerr*; and an excited *pyay* or *peep*, often in rapid series. Drum slows at end.

■ **BEHAVIORS** Famously associated with saguaros, but also found in riparian groves with standing deadwood. Flourishes in towns and even large cities, including Tucson and Phoenix.

■ **POPULATIONS** Only regularly occurring zebra-backed *Melanerpes* within its range in West. Mostly resident, but wanders slightly north and upslope fall to winter. Has withdrawn from former periphery of range in southeastern Calif. and far southern Nev.

Golden-fronted Woodpecker
Melanerpes aurifrons | GFWO | L 9¾" (25 cm)

■ **APPEARANCE** A bit more heavyset than closely related Gila and Red-bellied Woodpeckers. Golden-fronted is named for the small tuft of yellow at base of maxilla (upper mandible), but it is the extensive golden wash on the hind nape that attracts notice. Adult male has small red cap, lacking in female; juveniles of both sexes have red flecks on crown. Tail almost all-black, rump and uppertail coverts clean white; compare with Red-bellied.

■ **VOCALIZATIONS** Noisy; common call a short, raspy trill, *pr'r'r'r'r'*, less musical than corresponding call of Red-bellied Woodpecker. Drum has steady tempo.

■ **BEHAVIORS** In much of Tex. range, most strongly affiliated with mesquite. In limited western range, however, more associated with cottonwoods.

■ **POPULATIONS** Restricted in West to Big Bend region of Tex., especially Brewster Co., where fairly common. Unlike Red-bellied, rarely wanders; even in West Tex., rare away from Big Bend area.

Red-bellied Woodpecker
Melanerpes carolinus | RBWO | L 9¼" (23 cm)

■ **APPEARANCE** Extensive scarlet on crown and nape of male separates it from closely related Golden-fronted Woodpecker, but other plumages more similar. Compare especially juvenile Red-bellied and adult female Golden-fronted: Both have yellow tuft at base of maxilla and yellow-orange nape on otherwise gray face. Tail of Red-bellied extensively white in all plumages; Golden-fronted has mostly black tail and plain white rump.

■ **VOCALIZATIONS** All year, a muffled *chivf*, often doubled or trebled; also a bright, rolling *cheerr*, heard mostly late winter to summer, and a muffled whinny heard year-round. Drum steady.

■ **BEHAVIORS** Exceedingly generalist: pecks on wood, digs for ants in lawns, scarfs down fruit, engages in flycatching, even caches food. Birds out of range in West tend to find their way to feeders.

■ **POPULATIONS** An eastern woodpecker that for decades has been expanding westward, mostly to Great Plains, but also vagrants to Alta., Idaho, Nev., Ore., western Mont.

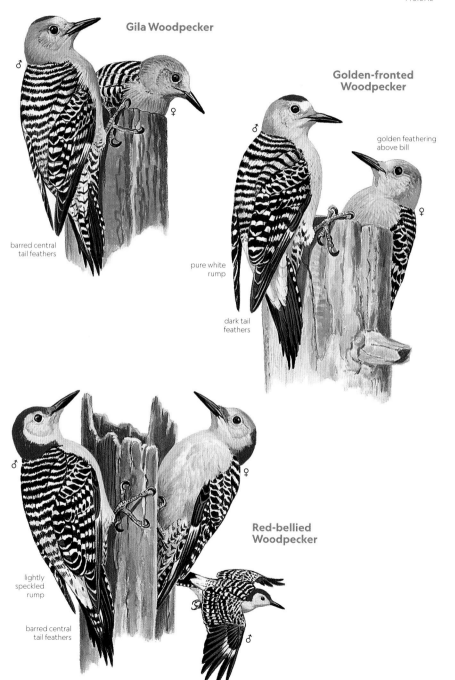

Yellow-bellied Sapsucker
Sphyrapicus varius | YBSA | L 8½" (22 cm)

- **APPEARANCE** Midsize woodpecker; extensively dark above but with large white wing patch that appears as vertical bar on perched bird. Adult male has red cap and red throat bordered broadly by black; female has red cap and white throat. Juvenile plumage, with little if any red, held well into winter; compare with juvenile Red-naped.
- **VOCALIZATIONS** Call a whiny, descending *pyeeer*, like distant Red-shouldered Hawk (p. 224). Drum distinctive: a few quick raps, then abruptly slowing, *ratta-tatta-tat ... tat ... tat ....*
- **BEHAVIORS** Nests in northern hardwood forests (maple, birch, aspen), where it drums noisily; winters in diverse forest types, where it is quiet and unobtrusive. Maintains feeding territories, both summer and winter, around sap wells, tidy rows of holes on tree trunk. These sap wells, drilled by all sapsucker species, ooze phloem, providing nourishment to the sapsuckers themselves as well as to other animals.
- **POPULATIONS** Closely related to Red-naped and Red-breasted Sapsuckers; amount of red on head increases in a progression from Yellow-bellied to Red-naped to southern *daggetti* Red-breasted to northern *ruber* Red-breasted. Yellow-bellied, breeding north to Alas. and wintering primarily in the Southeast and Mexico, is our most migratory woodpecker.

Red-naped Sapsucker
Sphyrapicus nuchalis | RNSA | L 8½" (22 cm)

- **APPEARANCE** Differs from similar Yellow-bellied in variable red swatch on nape of both sexes; red on throat of male Red-naped more extensive than on male Yellow-bellied, and most females have some red on throat (white on Yellow-bellied). White barring on back on both sexes arranged in two parallel columns (more diffuse on back of Yellow-bellied). Young acquire adultlike plumage by mid-autumn, months earlier than Yellow-bellied.
- **VOCALIZATIONS** Like Yellow-bellied, a descending squeal and rallentando drumming.
- **BEHAVIORS** Fairly common in high country in summer wherever there are aspens; on migration and winter, occurs in diverse woodlands.
- **POPULATIONS** The breeding sapsucker of the Interior West; both Red-naped and Yellow-bellied are regular, although scarce, on migration to western Great Plains.

Red-breasted Sapsucker
Sphyrapicus ruber | RBSA | L 8½" (22 cm)

- **APPEARANCE** Has much more red on head than other sapsuckers: Northern *ruber* almost entirely red-hooded, like Red-headed Woodpecker (p. 244); southern *daggetti*, with more black and white on head, also has more prominent white spotting on back. Sexes similar, unlike other sapsuckers.
- **VOCALIZATIONS** Call and drum like those of Yellow-bellied and Red-naped.
- **BEHAVIORS** More inclined to nest in conifer forests than hardwoods-loving Yellow-bellied and Red-naped; like other sapsuckers, found on migration and in winter wherever there are trees.
- **POPULATIONS** Northern *ruber* breeds mostly Ore. northward, southern *daggetti* mostly Calif. and far western Nev. Yellow-bellied, Red-naped, and *ruber* Red-breasted overlap as breeders around Prince George, B.C., a rare "triple point" for avian biogeography in the West. Their hybrid offspring migrate widely and add a challenge to fall and winter birding in the West.

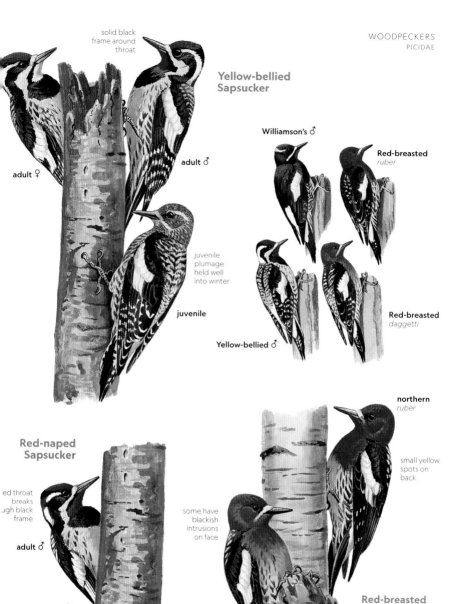

Williamson's Sapsucker
Sphyrapicus thyroideus | WISA | L 9" (23 cm)

■ **APPEARANCE** Similar in build to other sapsuckers (p. 248). Adult male mostly black; like other sapsuckers, has vertical white bar on wing and red on throat; the yellow belly is more prominent on Williamson's than on Yellow-bellied. Female very different, was for many years believed to be a separate species; head brown, flanks barred black and white, and belly yellowish, suggesting first-winter Yellow-bellied. Compare also with Gila Woodpecker (p. 246).

■ **VOCALIZATIONS** Call a harsh *shKYEEsssh*, dropping in pitch. Drum unique: 5–10 rapid raps, followed by a pause, then two to five more raps, then another pause, then another two to five raps.

■ **BEHAVIORS** Like Red-naped, a sapsucker of interior mountains; some habitat overlap, but Williamson's goes for drier woods, especially upper yellow pine (ponderosa, Jeffrey) zone, than Red-naped.

■ **POPULATIONS** Migratory, but not commonly seen on passage. Rare at best to coast.

American Three-toed Woodpecker
Picoides dorsalis | ATTW | L 8¾" (22 cm)

■ **APPEARANCE** Midsize woodpecker, about the same build as Hairy Woodpecker (p. 254). Has three toes and barred flanks like Black-backed Woodpecker; adult male of both species yellow-capped. Interior West female, with nearly white back, quite similar to female Hairy, but note barred flanks of American Three-toed (unmarked white on Hairy) and in good view, three, not four, toes.

■ **VOCALIZATIONS** Generally quiet; *plick* call like that of Hairy, but more muffled. Drum trails off, suggesting Pileated (p. 256) but not as powerful; both sexes drum.

■ **BEHAVIORS** Closely associated with fir forests, especially those that are some combination of old, diseased, and burned. Works one particular part of a tree for the longest time, often at eye level, flaking off bark and picking off arthropods.

■ **POPULATIONS** Scarce but widespread; resident, but some movement toward recently burned tracts. Northern subspecies *fasciatus* has more black on back than nearly white-backed *dorsalis* of central and southern Rockies.

Black-backed Woodpecker
Picoides arcticus | BBWO | L 9½" (24 cm)

■ **APPEARANCE** Slightly larger than American Three-toed; like that species, has only three toes (our other woodpeckers have four). Also like American Three-toed, has yellow crown (male) and barred flanks, but note solid black back of Black-backed, thought to be an adaptation for blending in with the charred trunks on which the bird forages.

■ **VOCALIZATIONS** Noisier than American Three-toed. Common call an odd, resonant cluck: *tock*, often with weak initial stutter, *t'tock*. Also gives rattles, squeals, and a grinding, ternlike snarl. Drum similar to that of American Three-toed; both sexes drum.

■ **BEHAVIORS** Both of the three-toed woodpeckers, genus *Picoides*, are attracted to burns, this one especially so. A single burned tree in an otherwise healthy forest may well have a Black-backed Woodpecker.

■ **POPULATIONS** Like American Three-toed, scarce but widespread; somewhat more prone to wandering, especially in search of burns, than American Three-toed.

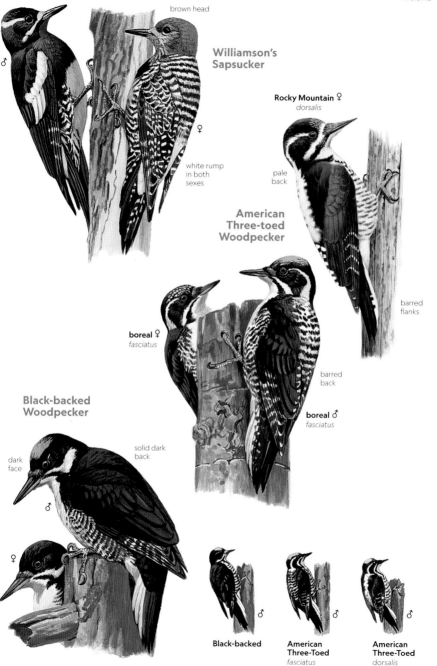

Downy Woodpecker
Dryobates pubescens | DOWO | L 6¾" (17 cm)

■ **APPEARANCE** Our smallest woodpecker; short-billed with unmarked white back. Adult male has red patch on hindcrown, lacking on female. Look-alike Hairy Woodpecker (p.254) has longer bill and unmarked outer tail feathers; outer tail feathers of Downy often flecked black. Compare also with Nuttall's and Ladder-backed Woodpeckers, closer relatives of Downy than Hairy.

■ **VOCALIZATIONS** Call a simple *pik;* also a rapid whinny of such notes, often given on landing. Drum, one to one and a half seconds in duration, a perfectly steady series of 15–25 raps.

■ **BEHAVIORS** Occurs anywhere there are trees: at plantings in our largest cities, as well as in unbroken tracts of forest wilderness. Especially in winter, feeds on mullein stalks; readily tends feeders.

■ **POPULATIONS** Complex variation in extent and hue of white in plumage. Many in Pacific Northwest have dirty gray-white ground color rather than gleaming white; this morphology pops up uncommonly elsewhere in West. "Eastern Downy Woodpecker," occurring to base of foothills of Rockies, has extensive white spotting in wings.

Nuttall's Woodpecker
Dryobates nuttallii | NUWO | L 7½" (19 cm)

■ **APPEARANCE** A bit larger and longer-billed than Downy Woodpecker. Similar to Ladder-backed Woodpecker; Nuttall's has more black on face, more extensive black below nape, and thinner white bars on black back. On Nuttall's, the white in the plumage is pure white; on Ladder-backed, the white in the plumage is imbued with a dirty brown-white hue. Juveniles of both Nuttall's and Ladder-backed have red on forecrown.

■ **VOCALIZATIONS** Call a sharp *pleek,* like Hairy Woodpecker's, often run together in excited series; also a faster series of duller notes in a trill. Drum steady, a bit faster than Downy's.

■ **BEHAVIORS** A lover of oak woodlands. In narrow zone of overlap with Ladder-backed Woodpecker, Nuttall's goes for cooler, wetter landscapes.

■ **POPULATIONS** Resident, but some wander upslope after breeding, and a few get well east into the Great Basin; multiple records for Nev. away from the Sierra. Hybridizes with Ladder-backed where ranges overlap.

Ladder-backed Woodpecker
Dryobates scalaris | LBWO | L 7¼" (18 cm)

■ **APPEARANCE** Just a hair smaller than Nuttall's Woodpecker. Like Nuttall's, a "zebra-back" in all plumages. Face and underparts of Ladder-backed have dirty brown-white wash; Ladder-backed also has less black on face and broader white bands on upperparts, including upper back (upper back solidly black on Nuttall's). Red on crown of male Ladder-backed more extensive than on Nuttall's; in close view, note buffy nasal tuft of Ladder-backed (clean white on Nuttall's). Compare also with Downy Woodpecker.

■ **VOCALIZATIONS** *Pik* note and whinny like Downy's, but not as flat-sounding. Drum fast, 30–50+ raps in one to one and a half seconds.

■ **BEHAVIORS** The desert *Dryobates;* occurs around mesquite, saguaros, and Joshua trees, but also around cottonwoods, willows, and elms in towns and riparian habitats alongside Downy.

■ **POPULATIONS** Sedentary. Body size and overall paleness variable, probably due more to individual differences than geographic trends.

WOODPECKERS
PICIDAE

Downy Woodpecker
- short stubby bill
- barred outer tail feathers

In many woodpecker species, juveniles of both sexes have red on forecrown as opposed to on hindcrown as in adult males.

juvenile

Nuttall's Woodpecker
- dark face with white borders
- white nasal tuft
- barring stops on upper back

Ladder-backed Woodpecker
- extensive reddish crown
- buffy nasal tuft
- barring extends to nape
- tail more barred than Nuttall's

Ladder-backed ♂   Nuttall's ♂

Hairy Woodpecker
Dryobates villosus | HAWO | L 9¼" (23 cm)

■ **APPEARANCE** Larger and longer-billed than Downy Woodpecker (p. 252), very similar in plumage. White outer tail feathers of Hairy usually unmarked; usually flecked black on Downy. Juvenile male has smudgy red forecrown, as in other *Dryobates* woodpeckers. Compare female Hairy with female American Three-toed (p. 250), especially in Interior West.

■ **VOCALIZATIONS** Call a sharp, shrill, far-carrying *pleek!* Also a rapid whinny of similar notes. Drum rapid, 25–40 raps in one to one and a half seconds.

■ **BEHAVIORS** Like Downy, a woodpecker of diverse forest types, but Hairy tends to stick to larger tracts. Attracted to burns, where look-alike female American Three-toed also occurs.

■ **POPULATIONS** Considered nonmigratory, but many wander widely after nesting; has a knack for finding small groves in arid lowlands. Geographic variation complex, roughly follows that of Downy Woodpecker. Many in Pacific Northwest are dirty brown-white below with mostly black wings; some in Rockies have almost entirely black wings. "Eastern Hairy," occurring west to foothills of Rockies, has extensive white spotting in wings.

White-headed Woodpecker
Dryobates albolarvatus | WHWO | L 9¼" (23 cm)

■ **APPEARANCE** About the same build as Hairy Woodpecker, but plumage unique. Both sexes mostly black but with striking white head; white wing flash in flight calls to mind Acorn Woodpecker (p. 244). Adult male has red on rear crown; adult female lacks red. Juvenile (both sexes) has red farther forward on crown than on adult male; this pattern is typical of most *Dryobates* woodpeckers.

■ **VOCALIZATIONS** Call like Hairy, but usually run together in very fast series of three to five or more notes. Drum variable; often has tentative beginning, then steady mid-section, then tapers off at end.

■ **BEHAVIORS** Like Williamson's Sapsucker (p. 250), a bird of montane pine forests. Feeds on tree sap, gleans arthropods from pine needles, and flakes bark for larvae. Attracted to burns; often feeds and even nests at eye level.

■ **POPULATIONS** Resident and nonmigratory. Has declined in heavily clear-cut regions, but also responds positively to forest management well informed by science.

Arizona Woodpecker
Dryobates arizonae | ARWO | L 7½" (19 cm)

■ **APPEARANCE** Fairly small woodpecker. Only woodpecker in West with solid olive-brown back, although appears simply black-backed in shade or low light; wings and tail also mostly dark. Underparts pale cream-white with extensive dark spots. Face dark and light like other *Dryobates* woodpeckers. Male has red nape, lacking on female; juvenile has red cap, farther forward than red on nape of adult male.

■ **VOCALIZATIONS** Call a sharp *pleenk* like Hairy's, a little weaker and squeakier. Tempo of drumming steady throughout.

■ **BEHAVIORS** Occurs in lower-elevation pine-oak woodlands in mountains; unobtrusive and hard to spot during breeding season. Hairy Woodpecker tends to be at higher elevations, Ladder-backed at lower.

■ **POPULATIONS** Resident and nonmigratory, although some go to lower elevations in cold weather fall to winter.

WOODPECKERS
PICIDAE

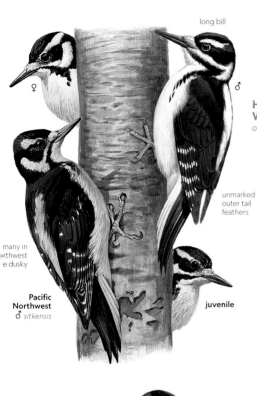

long bill

Hairy Woodpecker
orius

unmarked outer tail feathers

many in northwest are dusky

Pacific Northwest
♂ *sitkensis*

juvenile

white wing patch

White-headed Woodpecker

brown upperparts

Arizona Woodpecker

Hairy Rockies
♂ *orius*

Downy Rockies
♂ *leucurus*

Arizona ♂

255

Northern Flicker
Colaptes auratus | NOFL | L 12½" (32 cm)

■ **APPEARANCE** Large woodpecker with slightly decurved bill. In most of range, the only flicker: back golden brown with black bars; underparts white with black spots and breastband; feather shafts and undersurfaces of tail and wings have bright red or yellow tones; rump white. Male "Red-shafted Flicker" has unmarked brown nape and gray face with blobby red moustache; nape and face of female plain; both sexes have bright salmon flight feathers. Male "Yellow-shafted Flicker" has gray nape with red crescent and face buffy brown with blobby black moustache; female has red crescent on nape but is plain-faced; both sexes have bright yellow flight feathers. Compare with Gilded Flicker.

■ **VOCALIZATIONS** Year-round gives a bright *keeer* or *kee-yer*. Song a fast series of repeated *wick!* or *wake-up!* elements; drum long and steady.

■ **BEHAVIORS** Nests in tree cavities; common in towns and cities. Routinely drums on vents, gutters, and siding, and is a habitué of feeding stations. Our most terrestrial woodpeckers, flickers spend much time on lawns probing the soil for ants.

■ **POPULATIONS** "Red-shafted" breeds Interior West and Pacific Slope, "Yellow-shafted" from Great Plains eastward but also west across most of Canada to Alas. Intergrades, with blended characters, are frequent in West east of Rockies, much less so elsewhere.

Gilded Flicker
Colaptes chrysoides | GIFL | L 11½" (29 cm)

■ **APPEARANCE** A bit smaller than Northern Flicker. Flight feathers, with yellow ("gilded") linings, like "Yellow-shafted Flicker." Head and face marked like "Red-shafted": Male has gray face with blobby red moustache, lacking on female; both sexes have brighter yellow-brown crown than Northern. Gilded is also brighter overall and paler than Northern, with fainter barring above, a bulgier breastband, and spotting below more oblong.

■ **VOCALIZATIONS** Like Northern's, but main call sweeter, thinner, and higher, a simple *keer* or *keee*.

■ **BEHAVIORS** Like Northern's, although not as flamboyant. Characteristic of saguaro and Joshua tree forests; also in willows and cottonwoods along waterways.

■ **POPULATIONS** Nonmigratory; Gilded breeds at lower elevations than "Red-shafted" where ranges overlap, but some hybridization does occur.

Pileated Woodpecker
Dryocopus pileatus | PIWO | L 16½" (42 cm)

■ **APPEARANCE** Huge and crested; long-necked and long-billed. Body mostly black; wings flash white in flight. Both sexes emblazoned with fiery crest; male has red forecrown and red stripe on side of face; both marks are black on female.

■ **VOCALIZATIONS** Call a hurried series of *wuk* notes, speeding up in the middle; also a slower series of *wuk* or *wek* notes. Some calls flicker-like. Drum, resonant and powerful, trails off at end.

■ **BEHAVIORS** Classically a woodpecker of deep forests, but gets into urban districts with parks and wooded greenways. Spectacular to behold; obliterates stumps and trunks, wood chips flying everywhere.

■ **POPULATIONS** A natural engineer, the Pileated is important in nutrient cycling and the provisioning of nest cavities for owls, ducks, squirrels, and even martens.

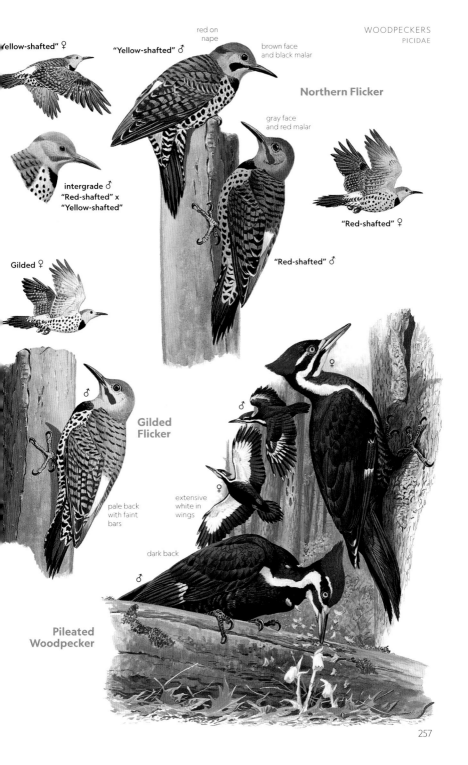

CARACARAS AND FALCONS | FALCONIDAE

Long assumed to be relatives of the hawks, eagles, and kin (pp. 218–231), the falcons were shown in the early 21st century to instead be more closely related to parrots (pp. 262–265) and passerines (pp. 266–453). Except for the unusual caracara, our falcons are fast-flying and pointy-winged.

Crested Caracara

Caracara plancus | CRCA | L 23" (58 cm) WS 50" (127 cm)

- **APPEARANCE** Caracaras are unfalconlike falcons, bulky and blocky; they are large-headed and big-billed, with long necks and legs. Wings rounded for a falcon and typically held flat. Head mostly pale, topped off with dark cap with wispy crest; bill and facial skin colorful. In flight, dark body contrasts with whitish tail and wing tips. Juvenile buffier overall than adult.
- **VOCALIZATIONS** In interactions with other scavengers, especially vultures (p. 216), a dry, froglike ratcheting.
- **BEHAVIORS** Found in open, arid country with perches: ranches, saguaro stands, streamside groves. Flies low to the ground; often seen standing or walking in the company of vultures at carrion.
- **POPULATIONS** Center of abundance in West around south-central Ariz., especially Pima Co. and Pinal Co. Casual far north and west of core range.

American Kestrel

Falco sparverius | AMKE | L 10½" (27 cm) WS 23" (58 cm)

- **APPEARANCE** Very small; about the same mass as a Mourning Dove (p. 76). Long-tailed and pointy-winged; Sharp-shinned Hawk (p. 220), only slightly larger, has rounded wings. Quite colorful: Chalk-blue wings of male contrast with bright russet back and wings; female rusty-winged. Both sexes have vertical black lines on face.
- **VOCALIZATIONS** Shrill *klee klee klee* or *killy killy killy*. This sound may induce scared rodents to urinate; the kestrel sees a bright UV trace in the urine, then pounces.
- **BEHAVIORS** Preys mostly on small rodents and large arthropods; usually seen singly in open country, especially along roadsides. Perched, habitually jerks its tail, a useful distinction from Merlin. Hovers ("kites") above prey, with body angled upward.
- **POPULATIONS** Despite well-publicized declines in East, seems to be holding its own in West. Accepting of urban landscapes, where it nests on buildings; has been helped by provisioning of nest boxes too.

Accidental to Hawaii

Merlin

Falco columbarius | MERL | L 12" (30 cm) WS 25" (64 cm)

- **APPEARANCE** Not much longer than American Kestrel, but bulkier. Low-contrast plumage overall: male bluish above, female brown above; both sexes have faint vertical stripe on face. Tail mostly dark with pale bars.
- **VOCALIZATIONS** Calls harsh and rapid, rising slightly, heard mostly in connection with territorial defense.
- **BEHAVIORS** Like the larger falcons (p. 260), catches birds on the wing. Flight fast and powerful on pointed wings, more suggestive of Peregrine (p. 260) than kestrel.
- **POPULATIONS** Three subspecies in West: pale *richardsonii* ("Prairie"), breeding northern Great Plains, wintering mostly interior; darker nominate *columbarius*, breeding Canada and Alas., scarce but widespread in winter in West; even darker *suckleyi* of Pacific Slope. Breeding range expanding south, especially in cities. Accidental to Hawaii.

Gyrfalcon

Falco rusticolus | GYRF | L 20–25" (51–64 cm) WS 50–64" (127–163 cm)

■ APPEARANCE Larger and bulkier than Peregrine and Prairie Falcons. Wings broad-based but pointed; tail, with many bands, also broad-based, somewhat tapered, and long. Three color morphs: white, gray, and dark. White morph distinctive, but beware occasional leucistic Red-tailed Hawk (p. 228). Gray morph and dark morph, especially juveniles, suggest huge Merlin (p. 258). Also compare gray morph with American Goshawk (p. 220).

■ VOCALIZATIONS Around nest, nasal squawks with quacking or honking quality. Usually silent in winter.

■ BEHAVIORS Found mostly on open Arctic tundra, nesting on cliffs near the sea; preys heavily on ptarmigan (p. 58). The few that make it well south each winter have the convenient habit of hanging around for a while.

■ POPULATIONS Most remain at high latitudes in winter, where Arctic warming is an emerging threat. Movements south in winter due in part to fluctuations in ptarmigan abundance.

Annual to Hawaii

Peregrine Falcon

Falco peregrinus | PEFA | L 16–20" (41–51 cm) WS 36–44" (91–112 cm)

■ APPEARANCE Sleek and powerful with long, pointed wings. Wing tips extend to tail tip on bird at rest. Adult barred below and bluish above with bold "helmet." Juvenile browner, with streaking, not barring, below. Wings from below uniformly dusky, lacking dark "wingpits" of Prairie.

■ VOCALIZATIONS Long, nasal squeals, rising in pitch, often in slow series: *rraaaahh ... rraaaahh ... rraaaahh ....* Typically heard around eyrie, especially right after fledging.

■ BEHAVIORS Famously the fastest animal on Earth. Flight speed deceptive; covers tremendous amounts of sky seemingly effortlessly. Nests widely in wilderness habitats like Arctic seacoasts and rugged interior mountains, but also on bridges and buildings in big cities. Migrants occur anywhere.

■ POPULATIONS Continues to recover from sharp decline in 20th century caused by pesticides. Northern breeders *(tundrius),* seen widely on migration, are paler than buff-washed *anatum,* a widespread breeder farther south; *pealei,* breeding near coast from B.C. north to Aleutians, even darker. Urban Peregrines may be "Pseudogrines," mixes of *anatum* and *tundrius.* Rare but annual in winter to Hawaii.

Prairie Falcon

Falco mexicanus | PRFA | L 15½–19½" (39–50 cm) WS 35–43" (89–109 cm)

■ APPEARANCE About the size of Peregrine, but longer-tailed and shorter-winged. Pale; sandy brown plumage matches arid landscapes favored by the species. In flight, black "wingpits" prominent. Juvenile a bit darker than adult and streaked below; adult more spotted below. Brown moustache weaker than Peregrine's, stronger than Merlin's; compare especially with female "Prairie Merlin."

■ VOCALIZATIONS Noisy at eyrie like Peregrine; call similar, but averages shorter and higher in Prairie. Usually silent in winter.

■ BEHAVIORS Nests on buttes in open country and cliffs in foothills. In winter, disperses to badlands, grasslands, and shrublands, where it feeds heavily on Horned Larks (p. 308); often in company of Ferruginous Hawks (p. 230).

■ POPULATIONS Only weakly migratory; winter range appears to be shifting north.

CARACARAS AND FALCONS
FALCONIDAE

dark-morph juvenile

gray-morph adult

gray-morph juvenile

wing coverts darker than remiges

Gyrfalcon

white-morph adult

long tail

juvenile
tundrius

Peregrine Falcon
anatum

adult

juvenile
tundrius

juvenile

adult
northern
tundrius

adult

adult
northern
Pacific coast
pealei

juvenile
pealei

Prairie Falcon

adults

black "wingpits" prominent in flight

PARAKEETS AND PARROTS | PSITTACIDAE

This mostly New World assemblage in the order Psittaciformes includes several groups popular in aviculture, among them amazons, conures, and macaws. All in the West are introduced from outside the region, although the Thick-billed Parrot (p. 461) formerly reached Southeast Arizona from Mexico. Dozens of species of parakeets, parrots, and kin have been documented in the wild in the West; treated here are generally those species with the largest established populations at the present time.

Nanday Parakeet
Aratinga nenday | NAPA | L 13¾" (35 cm)
Indigenous range centered on Paraguay; "Nanday" is the Guarani name for the species. Well established in Los Angeles region, especially Malibu and eastern Ventura Co. Slender and long-tailed, with head and bill black. Breast washed blue, "trousers" red. Flight feathers mostly black below, tinged blue above. Calls high, clipped, and grating: *kraah!* and *rrraaay!*

Mitred Parakeet
Psittacara mitratus | MIPA | L 15" (38 cm)
Indigenous to the east slope of the Andes from Peru to Argentina. Well established in Los Angeles, even right in the city. A large parakeet, close to the length of a small macaw. Green overall, including leading edge of wing, with extensive but splotchy red on face. Calls rough and braying. Compare especially with smaller Red-masked Parakeet.

Red-masked Parakeet
Psittacara erythrogenys | RMPA | L 13" (33 cm)
Indigenous to west slope of the Andes from Ecuador to Peru. Established in San Francisco Co. and Los Angeles Co.; present strongholds around Pasadena and Claremont; generally outnumbered there by similar Mitred Parakeet. Also established in Hawaii, with strongholds around Honolulu and western Hawaii I. Red-masked is smaller than Mitred, with solidly red face and red on leading edge of wing below. Calls more nasal than Mitred's, with laughing quality.

White-winged Parakeet
Brotogeris versicolurus | WWPA | L 8¾" (22 cm)
Indigenous to lowlands of northern S. Amer., where it occurs in stupendous flocks along Amazon R. Formerly established in and around San Francisco and Los Angeles, but has declined in recent years. Small, with white triangle on wing prominent in flight. Calls light and chirpy. Used to be lumped with paler and more yellow-bodied Yellow-chevroned Parakeet as a single species, called Canary-winged Parakeet.

Yellow-chevroned Parakeet
Brotogeris chiriri | YCPA | L 8¾" (22 cm)
Indigenous to lowlands of central S. Amer., south of range of White-winged Parakeet. Widespread and well established in Los Angeles region, including right in the city; outnumbers similar White-winged Parakeet. Yellow bar across green wings prominent in flight. Quite vocal; calls a bit sharper, shriller than those of White-winged. Where introduced in the same area, White-winged and Yellow-chevroned Parakeets sometimes hybridize, complicating field identification.

Red-crowned Parrot 🇪🇳
Amazona viridigenalis | RCPA | L 13" (33 cm)
Indigenous only to a small swatch of northeastern Mexico, but in West well established now in and around Los Angeles, San Diego, and Honolulu. The population in the Los Angeles area alone probably exceeds that of the indigenous population in Mexico. A typical amazon, stocky and short-tailed. Red crown extensive on male, reduced on female. Note also blue wash on hindnape; red secondaries are prominent on bird in flight. Calls include harsh, down-slurred whistle followed by pounding squawks. Red-lored Parrot, *A. autumnalis* (also pictured opposite), is like Red-crowned, but has yellow patch on face; it is uncommon around Los Angeles and San Diego.

Lilac-crowned Parrot
Amazona finschi | LCPA | L 13" (33 cm)
Indigenous to western Mexico. Well established in Los Angeles and San Diego regions; outnumbered in most places by Red-crowned, but Lilac-crowned is numerous in and around Pasadena. Crown and nape washed lilac; forehead dull crimson. Longer-tailed, paler, and yellower than Red-crowned. Gives pounding squawks like Red-crowned, mixed with harsh whistles, rising or falling. Introduced populations of Lilac-crowned and Red-crowned are hybridizing to some extent.

Yellow-headed Parrot
Amazona oratrix | YHPA | L 14½" (37 cm)
Indigenous to Mexico and Belize, where declining. Well established around San Diego and Los Angeles, with stronghold in and around Pasadena. Larger than Red-crowned and Lilac-crowned Parrots. Adult yellow-headed, juvenile less so. Calls mellower than those of Red-crowned and Lilac-crowned, suggesting a corvid (pp. 292–301) or even a human.

AUSTRALASIAN PARROTS AND LOVEBIRDS | PSITTACULIDAE
Indigenous to the Old World, this is the largest family in the parrot order, with nearly 200 species globally. Like psittacids (pp. 262–265), they are brightly colored, with hooked bills and raucous calls. Well-known representatives include the nectivorous lories, the highly social lovebirds, and the pet store "budgie" (Budgerigar, *Melopsittacus undulatus*).

Rose-ringed Parakeet
Psittacula krameri | RRPA | L 15¾" (40 cm)
Indigenous to Africa and South Asia. Widely established globally, including in West. Largest concentration in Calif. in and around Bakersfield; smaller numbers San Francisco, Los Angeles, and San Diego. Also well established in Hawaii on Kauai and Oahu. Tail, thin and blue, is quite long; bill red. Male has pink-and-black neck ring, lacking in female and young. Calls include a clipped *kyeer*, dropping in pitch.

Rosy-faced Lovebird
Agapornis roseicollis | RFLO | L 6¼" (16 cm)
Indigenous mostly to Namibia. Introduced and established in West in Hawaii, especially islands of Maui and Hawaii, and Phoenix metro area, where expanding. Small, chubby, and short-tailed. Face and breast peach-red; tail multicolored with much blue. Shrill, screeching calls include a rapid chatter and short clipped notes. Nests and roosts in palms and even saguaros.

PARAKEETS AND PARROTS | AUSTRALASIAN PARROTS AND LOVEBIRDS
PSITTACIDAE | PSITTACULIDAE

BECARDS | TITYRIDAE

The birds in this midsize Neotropical family are found mostly in woodlands, and they are notable for their curious names: tityra, purpletuft, mourner, xenopsaris, and others. Well over a third are classified as becards, with the Rose-throated being the only tityrid that reaches the West. Becards are large-headed and broad-billed, and they are capped or crested; they wait quietly on perches, and they build impressive globular nests. All birds from this point forward are in the gargantuan order Passeriformes, accounting for well over half of all the world's bird species.

Rose-throated Becard
Pachyramphus aglaiae | RTBE | L 7¼" (18 cm)

Scarce but present year-round in Ariz., mostly in Santa Cruz Co. Like other becards, stocky and short-tailed. In the West, male is dark gray above, paler gray below with rosy throat; female warm buff below, darker above with extensive dark cap. Call a weak *pip-s'WEEeeeee,* distinctive but easily overlooked.

TYRANT FLYCATCHERS | TYRANNIDAE

With 400+ species, the Tyrannidae are the largest of all bird families. They are staggeringly diverse in the Neotropics, and close to 40 species are annual north of Mexico. As their name implies, flycatchers are aerial insectivores; they are sit-and-wait predators that scan carefully for prey, fly out to get it, then return to a perch. Most have broad-based bills equipped with rictal bristles, whisker-like structures important in tactile sensation. Flycatchers are large-headed, and many are drab.

Northern Beardless-Tyrannulet
Camptostoma imberbe | NOBT | L 4½" (11 cm)

■ **APPEARANCE** Our smallest flycatcher and among the smallest passerines in the West; bill is short and thick, with orange base. Called "beardless" because it lacks rictal bristles. Upright posture, slight crest, and wing bars suggest a tiny *Contopus* flycatcher (p. 274). Pale lores and eyebrow aid in separation from similarly diminutive but unrelated juvenile Verdin (p. 302).
■ **VOCALIZATIONS** Song a series of four or five rich whistles, *pyee pyee pyee pyee.*
■ **BEHAVIORS** Found in dense thickets, often willow and mesquite; nest is globular with side entrance. Active for a flycatcher, with distinctive feeding method: instead of flycatching, gleans arthropods from foliage.
■ **POPULATIONS** Locally common in limited range in West. Increasingly found in winter; breeding range may be expanding north.

Sulphur-bellied Flycatcher
Myiodynastes luteiventris | SBFL | L 8½" (22 cm)

■ **APPEARANCE** One of our largest flycatchers. Streaked all over, with particularly bold marks on face; tail and rump rufous, underparts washed yellow.
■ **VOCALIZATIONS** Call explosive and squeaky, *PSYEE-tsyee-yik,* like a dog's rubber squeeze toy.
■ **BEHAVIORS** Occurs in broadleaf forests in canyons, with special affinity for sycamores (genus *Platanus*); often well up in canopy. Nests in tree cavities, including old flicker (p. 256) holes.
■ **POPULATIONS** Occurs only in warmer months in West: Most return after mid-May, and most are gone again by mid-Sept.

Dusky-capped Flycatcher
Myiarchus tuberculifer I DCFL I L 6¾" (17 cm)

■ **APPEARANCE** Our smallest *Myiarchus* flycatcher; birds in this genus are long-tailed and weakly crested, with black bills, gray bibs, washed-out yellow bellies, and rufous in the wings and tail. Dusky-capped is darker overall than larger Ash-throated, with a longer bill, richer yellow below, and, in close view, rufous in secondaries.

■ **VOCALIZATIONS** Call a simple whistle, *peeeer*, with sad quality; "Mournful Pierre" is a name sometimes given this bird.

■ **BEHAVIORS** Common but easily overlooked in broadleaf or broadleaf-conifer forests in foothills, canyons, and mountains. Nests in cavities, including nest boxes and woodpecker holes.

■ **POPULATIONS** Returns to Ariz. and N. Mex. breeding grounds in Apr., departs Oct. Very rare but annual, fall to winter, to Southern Calif.; a few wander each summer to West Tex.

Ash-throated Flycatcher
Myiarchus cinerascens I ATFL I L 7¾" (20 cm)

■ **APPEARANCE** Midway in size between smaller Dusky-capped and larger Brown-crested Flycatchers; proportionately smaller-billed than either species. Paler overall than other *Myiarchus* flycatchers; belly especially pale. In close view or photos, note solidly dark-tipped tail from below.

■ **VOCALIZATIONS** Call a short, rough, rolling *prrrp*; also a disyllabic *k'brick* or *k'breer*, not unlike Cassin's Kingbird (p. 270).

■ **BEHAVIORS** Common in arid woodlands, particularly those with junipers, but also gets into riparian habitats. Like other *Myiarchus* flycatchers, a cavity nester.

■ **POPULATIONS** Breeding range expanding north due in part to nest box provisioning. At midlatitudes in West (Colo., Utah, northern Nev.), present mostly late Apr. to mid-Aug.

Great Crested Flycatcher
Myiarchus crinitus I GCFL I L 8½" (22 cm)

Only *Myiarchus* flycatcher in much of the East; barely reaches our area in broadleaf groves in western Great Plains. Larger-billed than Ash-throated; yellow belly of Great Crested contrasts well with cold gray breast. In close view or review of photos, note white-edged tertials, mostly rufous undertail, and a bit of orange at base of otherwise black bill. Call a far-carrying, whistled *wheee-ip*, ending abruptly; also a buzzy, somewhat shrill *bzhrrrrrp*, often doubled or trebled. Uncommon breeder to eastern Colo.; farther north, gets west to around Edmonton, Alta.

Brown-crested Flycatcher
Myiarchus tyrannulus I BCFL I L 8¾" (22 cm)

■ **APPEARANCE** Larger overall and bigger-billed than Ash-throated Flycatcher. Also darker overall than Ash-throated, with darker gray throat and deeper yellow belly. From below, center of tail rufous to tip; rufous tail of Ash-throated is dark-tipped.

■ **VOCALIZATIONS** Song a pleasing, rolling, trisyllabic *prrrr da-rrrr (What we'll do!)*, often run out to five or seven syllables; also a short *whip*.

■ **BEHAVIORS** A flycatcher of woodlands, generally at lower elevations: from riparian groves to more open saguaro stands and also into lower elevations of montane forests.

■ **POPULATIONS** Range has been expanding slowly in West for close to a century. Reaches U.S. breeding grounds late Apr., gone by early Sept.

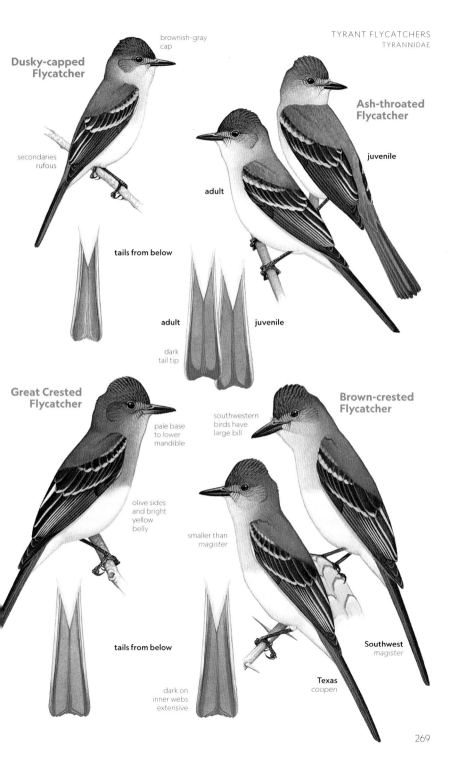

Tropical Kingbird
Tyrannus melancholicus | TRKI | L 9¼" (23 cm)
- **APPEARANCE** Large-billed, with a notched, grayish tail. Back olive green. Similar Western Kingbird has slighter bill, unnotched black tail with white edges, and back more gray-green.
- **VOCALIZATIONS** Calls short and sharp, with mechanical quality; *pleek* and *plick* notes uttered in series, often stuttering.
- **BEHAVIORS** Like all kingbirds, a noisy and conspicuous inhabitant of open habitat with perches; kingbirds, including this species, are aggressive toward other kingbirds, as well as other, sometimes much larger, bird species. Tropical is not averse to human-modified habitats: canals, other waterways, and golf courses.
- **POPULATIONS** Breeds in Ariz., where present May to Sept. Regular in small numbers on West Coast, especially Calif., fall to early winter. Casual inland well north of normal breeding range.

Cassin's Kingbird
Tyrannus vociferans | CAKI | L 9" (23 cm)
- **APPEARANCE** Same size and build as Western Kingbird, but darker overall. Head and breast of Cassin's dark gray, creating hooded impression; small white cheek patch contrasts well with otherwise dark gray hood. Tail of Cassin's lacks white edges of Western; instead, has indistinct pale band across tip of tail.
- **VOCALIZATIONS** Call, an emphatic *ch'BEW*, distinct from Western's, but similar to Ash-throated Flycatcher's (p. 268).
- **BEHAVIORS** Breeds in very open woodlands, especially pinyon-juniper "forests" in arid interior. In limited winter range in West, often in parks and watercourses near coast.
- **POPULATIONS** Breeding range oddly disjunct; especially at periphery of range (Wyo., Mont., Trans-Pecos), numbers vary from year to year. Some disperse north and east after breeding; migrates later in fall than Western.

Western Kingbird
Tyrannus verticalis | WEKI | L 8¾" (22 cm)
- **APPEARANCE** Kingbird of average size and build; across much of the West, the default kingbird. Yellow belly and white-edged (not white-tipped) tail instantly separate it from Eastern Kingbird (p. 272); note also contrast between pale gray-green back and nearly black tail. Compare with Cassin's and especially Tropical Kingbirds, both prone to vagrancy. In brief or distant view, Say's Phoebe (p. 282), with black tail and apricot belly and undertail coverts, suggests Western Kingbird; also, American Robin (p. 348), although larger and obviously different in plumage, can look weirdly similar in flight or perched at a distance.
- **VOCALIZATIONS** A constant noisemaker, starting up at the first hint of dawn. Gives short *rik!* and *reek!* notes, often run out into longer series: *reek! a-rik! arikaree!*
- **BEHAVIORS** One of the most common roadside sights in summer in the West; perches on signs, fences, and utility wires. Kenn Kaufman's classic *Kingbird Highway* takes its name from the species.
- **POPULATIONS** Present only in the warmer months in West; fall migration early, with most on the move by Aug. A kingbird in the West from early autumn to early spring is likely a species other than Western, even in places where Western is the predominant or only breeder.

TYRANT FLYCATCHERS
TYRANNIDAE

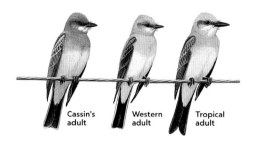

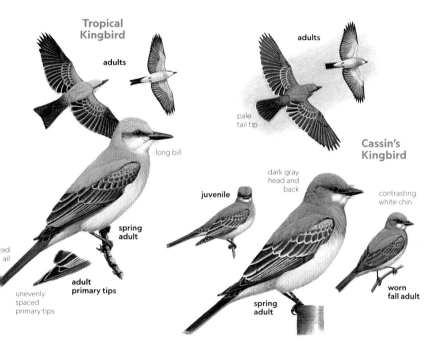

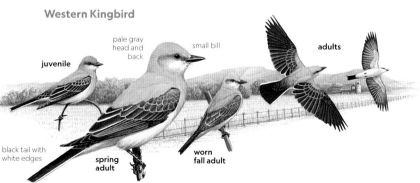

Thick-billed Kingbird
Tyrannus crassirostris | TBKI | L 9½" (24 cm)

■ **APPEARANCE** Largish kingbird with very large bill; tail weakly notched. Especially spring to summer when in worn plumage, adult roughly matches appearance of smaller, slighter-billed, darker Eastern Kingbird, but the two do not usually co-occur. Fresh adults (fall to winter) and juvenile can be quite yellow below; compare especially with Tropical Kingbird (p. 270).

■ **VOCALIZATIONS** Call a shrill, rising whistle: *sshREEET!* Also a falling stutter.

■ **BEHAVIORS** Nests at low elevations along rivers and canyon bottoms with widely scattered broadleaf trees; cottonwood, walnut, and most of all sycamore favored. Noisy and conspicuous, but nests and feeds in treetops.

■ **POPULATIONS** Mostly Mexican species that breeds north in our area to Ariz. and far southwestern N. Mex., where present May to Aug.; numbers increasing and range slowly expanding. Annual in small numbers fall to spring from coastal Southern Calif. to Lower Colorado River Valley.

Eastern Kingbird
Tyrannus tyrannus | EAKI | L 8½" (22 cm)

■ **APPEARANCE** A fairly small kingbird, but larger than superficially similar Eastern Phoebe (p. 282). Dark above, with wings and back dark gray, cap black; throat white. Black tail has broad white band at tip, prominent in flight. Adult has red patch, usually concealed, on crown. Juvenile browner above than adult and a bit grayer below; Eastern Phoebe smaller and drabber, lacks white tail tip.

■ **VOCALIZATIONS** Calls *dzzt* and *bzzeert*, shrill and buzzy; often run together in sputtering series, like a downed electrical wire.

■ **BEHAVIORS** Even among the kingbirds, *Tyrannus tyrannus* is notably pugilistic; drives off larger species—crows and even hawks—in midair assaults. In broad zone of overlap with Western Kingbird (p. 270), Eastern goes for wetter microhabitats: marshes, ponds, and rivers with wooded edges.

■ **POPULATIONS** Despite being "eastern," occupies vast breeding range in West, from Yukon to N. Mex. Even well west into the Pacific Northwest, outnumbers Western Kingbird in some places.

Scissor-tailed Flycatcher
Tyrannus forficatus | STFL | L 13" (33 cm)

■ **APPEARANCE** A kingbird with a very long tail. Outer tail feathers of adult male especially long; tail of adult female shorter, and tail of juvenile shorter still. Adult, with salmon and orangey highlights, distinctive, but juvenile is paler, with some resembling pale juvenile Western Kingbird.

■ **VOCALIZATIONS** Calls similar to those of Western Kingbird, but individual notes of Scissor-tailed a bit sharper, *pleep!* and *pleek!*

■ **BEHAVIORS** Nests in open habitats, arid and sunny, with utility wires and scattered trees, in same landscapes as Western Kingbird. Adult male watching from wire or snag unmistakable; splays tail out in flight, especially when chasing after insects.

■ **POPULATIONS** Relatively range-restricted, breeding mostly southern Great Plains; regular breeder in our area to eastern N. Mex.; first arrivals early Apr., many hang on till early Oct. Quite prone to vagrancy; annual to West Coast, especially Southern Calif.

TYRANT FLYCATCHERS
TYRANNIDAE

Thick-billed Kingbird — 1st fall, pale yellow belly, thick bill, spring/summer

Eastern Kingbird — white tail tip

Scissor-tailed Flycatcher — juvenile, adult ♂, salmon pink "wingpits", juvenile, adult ♂

Olive-sided Flycatcher
Contopus cooperi | OSFL | L 7½" (19 cm)

- **APPEARANCE** Larger than Western Wood-Pewee. Long-winged, short-tailed, and large-headed. Wing bars weak and eye ring absent; bill mostly dark. Most wear a "vest," with dark breast broken by vertical white stripe. White tufts poking out of back can be revealed or concealed in a matter of seconds.
- **VOCALIZATIONS** Song a strange, memorable *pip PREEeee preeer*, notoriously mnemonicized as *Quick! Three beers!* Calls include a single, sharp *peep* and a longer series, *pip pip pip*.
- **BEHAVIORS** Widespread in summer in boreal forests all over the West, but also down to sea level on Pacific coast, even in eucalyptus groves. Feeding method distinctive: perches atop prominent snag, sallies out considerable distance, nabs flying insect, and returns to exact same snag.
- **POPULATIONS** Late spring migrant, not reaching midlatitudes in interior until mid-May; following short breeding season, is on the move again by early Aug.

Greater Pewee
Contopus pertinax | GRPE | L 8" (20 cm)

- **APPEARANCE** About same length as Olive-sided Flycatcher, but body shape more attenuated: longer tailed than Olive-sided, with wispy crest. Lower mandible of Greater Pewee pale orange; underparts uniform olive-gray, not "vested" like Olive-sided.
- **VOCALIZATIONS** Song a couple of whistles, then a slurred note: *syeee syeee syeeEEEEyeee (Jo-sé Ma-RÍ-a)*. Also *pip* and *peep* notes like Olive-sided's.
- **BEHAVIORS** Nests and feeds well up in pines at middle to high elevations (7,000–10,000 ft/2,130–3,050 m); like other *Contopus* flycatchers, watches from exposed perches.
- **POPULATIONS** Arrives on N. Mex. and Ariz. breeding grounds early Apr., stays through Sept. Casual in summer to West Tex.; also fall to winter to Southern Calif.

Western Wood-Pewee
Contopus sordidulus | WEWP | L 6¼" (16 cm)

- **APPEARANCE** Smallest *Contopus* flycatcher in West. Compare with larger species in genus, but also with Willow Flycatcher (p. 276); Western Wood-Pewee shorter-tailed and longer-winged, with grayer and darker tones than Willow. Eastern Wood-Pewee (p. 461), very rare stray to western Great Plains, extremely similar; Western has fainter wing bars than Eastern, dirtier undertail coverts, browner breast, and bill darker beneath.
- **VOCALIZATIONS** All day long on breeding grounds, gives descending buzz ending in weak whistle, *pZZZzzeee*; dawn song, heard only on breeding grounds, mixes short whistles with buzzier phrases.
- **BEHAVIORS** One of the most characteristic and conspicuous species of western woods, from lowland broadleaf groves to spruce-fir forests near timberline. Like larger *Contopus* flycatchers, watches for prey from exposed, often dead, snags. Willow Flycatcher more likely to be in live, usually broadleaf, vegetation; is more active and more prone to twitching its tail.
- **POPULATIONS** Late spring migrant, especially in interior, with many still on the move well into June; most are headed back south by Aug. Completely vacates West in winter; Townsend's Solitaire (p. 344) accounts for many erroneous reports of winterers.

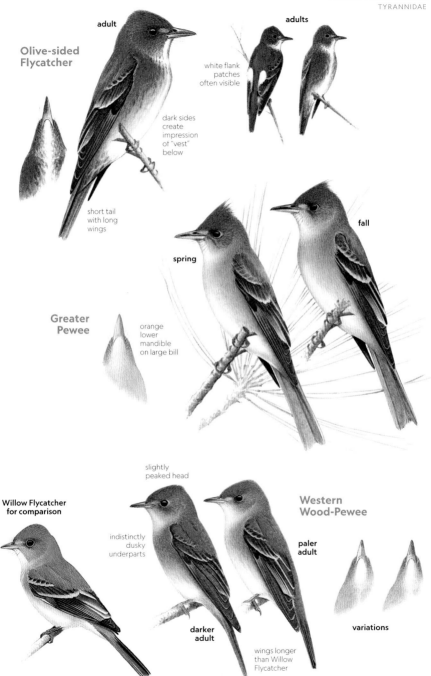

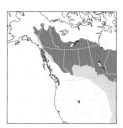

Yellow-bellied Flycatcher
Empidonax flaviventris | YBFL | L 5½" (14 cm)

■ **APPEARANCE** A midsize empid (genus *Empidonax*, pp. 276-281), Yellow-bellied is compactly shaped: short-tailed and long-winged; large-headed and small-billed. Most are extensively yellow below but with broad olive breastband. Eye ring also yellow; yellow wing bars contrast with otherwise black wings. Bill entirely orange-yellow below.
■ **VOCALIZATIONS** Song a rough *rrr-bunk*, similar to Least Flycatcher's (p. 278), but burrier. Call a whistled *prree-rreee*.
■ **BEHAVIORS** Nests on or near the ground in boreal bogs. On migration, finds its way to dense vegetation; vagrants in West often notably active.
■ **POPULATIONS** Very rare at best in West away from breeding grounds; the few records here south of Canada and west of the Rockies are from fall. Breeding range in Alas. may represent a recent expansion.

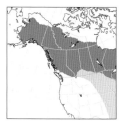

Alder Flycatcher
Empidonax alnorum | ALFL | L 5¾" (15 cm)

■ **APPEARANCE** Nearly identical to Willow Flycatcher; silent migrants of either species often reported as "Traill's Flycatcher," the name formerly given both species. Both are large empids, broad-billed with eye ring reduced or absent. Alder is slightly longer-winged and greener overall, especially on back and crown. Alder has a very thin eye ring; Willow usually lacks eye ring.
■ **VOCALIZATIONS** Song a harsh, low *rrree-beeeuh*, given 8-20 times per minute; lower-pitched with less variation in pitch than song of Willow. Call a high, bright *peep* like Hammond's Flycatcher's (p. 278).
■ **BEHAVIORS** Nests close to ground in dense thickets, often early successional with scattered small trees. Where breeding ranges of Alder and Willow overlap in West, the former tends to occur in wetter microhabitats.
■ **POPULATIONS** Breeding ranges of Alder and Willow largely nonoverlapping in West, but Alder may be expanding south into range of Willow. Migrates mostly to our east, with western limits of normal migration route poorly known; likely regular, especially late spring, from eastern Colo. north to eastern Mont.

Willow Flycatcher
Empidonax traillii | WIFL | L 5¾" (15 cm)

■ **APPEARANCE** Famously similar to Alder Flycatcher: Willow is browner overall than Alder, especially above, and slightly shorter-winged; Alder shows a very thin but complete eye ring, typically lacking on Willow. An overlooked ID challenge in the West is with Western Wood-Pewee (p. 274), genus *Contopus*; note the wood-pewee's longer wings, more slender build overall, and colder gray plumage.
■ **VOCALIZATIONS** Song an abrupt *peep-bew* or *fitz-bew*, given 5-15 times per minute; also a simpler *brrrew*, burry and rising. Call, often given by migrants, a liquid *whit*, distinct from *peep* of Alder.
■ **BEHAVIORS** Breeds and migrates widely in shrubby habitats. More than other empids, hunts from exposed snags in the manner of wood-pewees.
■ **POPULATIONS** Color varies from brownish subspecies *adastus* and *brewsteri* (widespread in West) to paler nominate *traillii* (reaching western Great Plains) to even paler, endangered *extimus* (Southwest). Late spring migrant, fairly early fall migrant.

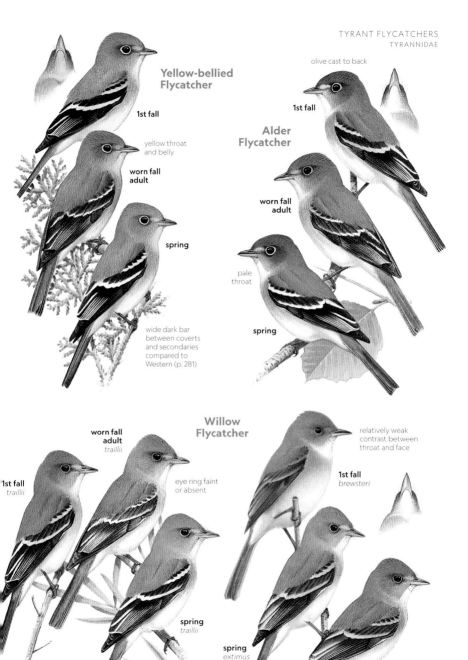

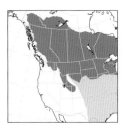

Least Flycatcher
Empidonax minimus | LEFL | L 5¼" (13 cm)

■ **APPEARANCE** Small, even for an empid (genus *Empidonax*, pp. 276–281); wings relatively short. Head large, tail short; bill short but broad. Boldly marked, with prominent eye ring and contrasting dark and light on wings; throat pale and lower mandible orange, variably tipped dark.

■ **VOCALIZATIONS** Song a simple, two-note *t'tek* (*Quebec*), rough and grating, often repeated rapidly, one per second. Call a dry *wit*, also often in quick succession.

■ **BEHAVIORS** Nests in broadleaf woods, typically early successional with sunny openings. Locally common on breeding grounds, but also absent from seemingly suitable swaths of habitat. Fussy and noisy; migrants call frequently and even sing.

■ **POPULATIONS** Migrates east of Rockies; can be fairly common in our area on migration on western High Plains. Breeds erratically west to foothills of Rockies; where it does occur, typically around lake edges and floodplains, can be locally common. Passage early in spring for an empid; migration protracted in fall, with some lingering late. Adults molt after fall migration, but young before, an unusual strategy.

Hammond's Flycatcher
Empidonax hammondii | HAFL | L 5½" (14 cm)

■ **APPEARANCE** Small like Least Flycatcher, but long-winged and small-billed; blocky head and short tail impart front-heavy look. Eye ring variable but usually prominent, often pinched back in rear, suggesting Western Flycatcher (p. 280). Molts right after nesting, so fall migrants are in bright, fresh plumage: Gray head contrasts with olive breast and back.

■ **VOCALIZATIONS** Song comprises three rough notes, widely spaced: *seelick ... srrbrrt ... srrurrt*. Call a sharp *peep*, suggestive of Olive-sided Flycatcher (p. 274) or even Pygmy Nuthatch (p. 326).

■ **BEHAVIORS** Breeds in conifer forests, from dry pinewoods in foothills to wet spruce-fir near treeline; nests and forages well up in canopy. Habitat generalist on migration.

■ **POPULATIONS** Broad-front migrant across West, but generally scarce east of Rockies, with peak spring migration at midlatitudes in May; fall migration more protracted, with some lingering late.

Dusky Flycatcher
Empidonax oberholseri | DUFL | L 5¾" (15 cm)

■ **APPEARANCE** A small empid; short-winged like Least, not quite as small-billed as Hammond's. Longer-tailed than both Least and Hammond's. Throat pale like Least's, but bill averages duskier below. Molts on wintering grounds, so many adults on fall migration are worn and dingy, the opposite of Hammond's.

■ **VOCALIZATIONS** Tripartite song usually has higher, squeakier elements than that of Hammond's. Call a dry *wit* like Gray Flycatcher (p. 280). On breeding grounds, a bright *dyeee* or *dyeeyick*.

■ **BEHAVIORS** Intermediate in nesting habitat preference between Hammond's and Gray; often on shrubby hillsides of lower slopes, also in open woods. Migrants favor brushy habitats.

■ **POPULATIONS** Migrates a bit later in spring than Hammond's and, especially, Gray Flycatchers; some get east to High Plains on migration. Like Gray, essentially a species of the Interior West; scarce at best along West Coast.

Gray Flycatcher
Empidonax wrightii | GRFL | L 6" (15 cm)

■ **APPEARANCE** Long-billed, long-tailed, and proportionately short-winged empid (genus *Empidonax,* pp. 276–281). Head small and smoothly rounded. Plumage low-contrast and grayish overall.

■ **VOCALIZATIONS** Song tripartite like those of Hammond's and Dusky Flycatchers (both on p. 278), with middle element squeakier than first and third elements. Call a toneless *whit* like Dusky's.

■ **BEHAVIORS** Nests in sagebrush and sagebrush-juniper landscapes; migrants stay close to ground at edges of clearings and waterways. Slowly dips tail like Eastern Phoebe (p. 282); other empids flick tail in jerky movements.

■ **POPULATIONS** An early spring migrant, arriving at midlatitudes by late Apr. Breeding range is expanding northeast.

Western Flycatcher
Empidonax difficilis | WEFL | L 5½–5¾" (14–15 cm)

■ **APPEARANCE** A midsize empid; weakly crested. Rather bright for an empid, with olive tones overall; fairly large bill is orange below. Wing bars and eye ring are tinged yellow; eye ring pinches back behind eye, creating "teardrop." Similar to Yellow-bellied Flycatcher (p. 276), a casual vagrant at best within range of Western.

■ **VOCALIZATIONS** Male "position note," given on territory, varies geographically: Pacific Slope population gives *swleee,* slurred and rising; Interior West population gives *peee psEEE (bopeep!),* with two discrete elements. Song tripartite like Hammond's, Dusky, and Gray Flycatchers, but elements higher and squeakier. Call, given year-round, a high *pseet.*

■ **BEHAVIORS** Breeds in shady groves, broadleaf or conifer, often near water; nests on sheltered banks, rock faces, and outbuildings. Habitat generalist on migration.

■ **POPULATIONS** Until 2023 treated as two species, Cordilleran Flycatcher (Interior West) and Pacific-slope Flycatcher (west of Cascade-Sierra axis). Interior West birds a bit brighter, more yellow-green, but much overlap; Channel Is. subspecies *insulicola* has brighter yellow breast. Annual vagrant to East, with most there presumed to refer to "Pacific-slope" subspecies.

Buff-breasted Flycatcher
Empidonax fulvifrons | BBFL | L 5" (13 cm)

■ **APPEARANCE** Our smallest empid; short-billed. Eye ring, like that of Western Flycatcher, pinches back in "teardrop"; bill orange beneath, also like Western. Most have bright cinnamon wash beneath; even the upperparts, especially nape, are washed cinnamon. Some are duller, more olive; these suggest Western Flycatcher, larger overall and larger-billed.

■ **VOCALIZATIONS** Call a muffled, rising *tchit,* similar to *auduboni* Yellow-rumped Warbler's (p. 438). Song a series of snappy *p'd'dyee* elements, suggestive of Western Tanager (p. 444) call, but higher, squeakier.

■ **BEHAVIORS** Strongly associated with dry pine forests, 6,000–9,000 ft (1,830–2,740 m), where it nests and forages; lightly burned tracts, with mosaic of live and dead trees, favored. Scarce in West, but loose colonies sometimes establish.

■ **POPULATIONS** Present on Ariz. breeding grounds early Apr. to early Sept. Formerly more regular in southwestern N. Mex. and West Tex., but now mostly gone there.

TYRANT FLYCATCHERS
TYRANNIDAE

small-headed

winter

spring

Gray Flycatcher

dips long tail smoothly downward

all-orange lower mandible

1st fall

worn fall adult

Western Flycatcher

spring

dark bar between secondaries and wing bars small or absent; compare with Yellow-bellied

pale 1st fall

relatively long tail

fresh

large rounded head with conspicuous eye ring

Buff-breasted Flycatcher

worn summer adult

281

Black Phoebe
Sayornis nigricans | BLPH | L 6¾" (17 cm)
- **APPEARANCE** Like other phoebes, a medium-small flycatcher with a long tail. Stunningly pied in all plumages: Adult's snow-white belly contrasts with black upperparts; juvenile similar, but with buff fringing above.
- **VOCALIZATIONS** Song squeaky, *ts'wee! t'wee*. Call a flat, sharply down-slurred *swip*.
- **BEHAVIORS** Prominent denizen of coastal and desert waterways. Forages close to ground; catches airborne insects, but also snags arthropods from the water's surface. Dips tail while perched. Often nests on bridges and buildings, but also natural substrates like streambanks.
- **POPULATIONS** Generally resident, but some disperse after breeding, north as well as south. Breeding and wintering ranges expanding north overall.

Eastern Phoebe
Sayornis phoebe | EAPH | L 7" (18 cm)
- **APPEARANCE** About the same size and shape as Black Phoebe. Dark overall, but not as pied as Black; Eastern is muddy brown above, off-white below. Juvenile and freshly molted adult, commonly seen late summer to fall, are yellowish below, suggesting Say's Phoebe. Compare also with genera *Contopus* (p. 274) and *Empidonax* (pp. 276–281).
- **VOCALIZATIONS** Song alternates between emphatic *fee-bee!* and rolling *frree-brree* elements. Call a bright *chip!*
- **BEHAVIORS** Habits similar to Black Phoebe's, but not as tightly associated with aquatic habitats; nests around bridges, cabins, and outbuildings. Stereotyped tail-dipping distinctive.
- **POPULATIONS** Breeding range expanding west in Great Plains to Rockies. Hardy; like Say's Phoebe, migrates early in spring.

Say's Phoebe
Sayornis saya | SAPH | L 7½" (19 cm)
- **APPEARANCE** A bit larger than Black and Eastern Phoebes, and not as compactly built. Apricot underparts contrast with dusky plumage overall; tail blackish. With dark tail, colorful belly, and generally gray plumage, can be confused with Western Kingbird (p. 270).
- **VOCALIZATIONS** Call a down-slurred *see-ur*, given year-round. Song consists of *ch'peee* and *ch'pee-dar* phrases, slowly alternated.
- **BEHAVIORS** Not averse to water, but generally in more arid habitats than Black and Eastern; often finds its way to barns, sheds, and homes. Like Black and Eastern, dips tail, but not as incessantly.
- **POPULATIONS** The most highly migratory phoebe; returns early in spring, departs late in fall. Winter range expanding north.

Vermilion Flycatcher
Pyrocephalus rubinus | VEFL | L 6" (15 cm)
- **APPEARANCE** Phoebe-like in build, but smaller. Adult male intense vermilion and dark brown; extent of vermilion reduced on immature male. Drabber females, especially immatures, told by salmon or yellow underparts, streaky breast, and well-patterned face.
- **VOCALIZATIONS** Male song, given in flight, is rapid and stuttering. Call a rising *psee!*
- **BEHAVIORS** Habits phoebe-like: dips tail; forages in open country, often around water, with low perches.
- **POPULATIONS** Mostly sedentary, but some wander north, especially fall to early winter; scattered breeding records well north of mapped range, too.

VIREOS | VIREONIDAE
These small birds, found mostly in wooded habitats, are related to elepaios (p. 290), shrikes (p. 292), and corvids (pp. 292–301). Many vireos are drab and disinclined to venture out into the open, but they have loud and often incessant songs. With hook-tipped bills, they glean arthropods from vegetation.

Bell's Vireo
Vireo bellii | BEVI | L 4¾" (12 cm)

■ **APPEARANCE** Tiny but long-tailed; smallest vireo in the West. A particularly drab species: dirty olive overall, with faint eyeline, indistinct eye ring, and thin wing bars.

■ **VOCALIZATIONS** Song fast and grating, with the final note different from phrase to phrase: *ch'reedu-j'riddy-jurry-zheer?... ch'reedu-j'riddy-jurry-chroo!*

■ **BEHAVIORS** Sticks to tangles and thickets in brushy habitats, often near water. Active and fidgety, flipping its tail about, but hard to see; listen for the distinctive song.

■ **POPULATIONS** Varies from gray *pusillus* ("Least Bell's Vireo," endangered) of West Coast to more olive *arizonae* ("Arizona") of Ariz. to brighter *medius* (N. Mex, West Tex.) and even brighter nominate *bellii* (Midwest, breeds west to Colo.).

Gray Vireo
Vireo vicinior | GRVI | L 5½" (14 cm)

■ **APPEARANCE** Slender and long-tailed. Pale gray above and below. Plumbeous Vireo (p. 286) is more solidly built, with larger bill, shorter tail, and longer wings; Plumbeous also more boldly marked, with "spectacles" and wing bars more prominent than on Gray. Compare also with smaller Bell's Vireo, especially subspecies *pusillus*.

■ **VOCALIZATIONS** Song of simple phrases, *ch'wee* and *churrit*, brighter and less burry than Plumbeous's, and delivered less slowly.

■ **BEHAVIORS** Nests in open, arid pinyon-juniper country in foothills; cocks tail like Bewick's Wren (p. 332), also found in "PJ" woods. Plumbeous Vireo usually in ponderosa pines, farther upslope. Recently discovered wintering in groves of elephant tree, *Bursera microphylla*, in Anza-Borrego Desert of Southern Calif.

■ **POPULATIONS** Scarce, even in perfect habitat. Famously undetected on migration, with very few credible records of migrants.

Hutton's Vireo
Vireo huttoni | HUVI | L 5" (13 cm)

■ **APPEARANCE** Small and compact. Greenish overall with prominent wing bars, broken eye ring, and pale lores. More frequently confused with Ruby-crowned Kinglet (p. 322) than other vireos: Hutton's has blue-gray legs and thicker bill with hooked tip; kinglet has black bar, absent on Hutton's, below lower wing bar. Compare also with Cassin's Vireo (p. 286).

■ **VOCALIZATIONS** Song of short phrases repeated rapidly and endlessly. Phrases may be falling or rising, sweet or shrill, but the key is that they're always the same, unlike other vireos with simple, but alternating, songs.

■ **BEHAVIORS** Found year-round in evergreen trees, especially oak groves, but also oak-conifer woods. Fidgety, like Ruby-crowned Kinglet.

■ **POPULATIONS** Mostly nonmigratory, although a few disperse following breeding; all our other vireos are migratory. Coastal subspecies (nominate *huttoni*) smaller and brighter than interior *stephensi*, but no range overlap.

VIREOS
VIREONIDAE

Cassin's Vireo
Vireo cassinii | CAVI | L 5" (13 cm)

■ **APPEARANCE** Medium-large vireo, with sturdy build. Intermediate in hue between colorful Blue-headed and relatively monochrome Plumbeous Vireos; all three have wing bars and "goggles." Markings on face and throat of Cassin's more blended than on high-contrast Blue-headed.

■ **VOCALIZATIONS** Song comprises short, slurred, burry phrases, *zzzhreeeya ... shreeeyup*; intermediate in timbre between sweeter Blue-headed's and gruffer Plumbeous's, but much overlap.

■ **BEHAVIORS** Feeds methodically well up in trees, broadleaf or conifer; more generalist than pine-loving Plumbeous.

■ **POPULATIONS** Cassin's, Blue-headed, and Plumbeous Vireos, closely related, were treated as one species, the Solitary Vireo, until 1997; their breeding ranges are almost wholly disjunct, but they overlap on migration. Fall migration of Cassin's more easterly than spring migration; regular late summer to early fall east to western Great Plains and West Tex.

Blue-headed Vireo
Vireo solitarius | BHVI | L 5" (13 cm)

■ **APPEARANCE** Size and shape like Cassin's. "Helmet" blue-gray, back greenish, flanks yellow, and throat bright white; wing bars and "goggles" prominent; Blue-headed is deeper-hued than Cassin's, with strong contrast on face and throat.

■ **VOCALIZATIONS** Song comprises rich and sweet phrases, slowly delivered: *siri?* and *swee-uh-wee*, interspersed with burrier notes. Song slower and lower than Red-eyed Vireo's (p. 288), a bit sweeter than that of Cassin's.

■ **BEHAVIORS** Like the other "Solitary Vireos," nests and forages well up in trees, typically in extensive tracts of conifers or northern hardwoods; movements slow and methodical.

■ **POPULATIONS** Overlaps slightly with Cassin's on migration on western Great Plains; Cassin's averages later in spring and earlier in fall.

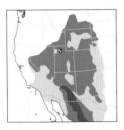

Plumbeous Vireo
Vireo plumbeus | PLVI | L 5¼" (13 cm)

■ **APPEARANCE** A bit larger than the two other "Solitary Vireos," with bigger bill. Spring and summer, gray above with prominent white wing bars and "goggles"; gray-white below. Freshly molted birds in fall tinged yellow-olive, resembling worn Cassin's that haven't yet molted.

■ **VOCALIZATIONS** Song like those of Cassin's and Blue-headed, but burrier. Cassin's sings faster on average than Plumbeous, but much overlap; ID based only on song unadvisable.

■ **BEHAVIORS** Like Cassin's and Blue-headed, feeds methodically in trees. Breeds in conifer forests, especially pine; migrants in any woodlot type.

■ **POPULATIONS** Migrates later in fall than Cassin's. Breeding ranges of all three "Solitary Vireos" have expanded in recent decades, due in part to changing forestry practices.

Yellow-green Vireo
Vireo flavoviridis | YGVI | L 6" (15 cm)

Closely related to Red-eyed Vireo, replacing it as a breeder in Mid. Amer. Large and long-billed like Red-eyed, with "mushier," less contrasting head pattern. Most are indeed imbued with a bright yellow-green wash, but beware the effects of lighting and feather wear. Song like Red-eyed's, but phrases faster, sounding like slurred chirps. Regular to coastal Calif. in very small numbers Sept. to Oct.; casual in summer to Ariz.

VIREOS
VIREONIDAE

Cassin's Vireo
- gray head blends into greenish back
- yellow edges to flight feathers
- worn Cassin's, especially spring and summer, are grayer overall
- ♂

Blue-headed Vireo
- sharp contrast between blue-gray head and green back
- white throat distinct from blue-gray head
- ♀
- ♂

Plumbeous Vireo
- conspicuous "goggles"
- gray edges to flight feathers

All vireos are more vividly colored after fall molt, which complicates field ID of the three "Solitary Vireos": Fresh Plumbeous approaches worn Cassin's, and fresh Cassin's approaches worn Blue-headed.

Yellow-green Vireo
- dull cap and diffuse stripes on face
- large bill
- adult
- immature

Red-eyed Vireo
Vireo olivaceus | REVI | L 6" (15 cm)

- **APPEARANCE** Large, long-billed vireo. Olive above; pale gray below with variable yellow wash. Adult has red eye and black stripes on face; juvenile has brownish eye, richer underparts, and weaker facial pattern, suggesting Philadelphia and Warbling Vireos. Compare also with closely related Yellow-green Vireo (p. 286), rare in West.
- **VOCALIZATIONS** Individual songs comprise pure-tone phrases of two to four syllables, separated by pauses of one-half to two seconds: *cheerio ... cheery ... j'reery ... j'cheer-cheery ....* Compare with songs of Blue-headed (p. 286) and especially Philadelphia Vireos. Call an unmusical *gwaaah* like Warbling Vireo's.
- **BEHAVIORS** Nests and forages high in broadleaf trees. Movements sluggish, but sings continually; known as the "Preacher Bird" for habit of droning on endlessly.
- **POPULATIONS** Periphery of breeding range unstable from year to year. Rare but regular late-spring vagrant throughout West (well into June); fall migration more protracted.

Philadelphia Vireo
Vireo philadelphicus | PHVI | L 5¼" (13 cm)

- **APPEARANCE** Slighter than Red-eyed Vireo, with shorter bill. Yellowish below with olive topside and grayish crown. Note dark line through eye (transocular); Warbling Vireo shows dark behind, but not in front of, eye. Yellow of Philadelphia below most intense at center of breast; on Warbling, yellow most intense on flanks. Tennessee Warbler (p. 426), especially breeding female, similar but has thin bill and white undertail coverts.
- **VOCALIZATIONS** Song like Red-eyed's; phrases of Philadelphia's average higher and more widely spaced, but much overlap. Deliveries differ: Philadelphia has limited repertoire and repeats song phrases; Red-eyed has larger repertoire, delivers song phrases seemingly randomly.
- **BEHAVIORS** Like Red-eyed, nests and forages in canopy of broadleaf forests, often in earlier successional stands than Red-eyed. Breeders defend territories against Red-eyeds.
- **POPULATIONS** Rarely seen in West away from breeding grounds; vagrants seen more often in fall than spring.

Warbling Vireo
Vireo gilvus | WAVI | L 5½" (14 cm)

- **APPEARANCE** Smaller and shorter-billed than Red-eyed Vireo. Pale olive overall, darker above than below. Most are weakly capped with a broad but diffuse white eyebrow. Philadelphia Vireo brighter and bolder, but fresh Warbling in fall can be rather yellow; even a dull Philadelphia has yellow wash on center of breast and dark line in front of eye, lacking on Warbling. Compare also with Red-eyed.
- **VOCALIZATIONS** Most in the West sing lazy, burry song, rising and falling, reminiscent of Blue Grosbeak (p. 448). Eastern population, breeding west to foothills of Rockies, sings a frenetic, herky-jerky series, around two seconds long, with a sharp and rising terminal note. Call a peevish *waaah* or *gwaaay*, harsh and unmusical.
- **BEHAVIORS** Common breeder wherever there are broadleaf trees, both extensive tracts in mountains and small groves by waterways in lowlands. Sings constantly, but quite hard to see.
- **POPULATIONS** Eastern (nominate *gilvus*) and western (*swainsoni*) groups may be distinct species; *swainsoni* smaller-billed and drabber overall.

MONARCH FLYCATCHERS | MONARCHIDAE

Despite having the name "flycatcher," the 100+ species in this family are not closely related to either the tyrant flycatchers (pp. 266–283) or the Old World flycatchers (p. 350). Widespread in Africa, Asia, and Oceania, the monarch flycatchers reach the West only in Hawaii, where they are represented by the elepaios, endemic to the main islands.

Endemic to Hawaii I.

Hawaii Elepaio
Chasiempis sandwichensis | HAEL | L 5½" (14 cm)

■ **APPEARANCE** Like all elepaios, small and compact, with sturdy legs and slight bill; tail, often cocked and fanned, broadens toward tip. All elepaios are clad in washed-out shades of gray and brown, with complex age-related variation; adult plumage, acquired after two years, features prominent white spotting and barring on wings, plus dark tail and white rump. Adult of this species brownish above; lores mostly brown and underparts gray-white with brown streaking. Also, adults sexually dimorphic; males have variable black on throat, females have clean white throat. Compare with juvenile Apapane (p. 364), washed in buff and black and with tail often cocked; but Apapane has decurved bill and lacks bold white patches.

■ **VOCALIZATIONS** Song thin and whistled, with a feeble introductory element and then a strongly slurred one *(e-le-PAI-o)*.

■ **BEHAVIORS** Active, inquisitive. Employs diverse foraging tactics: gleaning foliage, probing tree trunks, and flycatching. Found in forests, generally at higher elevations; in their actions, habitats, and general build, all elepaios suggest wrens, absent from Hawaii.

■ **POPULATIONS** Indigenous and endemic to Hawaii I., where it is the only elepaio. Sedentary and nonmigratory; thus, distinguished from other elepaios by range alone. Despite being restricted to Hawaii I., only one-fifth the area of San Bernardino Co., Calif., exhibits extensive geographic variation in color and pattern, likely reflecting local adaptation more than discrete genetic differences.

Kauai Elepaio
Chasiempis sclateri | KAEL | L 5½" (14 cm)

Only elepaio on Kauai; never occurs elsewhere. Like other elepaios, exhibits delayed plumage maturation, not acquiring adult plumage until age two. Adult has gray upperparts and white lores; underparts gray-white with blurry rufous breastband. Song similar to Hawaii Elepaio, but opening element rougher; total number of phrases per song averages fewer in Kauai Elepaio. As with Hawaii Elepaio, could be confused with juvenile Apapane (p. 364); another challenge on Kauai is juvenile White-rumped Shama (p. 350), with roughly similar color scheme, long tail, and white rump. Like Hawaii Elepaio, accepting of diverse forest types, but most often seen in higher-elevation tracts with indigenous trees, especially ohia, *Metrosideros polymorpha*.

Oahu Elepaio EN
Chasiempis ibidis | OAEL | L 5½" (14 cm)

The only elepaio on Oahu. Range completely disjunct from that of the other elepaios, but, as with Kauai Elepaio, compare with juvenile Apapane (p. 364) and juvenile White-rumped Shama (p. 350). Adult has bold white on wings and rump; note also brown upperparts and white lores of adult, combined with gray-white underparts streaked brownish. Song midway in timbre and complexity between Hawaii and Kauai Elepaios. Rarest of the elepaios; rat predation is an especially acute threat.

MONARCH FLYCATCHERS
MONARCHIDAE

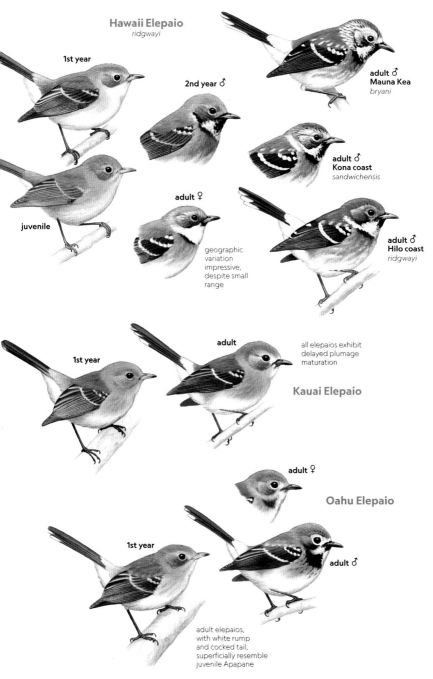

SHRIKES | LANIIDAE

These monochrome songbirds have the macabre habit of impaling their prey—large insects, lizards, even birds—on thorns and especially barbed wire fences. Their hooked bills and "bandit's masks" contribute to their fearsomeness.

Loggerhead Shrike
Lanius ludovicianus | LOSH | L 9" (23 cm)

- **APPEARANCE** Shorter-billed, shorter-tailed, and smaller overall than Northern Shrike. Black mask thicker on Loggerhead, especially around base of bill. Adult clean gray, black, and white; juvenile weakly barred below and scaly above. Compare also with Northern Mockingbird (p. 340).
- **VOCALIZATIONS** Calls include harsh *raaah* or *rraack*, but also sweeter *ch'peee*. Song consists of short trills and soft whistles.
- **BEHAVIORS** Solitary. Shrikes watch from wires or low perches, then go after their prey in sudden sorties just above the ground.
- **POPULATIONS** Partial migrant; returns in spring as the Northerns are leaving, departs in fall before the first Northerns arrive. Endangered subspecies *mearnsi*, particularly dark above, endemic to San Clemente I. off Calif.

Northern Shrike
Lanius borealis | NSHR | L 10" (25 cm)

- **APPEARANCE** Larger, longer-tailed, and longer-billed than Loggerhead Shrike. Paler overall than Loggerhead, with dark face mask thinner. Many sightings in winter are of juveniles, weakly barred below. Compare also with Northern Mockingbird (p. 340).
- **VOCALIZATIONS** Like Loggerhead, but generally silent in winter when most are found. One common call is a startling, retching *RRRAAAA*. Extraordinary is the species' practice of "acoustical luring," singing softly and ventriloquially to catch the attention of small birds—which it then eats.
- **BEHAVIORS** Nests on taiga; winters in open country with perches—including landscapes vacated for the winter by Loggerhead. Hunting behavior like Loggerhead.
- **POPULATIONS** Widespread but scarce in winter; arrives on wintering grounds in late fall, heads north by late winter. Weakly irruptive in West.

JAYS AND CROWS | CORVIDAE

Famous for their powers of cognition, the birds in this family are large to very large songbirds with sturdy feet and thick bills. Their calls are simple but varied. Many in the West are sedentary, but some are prone to seasonal irruptions.

Canada Jay
Perisoreus canadensis | CAJA | L 11½" (29 cm)

- **APPEARANCE** Slim and long-tailed; has small bill, a cold-weather adaptation. Adult has black hindcrown, large white cheek, and mostly gray plumage otherwise; suggests a huge chickadee.
- **VOCALIZATIONS** A variety of chuckles, whistles, and sighs; they carry well but sound soft.
- **BEHAVIORS** The infamous "Camp Robber" of the boreal forest. Singles or small groups sneak into day-use areas, snag a bit of tofu or s'mores, then carry on.
- **POPULATIONS** Nearly sedentary, with irruptions infrequent and unambitious. Amount of black on crown varies regionally: most extensive on Pacific Slope, whitest in southern Rockies, intermediate in much of Canada and Alas.

Pinyon Jay
Gymnorhinus cyanocephalus | PIJA | L 10½" (27 cm)
- **APPEARANCE** Stocky and short-tailed; bill is long and straight with nostrils exposed, unfeathered. Adult sky blue all over, deepest on head; juvenile duskier.
- **VOCALIZATIONS** Call an arresting *hya!* or *howay,* ringing and nasal; carries well. Flocks in chorus, especially at first light, are manic and mirthful.
- **BEHAVIORS** A pinyon pine specialist; *Gymnorhinus* means "naked nose," indicating an adaptation for feeding on sticky pinyon rosin. Feeds on ground as well as in trees. Gregarious; flocks very noisy as they fly from one stand of trees to another.
- **POPULATIONS** Irruptive; individuals or very small flocks sometimes range well out onto lowlands fall to winter. Irregular to West Tex., sometimes in large flocks.

Steller's Jay
Cyanocitta stelleri | STJA | L 11½" (29 cm)
- **APPEARANCE** Robust jay with long crest and relatively short tail. Head, throat, and back mostly blackish, suffused with deep brown. Rest of body deep blue; relative intensity and hue highly dependent on lighting; can appear almost black in shady forest. Closely related Blue Jay, slimmer and paler, has ample white in tail and wings; Blue Jay also has black breastband. Scrub-jays (p. 296), longer-tailed and lankier overall, lack crests.
- **VOCALIZATIONS** Calls, including a retching *RRRAAA* and a clanging *shook shook shook shook,* are notably harsh; but also gives a lovely "whisper song," airy and gurgling.
- **BEHAVIORS** Like Blue Jay, its eastern counterpart, bold and boisterous; tends feeders and raids campsites, but also fends for itself far from human activity.
- **POPULATIONS** Geographic variation quite complex, with well over a dozen subspecies. In southern Rockies, has white streaks on face, relatively pale and bright plumage overall, and especially long crest. Darker near coast, especially on Haida Gwaii.

Blue Jay
Cyanocitta cristata | BLJA | L 11" (28 cm)
- **APPEARANCE** Blue and crested like Steller's Jay, but Blue Jay is slimmer. Complexly patterned in blues and whites, blacks and grays; note especially black breastband and blue-and-white wings and tail. In flight, white in tail and trailing edge of wing obvious even from afar. Scrub-jays are longer-tailed with uncrested, smoothly rounded heads. Juvenile Blue Jay can be quite gray, but nevertheless shows some blue, combined with "necklace" and crest.
- **VOCALIZATIONS** Classic call is a nasal shriek, descending slightly, *aaaayyy,* often doubled or trebled. Also a variety of whistles, squeaks, and superb imitations of hawks. Infrequently heard "whisper song" a run-on jumble of slurred whistles.
- **BEHAVIORS** Like Steller's Jay, found around humans as well as in remote habitats. Noisy and aggressive like Steller's; singles or small groups often mob roosting hawks and owls. Generally in broadleaf groves, but gets around.
- **POPULATIONS** Fairly sedentary, but some perform a "stealth" migration through regions where others are resident. A few wander in winter to parts of northwestern U.S. where breeding does not normally occur. Hybridizes very rarely with Steller's.

Island Scrub-Jay
Aphelocoma insularis | ISSJ | L 12" (30 cm)
Endemic to Santa Cruz I., off Southern Calif., where it is the only scrub-jay. Like other scrub-jays, long-tailed and uncrested, with thin white eyebrow and incomplete breastband. Island is larger overall than other scrub-jays, with bigger bill; undertail coverts washed blue (gray or gray-white on other scrub-jays). Calls like those of California Scrub-Jay, but gruffer and not as ascending. Widespread within limited and isolated range in oak, pine, and chaparral, with most sightings by birders on day trips at and near boat docks.

California Scrub-Jay
Aphelocoma californica | CASJ | L 11" (28 cm)
- **APPEARANCE** Almost identical in build to Woodhouse's Scrub-Jay, but bill of California averages thicker. California is more contrastingly marked, but beware effects of lighting and feather wear; note especially the bright blue, broken breastband of California; breastband blurry and inconspicuous on Woodhouse's.
- **VOCALIZATIONS** Common call a shrill, nasal, rising *shreee!* or *scraay-aay!* Also a series of pounding notes like those of Steller's Jay (p. 294), but individual elements clearer, more ringing, more whistled: *yeet! yeet! yeet! yeet!*
- **BEHAVIORS** Common in oak woodlands, as well as chaparral both coastally and in foothills. Flourishes in towns and suburbs; tamer and bolder than Woodhouse's.
- **POPULATIONS** Nonmigratory, but range expanding north. Comes into contact with Woodhouse's at Calif.-Nev. border south of Reno, but otherwise no range overlap.

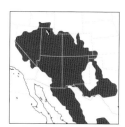

Woodhouse's Scrub-Jay
Aphelocoma woodhouseii | WOSJ | L 11" (28 cm)
- **APPEARANCE** In terms of shape, differs from California Scrub-Jay only in regard to bill structure: Woodhouse's is thinner-billed. Woodhouse's is also paler and less contrastingly marked than California, with only a trace of a blue breastband. Juveniles of all scrub-jays dusky, especially this species.
- **VOCALIZATIONS** Calls match those of California but are less explosive.
- **BEHAVIORS** Widespread but often uncommon in arid woodlands, especially pinyon-juniper. More of a recluse than California, but also not above visiting feeding stations.
- **POPULATIONS** Mostly resident, but a few wander fall to winter, especially in south of range.

Mexican Jay
Aphelocoma wollweberi | MEJA | L 11½" (29 cm)
- **APPEARANCE** A large jay; more heavyset than scrub-jays, with shorter tail. Less boldly patterned than scrub-jays; blue above and gray below. Lacks thin white eyebrow and variable breastband of scrub-jays. Compare also with Pinyon Jay (p. 294).
- **VOCALIZATIONS** Call a short, clipped, rising *shree!* or *shrayt*, often in excited chatter.
- **BEHAVIORS** Highly social; extended families make the rounds in woodlands in foothills and canyons, where they search for pine nuts and acorns.
- **POPULATIONS** Two subspecies in West: *arizonae*, pale and washed-out, in Ariz. and N. Mex.; and darker, more boldly marked, slightly smaller *couchii* in Chisos Mts., Tex. Even though family groups are always on the move, they do not wander far, and irruptions are almost unheard of in the species.

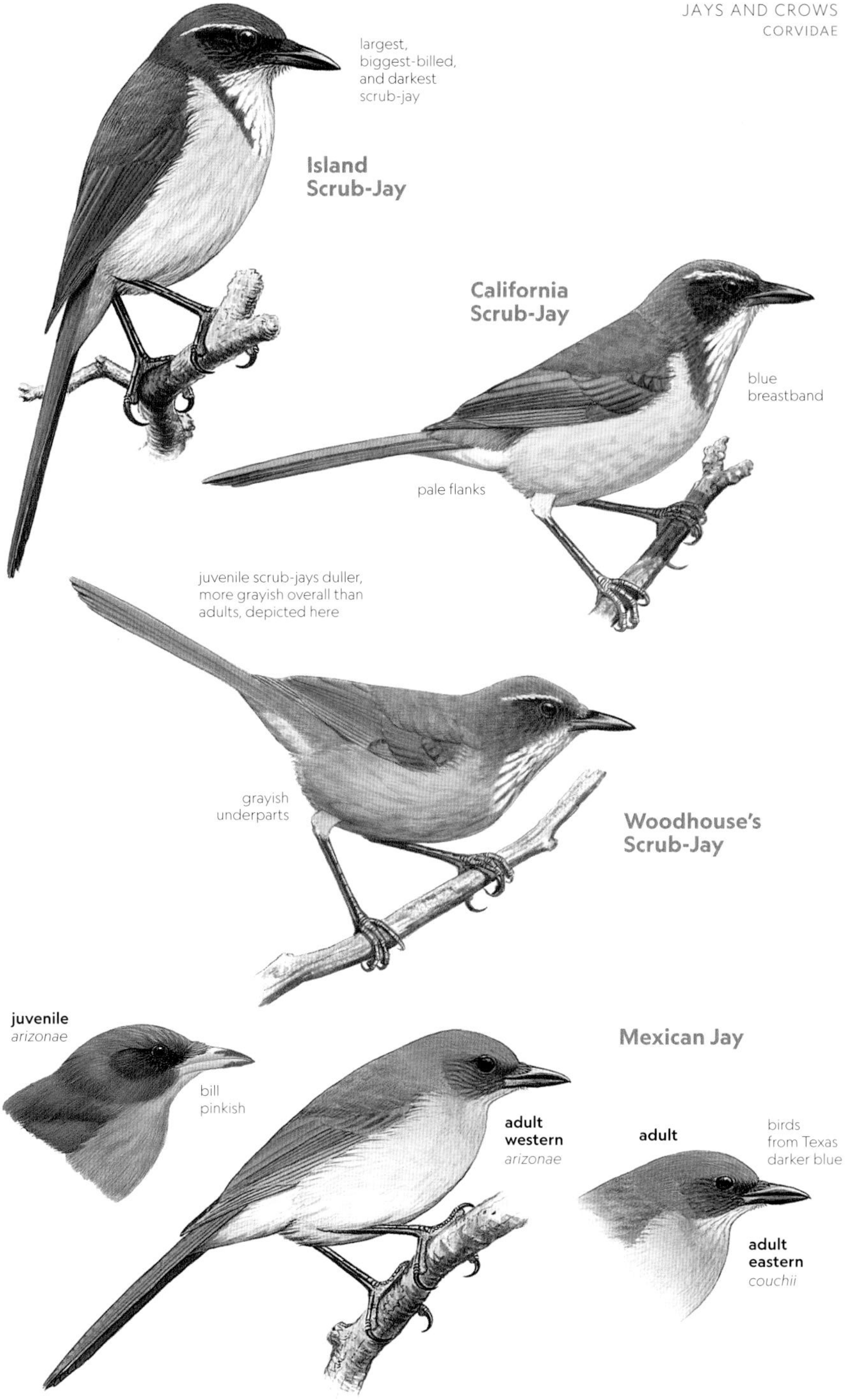

JAYS AND CROWS
CORVIDAE

Island Scrub-Jay
largest, biggest-billed, and darkest scrub-jay

California Scrub-Jay
blue breastband

pale flanks

juvenile scrub-jays duller, more grayish overall than adults, depicted here

grayish underparts

Woodhouse's Scrub-Jay

Mexican Jay

juvenile
arizonae

bill pinkish

adult western
arizonae

adult

birds from Texas darker blue

adult eastern
couchii

Clark's Nutcracker
Nucifraga columbiana | CLNU | L 12" (30 cm)

- **APPEARANCE** Stocky overall, with long bill and short tail; built like a small crow. Body soft gray. At rest, wings and tail appear mostly black; puffy white undertail coverts prominent. In flight, wings and tail flash white.
- **VOCALIZATIONS** Far-carrying cry rasping and wavering, *RRRAAA-a-a-a-a*, rising slightly; Steller's Jay (p. 294) sometimes gives a quite similar call. Also shorter notes, rising or falling, more whistled or nasal, jaylike.
- **BEHAVIORS** Widespread at high elevations, usually above foothills; gets up to timberline and beyond. Occurs in small groups or singly in mountain meadows and forests; also campsites, roadsides, and visitor centers.
- **POPULATIONS** Most are resident, flourishing in winter even at high elevations with deep snowpack. But a few wander to foothills and low valleys every year after breeding, and larger-scale irruptions, with dispersal far from mountains, are occasionally noted.

Black-billed Magpie
Pica hudsonia | BBMA | L 19" (48 cm)

- **APPEARANCE** Large and slender, with very long tail. Bill, stout and strong, is black. Plumage appears black and white in many views, but in good light shows metallic blue on wings and glistening green on tail.
- **VOCALIZATIONS** Call a rising *maaag?* or shorter *mag*, often run together in series: *mag mag mag mag*. Elaborate song is a jumble of whistles, rattles, and clacks interspersed with *mag* notes.
- **BEHAVIORS** Bold and sometimes aggressive, but also wary and flighty; often seen at roadkill. Usually occurs singly or in family groups, but also forms roosts of scores to occasionally 100+ birds. Generally restricted to cold and dry climes, from shrublands to forests to cities. Has one of the most distinctive nests of any bird in the West: a bulky, spheroid, and very large construction of twigs and small sticks.
- **POPULATIONS** Common in much of range; numbers crashed in many places during West Nile virus epidemic of early 2000s but have recovered everywhere.

Yellow-billed Magpie
Pica nuttalli | YBMA | L 16½" (42 cm)

- **APPEARANCE** A bit smaller than Black-billed Magpie, but nevertheless a large bird. Black-and-white patterning and iridescent highlights in wings and tail essentially identical to Black-billed, but bill bright yellow. Many, especially juveniles, show irregular yellow bare skin beneath eye. Compare also with California Scrub-Jay (p. 296), quite long-tailed; in fleeting glimpse, the two are confusable.
- **VOCALIZATIONS** Call repertoire like that of Black-billed: long notes given singly and shorter notes in series, with most calls rising in pitch.
- **BEHAVIORS** Locally common within limited range in oak savanna at low elevations; also around ranches and residential neighborhoods.
- **POPULATIONS** Nonmigratory; never gets into normal range of Black-billed Magpie. Reports of Yellow-billed Magpies out of range often refer to Black-billed Magpies with yellow-stained bills or simply carrying yellow food. Range has contracted somewhat, and recovery from West Nile virus has been slow.

American Crow
Corvus brachyrhynchos | AMCR | L 17½" (44 cm)

- **APPEARANCE** Well-built and stocky; entirely black. Bill strong and sturdy, but not as massive as that of ravens. In flight, wings and tail appear rounded; ravens have longer wings and ample tails, often fanned. Hawaiian Crow, *C. hawaiiensis* (p. 473), critically endangered, is considered extinct in the wild.
- **VOCALIZATIONS** Well-known call a straight-up *caw*, often in series. Also bill-clicks and spooky clattering.
- **BEHAVIORS** Famously clever; resourceful, quickly adapting to new resources and environments. Was found especially in farm country until mid-20th century but has become more of a habitat generalist in recent decades, with many now in large cities. Crows never soar; ravens often do.
- **POPULATIONS** Partially migratory inland. Resident crows along and near the coast from Salish Sea north to Alas. were until recently considered a separate species, Northwestern Crow, *C. caurinus*, smaller and hoarser.

Chihuahuan Raven
Corvus cryptoleucus | CHRA | L 19½" (50 cm)

- **APPEARANCE** Closer in heft to American Crow than to Common Raven but proportioned more like the latter. However, Chihuahuan isn't quite as long-winged and long-tailed as Common, and its bill is less massive. In close view or photographic review, note that the nasal bristles ("nose hairs") of Chihuahuan extend farther out on the bill than on Common. Completely black except for white-based throat feathers (gray-based on Common), but that character is variable, highly dependent on lighting, and usually invisible.
- **VOCALIZATIONS** Call harsh and nasal, a bit like a Mallard's quack: *rraack* or *rruuhh*.
- **BEHAVIORS** Occurs at lower elevations than Common, especially desert shrublands and shortgrass prairie. Bulky stick nests on roadside utility poles in creosote bush–dominated lowlands are often the handiwork of a pair of Chihuahuan Ravens.
- **POPULATIONS** Resident in much of range, but wanders northward, sometimes in sizable flocks, in winter.

Accidental to Hawaiian Is.

Common Raven
Corvus corax | CORA | L 24" (61 cm)

- **APPEARANCE** Enormous; largest passerine on Earth, exceeding Red-tailed Hawk (p. 228) in body length, body mass, and wingspan. Wings and tail long; bill long and massive. Nasal bristles don't extend as far out on bill as on Chihuahuan Raven. Throat feathers long and unkempt; in bright glare, the all-black throat feathers gleam brilliant white, suggesting Chihuahuan.
- **VOCALIZATIONS** Call a powerful, far-carrying, wavering croak: *kr'r'r'ruck*. Also short whistles and rattles, including notes like those of American Crow and Chihuahuan Raven.
- **BEHAVIORS** Habitat generalist; found year-round in hot canyons, on Arctic and alpine tundra, along seacoasts, and around big cities. Gets into deserts and arid grasslands where Chihuahuan occurs. Habitually soars with tail splayed out; Chihuahuan also soars, but not as often.
- **POPULATIONS** Range-restricted California clade and widespread Holarctic clade were formerly separate species, but have conjoined in recent geologic time, reversing their earlier separation. Accidental to Hawaiian Is.

JAYS AND CROWS
CORVIDAE

VERDIN | REMIZIDAE
All but one species, the Verdin, in this family are found only in the Old World. Remizids are very small, with fine bills, small tails, and short wings. Their closest relatives are the chickadees and titmice.

Verdin
Auriparus flaviceps | VERD | L 4½" (11 cm)
- **APPEARANCE** Tiny, with pointed bill. Adult male has yellow head and red "shoulders"; yellow and red reduced on adult female. Juvenile, plain gray, is recognized by bill structure and small size; compare with Northern Beardless-Tyrannulet (p. 266) and especially Lucy's Warbler (p. 428).
- **VOCALIZATIONS** Call a descending *beef* or *byeev*. Also a ringing *byeeo*, doubled or trebled.
- **BEHAVIORS** A desert insectivore; feeds actively in shrubs and small trees, especially mesquite. Nest distinctive, a spheroid construction of thorns and cactus spines.
- **POPULATIONS** Nonmigratory, but range has expanded in some regions with desertification of former grasslands in West.

CHICKADEES AND TITMICE | PARIDAE
Familiar feeder visitors in much of the West, the parids are small and short-winged, active and acrobatic. They are sociable, feeding in little flocks that often attract other songbirds. Our species are clad in grays, blacks, and whites, and they are found mostly in woodlands.

Black-capped Chickadee
Poecile atricapillus | BCCH | L 5¼" (13 cm)
- **APPEARANCE** A little ball of a bird, with classic chickadee plumage: black-capped and black-bibbed, with a huge white cheek patch. Note also buffy flanks and frosty white panel on gray wings.
- **VOCALIZATIONS** Gets its name from its *chick-a-dee-dee-dee* call, rough and rapid. Song in much of range a sweet, thin, whistled two-note *dee-dyeee*, the first note higher. In Pacific Northwest, song runs out to three or four notes, suggesting Mountain Chickadee or even White-throated Sparrow (p. 400).
- **BEHAVIORS** Where Mountain and Black-capped Chickadees co-occur, the latter is more inclined to broadleaf, riparian, and lowland habitats.
- **POPULATIONS** Mostly nonmigratory, but some wander fall to winter. Hybridizes with Mountain Chickadee in some regions, for example, south-central Wyo. and central N. Mex.

Mountain Chickadee
Poecile gambeli | MOCH | L 5¼" (13 cm)
- **APPEARANCE** Almost identical in size and shape to Black-capped; Mountain is grayer overall. Black cap of Mountain is broken by bold white eyebrow, but beware effects of molt and wear: Mountain's eyebrow can be inconspicuous, and aberrant Black-caps can have white-flecked cap.
- **VOCALIZATIONS** *Chick-a-dee-dee-dee* call hoarser than Black-capped's. Song a series of three or four thin whistles, distinct from Black-capped's in most regions of range overlap, but songs converge in Pacific Northwest.
- **BEHAVIORS** Common in coniferous forests; flourishes in winter where there is deep snow cover, relying on superb spatial memory to find previously hidden food.
- **POPULATIONS** A few go to lowlands fall to winter. Rocky Mountains group *gambeli* has warmer flanks and broader eyebrow than Pacific group *baileyae*.

Mexican Chickadee
Poecile sclateri | MECH | L 5" (13 cm)

- **APPEARANCE** Same build as Black-capped and Mountain Chickadees (p. 302). Head and face patterned like Black-capped's, but darker gray overall; black on throat and gray on flanks extensive.
- **VOCALIZATIONS** Short, gurgling outbursts and longer, jangling series; the "chickadee" call begins with a shrill trill followed by harsh dee notes. Lacks a song corresponding to the whistled utterances of Black-capped and Mountain.
- **BEHAVIORS** Occurs from mid-elevation ponderosa pine forests to higher spruce-fir tracts; wanders lower in winter to oak-sycamore woods.
- **POPULATIONS** Restricted in West at present time almost entirely to Chiricahuas, Ariz.; very scarce now in Animas Mts., N. Mex. The only chickadee in its range.

Chestnut-backed Chickadee
Poecile rufescens | CBCH | L 4¾" (12 cm)

- **APPEARANCE** Small and chubby. Patterned like Black-capped, but with dark brown cap, rich chestnut on back, and variable chestnut on flanks.
- **VOCALIZATIONS** "Chickadee" call a couple of sharp notes followed by a harsh buzz. Often gives only the opening notes, in series or singly.
- **BEHAVIORS** Found in diverse forest types, but most common in humid evergreen tracts; also in towns and cities where sufficiently wooded.
- **POPULATIONS** Extent of chestnut on flanks varies with latitude; from San Francisco Bay Area southward, flanks mostly gray. Range expanding south and inland in Calif.

Boreal Chickadee
Poecile hudsonicus | BOCH | L 5½" (14 cm)

- **APPEARANCE** Same size, shape, and basic plumage pattern as Black-capped Chickadee, but browner overall; many have disheveled look. Cap chocolate, flanks orangey-brown, wings plain gray; white in cheek reduced compared to Black-capped. Compare also with Chestnut-backed and Gray-headed Chickadees.
- **VOCALIZATIONS** "Chickadee" call of Boreal especially nasal and tinny. Lacks whistled song of Black-capped and Mountain, but gives a short gurgle, often ending in a loose trill.
- **BEHAVIORS** In broad range of overlap with Black-capped, Boreal is more specialized, tending to stick to conifer woods, especially spruce and fir. Absent from humid coastal forests where Chestnut-backed occurs.
- **POPULATIONS** Weakly irruptive at best in West, with just a small push south from southern edge of core range.

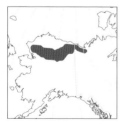

Gray-headed Chickadee
Poecile cinctus | GHCH | L 5½" (14 cm)

- **APPEARANCE** Our largest chickadee; its size is an adaptation to the cold climes where the bird occurs. Similar to Boreal Chickadee, but cap brown-gray on Gray-headed, flanks not as dark, and white on cheek more extensive.
- **VOCALIZATIONS** "Chickadee" call nasal and tinny. Also gives a jumble of thin, sweet, slurred whistles.
- **BEHAVIORS** More inclined to broadleaf stands (birch, willow) than Boreal, but also gets into spruce; favors river edges and floodplains.
- **POPULATIONS** Nonmigratory and apparently not irruptive. Widespread at high latitudes in Old World, where range retracting; range may be shrinking here too.

CHICKADEES AND TITMICE
PARIDAE

Mexican Chickadee
- extensive black bib
- gray flanks

Chestnut-backed Chickadee
- rich chestnut back and sides
- northern *rufescens*
- coastal central California *barlowi*
- gray sides

Boreal Chickadee
- juvenile
- worn summer
- orangey-brown flanks
- gray wash to cheeks
- fresh fall

Juveniles of all chickadees similar to adults but slightly duller and with "fluffier" feathering.

Gray-headed Chickadee
- pale edges to coverts and secondaries
- pure white cheeks
- slightly paler flanks than Boreal

Bridled Titmouse
Baeolophus wollweberi | BRTI | L 5¼" (13 cm)

■ **APPEARANCE** Like other titmice, crested; black-and-white patterning ("bridled") on face vaguely suggests Mountain Chickadee (p. 302), the same size but uncrested.

■ **VOCALIZATIONS** Song a rapid series of sharp notes, monosyllabic or disyllabic, all the same. Main call a stuttering, two-part chatter, starting with sharp notes, ending with huskier notes; like the "chickadee" call of chickadees (pp. 302–305), but with more notes.

■ **BEHAVIORS** Found in oak and mixed oak-conifer woods; tends to occur at lower altitudes than Mexican Chickadee (p. 304) and higher altitudes than Juniper Titmouse, but elevational overlap with both. Sociable; practices cooperative breeding, wherein nonreproductives help raise close kin.

■ **POPULATIONS** Considered nonmigratory, but some move downslope starting in late Sept., heading back up again by early Apr.

Oak Titmouse
Baeolophus inornatus | OATI | L 5" (13 cm)

■ **APPEARANCE** Almost identical to Juniper Titmouse; Oak Titmouse is slightly smaller, with browner hues.

■ **VOCALIZATIONS** Songs varied, but many are ringing and repetitious: *leeda leeda leeda* and *d'lee d'lee d'lee*. Also a rapid "chickadee" call: *tsicka tsicka tsicka zzzzz*.

■ **BEHAVIORS** Noisy and energetic. Roams sun-kissed oak woods in small flocks; also in towns and cities.

■ **POPULATIONS** Nonmigratory; almost no range overlap with Juniper Titmouse, also sedentary. Oak Titmice in north of range average paler and smaller-billed than those farther south.

Juniper Titmouse
Baeolophus ridgwayi | JUTI | L 5¼" (13 cm)

■ **APPEARANCE** Slightly larger than Oak Titmouse and even drabber; Juniper Titmouse is just plain gray all over.

■ **VOCALIZATIONS** Repertoire like that of Oak Titmouse, but songs and calls faster, not as musical.

■ **BEHAVIORS** Found mostly in arid, rocky juniper and pinyon-juniper woods in foothills and badlands. Spirited and noisy like Oak Titmouse.

■ **POPULATIONS** Widespread in suitable habitat, but generally uncommon. Oak and Juniper Titmice were until 1997 treated as a single species, the Plain Titmouse.

Black-crested Titmouse
Baeolophus atricristatus | BCTI | L 5¾" (15 cm)

■ **APPEARANCE** Largest titmouse in the West; long-crested. Crest of adult solid black, cut off sharply from pale gray forehead. Rest of plumage mostly gray, darker above than below; flanks washed buff. Crest of juvenile concolor with rest of upperparts; compare with Juniper Titmouse.

■ **VOCALIZATIONS** Song more ringing, more musical than any other titmouse in West: five or more *deer!* or *peedah!* notes in rapid succession. Also a very nasal scold, often run out to several notes.

■ **BEHAVIORS** Found in oak-mesquite woodlands and shrublands, getting up into pinyon-juniper foothills, especially where there is an oak admixture.

■ **POPULATIONS** Nonmigratory; range tracks that of mesquite, generally expanding. In range in West, usually the only member of its family, but barely overlaps Juniper Titmouse near Davis Mts., Tex.

CHICKADEES AND TITMICE
PARIDAE

LARKS | ALAUDIDAE

A mostly Old World assemblage numbering around 100 species, the larks are terrestrial birds of open habitats. Despite superficial behavioral and morphological similarities with pipits (p. 358) and longspurs (pp. 380–383), they are part of a group that includes the swallows (pp. 310–315) and even the Bushtit (p. 316). Larks are prized in many cultures for their songs, often delivered in flight.

Established on main islands; also some natural vagrants

Eurasian Skylark
Alauda arvensis | EUSK | L 7¼" (18 cm)

■ **APPEARANCE** About the same size as the largest New World sparrows (pp. 386–409). Streaky all over with short tail, slight bill, and underwhelming crest. Bill and legs orangey; eye black and beady. Tail and trailing edge of wing edged white; hard to see on bird at rest, more prominent in flight.

■ **VOCALIZATIONS** This is the bird made famous by Percy Bysshe Shelley ("To a Skylark"). The song, given in towering flight, is a nonstop issuance of short buzzes and whistles. Flight call a rough *chrup* or *churrup,* buzzy and descending.

■ **BEHAVIORS** Always in open, often desolate, country; favored fields are dry and weedy, with little if any woody vegetation. Except when in display, keeps close to the ground; walks, does not hop.

■ **POPULATIONS** Introduced and widely established on main Hawaiian islands (nominate *arvensis*), although rare on Kauai; also introduced and declining in Salish Sea region (nominate *arvensis*), where restricted now mostly to northern outskirts of Victoria, B.C. Asian subspecies *pekinensis* annual in small numbers to Bering Sea region and casual farther south along coast, mostly in winter, to Calif; *pekinensis* also accidental to Hawaiian Is.

Horned Lark
Eremophila alpestris | HOLA | L 6¾–7¾" (17–20 cm)

■ **APPEARANCE** A bit larger than American Pipit (p. 358) and Lapland Longspur (p. 380), two species with which it often co-occurs. Variable sandy brown above, paler below; tail, mostly black, has thin white edges. Facial pattern of adult, yellow and black with fine "horns," unique but variable. Juvenile, spotted above and weakly streaked below, can be confused with Sprague's Pipit (p. 358); on pipit, note pink legs, brighter bill, and extensive white on tail.

■ **VOCALIZATIONS** Flight call a descending *ts'tseee,* sweet and jingling, not unlike American Tree Sparrow's (p. 394); flight call of American Pipit more strongly descending, less musical. Song, given perched or in flight, a couple of stutters, then a short jumble: *pik plek p'TEE'd'DEEDle'eedle'ek;* sings on moonlit nights, spring to early summer.

■ **BEHAVIORS** Always in open habitats, often extreme or degraded: alpine tundra, salt flats, barren agricultural fields in winter, and military test sites. Occurs in compact flocks, sometimes quite large, in nonbreeding season; flocks flush frequently. Birds on the ground meander by walking, not hopping, their bodies low to the ground.

■ **POPULATIONS** Occurs widely in Northern Hemisphere, with 40+ subspecies. A sampling from the West includes but is not limited to: northern *arcticola* and *hoyti,* large and pale; *sierrae* and *strigata* of Pacific Northwest and Northern Calif., with deeper yellow on face and breast; *rubea,* mostly Central Valley of Calif., reddish above. Migratory strategies vary among subspecies; most or all breeding north of southern Canada withdraw south in winter.

LARKS
ALAUDIDAE

REED WARBLERS | ACROCEPHALIDAE
Numbering close to 60 species worldwide, the reed warblers are, in a word, drab. Most are brown above and paler gray below, with a long, thin bill and an indistinct eyebrow. Many, but not all, occur in wetlands, where they are more often heard than seen.

Millerbird 🅲🆁
Acrocephalus familiaris | MILL | L 5" (13 cm)
Found only on Nihoa and Laysan; the birds on Laysan are recent transplants from Nihoa. A distinctive population, perhaps a separate species, occurred on Laysan but went extinct in 1923. The Millerbird is a classic reed warbler, gray-brown with a hint of an eyebrow and the bill long and thin. The song is a jumble of short whistles and scratchy notes. Where vegetation occurs on Nihoa and Laysan, Millerbirds occur. Critically endangered, with total population only in low hundreds; presumably has always been very rare, owing to very limited distribution.

SWALLOWS | HIRUNDINIDAE
The birds in this globally distributed family are aerial insectivores with small bills and long, pointed wings. Despite behavioral similarities with swifts (pp. 82–85), swallows are unambiguously songbirds: They habitually perch on wires (our swifts never do), and their flight is languid (swifts have unique twittering flight); many are gifted songsters and brightly colored.

Tree Swallow
Tachycineta bicolor | TRES | L 5¼" (15 cm)
- **APPEARANCE** Midsize swallow with shallowly notched tail. Adult steely green or blue above, underparts snow-white; lacks intrusions of white onto face and rump of Violet-green Swallow. Juvenile is solid brown above; many have diffuse breastband, suggesting Bank Swallow (p. 312).
- **VOCALIZATIONS** Song a short jumble of *slweet* and *tlit* notes, with liquid or gargling quality. Frequently calls in flight, giving one to a few of the song elements as it goes.
- **BEHAVIORS** Usually found around freshwater, especially wooded ponds and river edges. Nests in cavities; readily takes to bird boxes. In winter, supplements diet of flying insects with waxy berries.
- **POPULATIONS** Returns earlier in spring than most swallows in West. Unlike Tree Swallows in the East, which are famous for lingering late in fall, many in the West head south by late summer.

Violet-green Swallow
Tachycineta thalassina | VGSW | L 5¼" (13 cm)
- **APPEARANCE** A bit smaller than Tree Swallow, and more colorful. Magnificent in good light, with purple rump, splashes of blue, and every shade of green. White beneath wraps up above eye and onto rump; also intrudes to axillaries ("wingpits"). White-throated Swift (p. 84), only distantly related, has similar pattern.
- **VOCALIZATIONS** Call repertoire matches Tree Swallow's, but notes scratchier, less gurgling.
- **BEHAVIORS** Favors drier habitats than Tree Swallow, especially pinewoods. Usually nests in natural cavities, not bird boxes, and often in small colonies.
- **POPULATIONS** Arrives a bit later in spring than Tree Swallow; stages in large numbers as early as midsummer, then begins southbound migration.

REED WARBLERS | SWALLOWS
ACROCEPHALIDAE | HIRUNDINIDAE

Bank Swallow
Riparia riparia | BANS | L 4¾" (12 cm)

- **APPEARANCE** Our smallest swallow; compact overall, with narrow wings and slight tail, gently forked. Adult brown above and mostly white below, with broad brown breastband sharply set off from gleaming white throat; Northern Rough-winged Swallow, brown above and paler gray-brown below, has dusky throat and lacks breastband. Juvenile Bank has thin whitish wing bars and scalloping above; juvenile Northern Rough-winged has thick cinnamon wing bars. Compare also with juvenile Tree Swallow (p. 310), dusky brown above and often with blurry breastband.
- **VOCALIZATIONS** Call a harsh buzz, typically given in flight, in rapid series, accelerating slightly: *bzhh bzhh bz bz'bz'bz'bz'bz*. Also just a single *bzhh*.
- **BEHAVIORS** Nests colonially in earthen embankments along rivers and pond edges; also uses quarries and roadcuts. Wingbeats choppy; wingbeats of Northern Rough-winged more languid.
- **POPULATIONS** Worldwide range vast, but summer distribution in West quite local, reflecting habitat availability; some colonies are very large, but many are small.

Northern Rough-winged Swallow
Stelgidopteryx serripennis | NRWS | L 5" (13 cm)

- **APPEARANCE** Slightly larger and longer than Bank Swallow, with broad, blunt-tipped tail. Our drabbest swallow: Adult dull brown above and sandy gray-brown below, becoming whiter toward belly and undertail coverts. Juvenile like adult, but with cinnamon wing bars and overall rufous suffusion on wings. The "rough wing," visible in good digital photographs, indicates serrations on the outer primaries, their function unknown.
- **VOCALIZATIONS** Call a flatulent *rrrpp*, often in fast series. Not as harsh as Bank Swallow's; suggests flight call of Dickcissel (p. 452).
- **BEHAVIORS** Like Bank Swallow, nests in embankments; nests singly or in very small colonies, and does not excavate its own nest sites. Usually forages over water. Wingbeats languid; wingbeats of Bank Swallow choppier.
- **POPULATIONS** Departs most of breeding range in West by Sept. Unlike many aerial insectivores, appears to be in general good health; increasingly seen in winter in Southern Calif. and southern Ariz.

Purple Martin
Progne subis | PUMA | L 8" (20 cm)

- **APPEARANCE** Our largest swallow; long-winged and with tail deeply notched. Adult male dark purple-blue, appearing black in poor light. Adult female has scaly gray underparts; juvenile like female but paler.
- **VOCALIZATIONS** Main call a down-slurred whistle, low and rich; also a short, slow trill, descending. Song a pleasing gargle.
- **BEHAVIORS** Although Purple Martins in the East nest only in humanmade structures, those in the Interior West often use natural cavities. Use of nest boxes (called "apartments"), as well as bridges and buildings, increasing on Pacific Slope from B.C. to Northern Calif.
- **POPULATIONS** Returns very early to southern U.S. states, then slowly fans northward; southbound migration also early. Most in our area are western *arboricola* subspecies, but nominate eastern subspecies *subis* reaches Canadian Rockies; *subis* gathers in impressive flocks on breeding grounds before migrating south in late summer, but *arboricola* generally does not.

Accidental to Hawaiian Is.

Barn Swallow
Hirundo rustica | BARS | L 6¾" (17 cm)

- **APPEARANCE** Fairly large swallow; tail very long and deeply forked. Adult orange below and lustrous blue above; color scheme broadly similar to that of Cliff and Cave Swallows. Juvenile whiter below than adult, with shorter tail, but even recently fledged young show hints of telltale tail shape.
- **VOCALIZATIONS** Calls, frequently given in flight, simple *vree* and *frit* notes, rising sharply, often in short, fast series. Song, given by both sexes, a lengthy outpouring of chirpy phrases ending in a twangy, rising note and dry rattle.
- **BEHAVIORS** Tolerant of humans; classically associated with farm country but may be found almost anywhere. Conspicuously tends open-cup earthen nests under bridges, verandas, and lampposts.
- **POPULATIONS** Widespread and still common, but an emerging threat is rapid reduction of the aerial prey base of this and other swallows. Departs Interior West later in fall than other swallows. Two white-bellied Eurasian subspecies are regular in very small numbers to western Alas.; orange-bellied N. Amer. subspecies and both Eurasian subspecies accidental to Hawaiian Is.

Accidental to Hawaii

Cliff Swallow
Petrochelidon pyrrhonota | CLSW | L 5½" (14 cm)

- **APPEARANCE** Heavyset swallow with broad, blunt-tipped tail; bluish above and orangish below like Barn Swallow. Bright buff-orange rump especially prominent in flight; forehead pale cream, known as "headlights" for being so conspicuous on bird flying toward observer. Juvenile pale-rumped like adult but has dark forehead; compare with Cave Swallow.
- **VOCALIZATIONS** Flight call breathy and nasal, typically down-slurred: *reer* or *vreen*. Complex song comprises grinding and cranking notes, interspersed with flight calls.
- **BEHAVIORS** Nests in densely packed colonies in culverts and under bridges; individual nests, made of mud and straw, are domed with side entrances. Gregarious, even for a swallow.
- **POPULATIONS** Mostly Mexican subspecies *melanogaster*, with darker forehead, reaches Southeast Ariz., west of where Cave normally occurs. Accidental (subspecies unknown) to Hawaii.

Cave Swallow
Petrochelidon fulva | CASW | L 5½" (14 cm)

- **APPEARANCE** Similar to Cliff Swallow, but not as heavyset and head pattern reversed: Adult Cave has dark forehead and pale throat; adult Cliff has pale forehead and dark throat. Juvenile duskier on head; juvenile Cliff can be quite similar, but note richer rump and paler throat of juvenile Cave.
- **VOCALIZATIONS** Calls varied, including rising notes (like Barn Swallow) and falling notes (like Cliff Swallow). Notes tend to be purer and less nasal than those of Cliff, but much variation.
- **BEHAVIORS** Nest similar to Cliff Swallow's, but more open. Still nests in caves (including, fittingly, at Carlsbad Caverns N.P., N. Mex.), but also nests alongside Cliff Swallows under bridges and in culverts.
- **POPULATIONS** Did not occur in West until 20th century. Rapidly expanded north to West Tex. and as far northwest as Sierra Co., N. Mex., but expansion has slowed or stalled in 21st century.

BUSHTITS | AEGITHALIDAE

Also known as long-tailed tits, the aegithalids are a mostly Old World assemblage. They are active and flocky, with complex social systems. Recent research indicates a close relationship with bush-warblers and leaf warblers, also Old World families.

Bushtit
Psaltriparus minimus | BUSH | L 4½" (11 cm)

■ **APPEARANCE** Tiny and drab; a dirty cotton ball with a toothpick for a tail. Eye yellow in adult female, black in adult male. Interior *plumbeus* plain gray except for weakly contrasting brown auriculars ("ears"); coastal subspecies (nominate *minimus*) has weakly contrasting brown cap.

■ **VOCALIZATIONS** Notes short and sharp: *spik, speek*. Also a ringing *lilililili* and a descending *WHEE-whee-whee-whi-whi-whi*.

■ **BEHAVIORS** Energetic flocks roam shrubby habitats, whether in arid wildlands or suburban neighborhoods with suitable plantings.

■ **POPULATIONS** Some *plumbeus* males, especially young in late summer, have black auriculars; these "Black-eared Bushtits," common in Mexico, were once considered a distinct species.

BUSH-WARBLERS | CETTIIDAE

Like bushtits, the bush-warblers are small, often drab, mostly long-tailed inhabitants of shrubby habitats. Also known as cettiids, they are best represented in Africa and Asia.

Established on main islands

Japanese Bush-Warbler
Horornis diphone | JABW | L 5½" (8 cm)

■ **APPEARANCE** Small and relatively long-tailed; brown and gray overall, with pale eyebrow. Akikiki (p. 362), rare and local, has very different shape; Millerbird (p. 310), also rare and local, does not overlap in range.

■ **VOCALIZATIONS** Song a monotone whistle followed by brief gurgle: *eeeeeeeee gli'g'g'glii*. Call a rough smacking sound, sometimes in series.

■ **BEHAVIORS** Hard to see, especially late summer to early winter when not singing; sticks to dense vegetation, especially undergrowth.

■ **POPULATIONS** Beloved harbinger of spring in Japan; introduced to Oahu 1930s, has since spread to the other main islands.

LEAF WARBLERS | PHYLLOSCOPIDAE

The dreaded "phylloscs" are an extreme field ID challenge for birders in the Old World. Taxonomy is vexing, with at least 80 species in the family, all in the genus *Phylloscopus*. They are fidgety foliage-lovers, and many are washed greenish or yellowish.

Arctic Warbler
Phylloscopus borealis | ARWA | L 5" (13 cm)

■ **APPEARANCE** Superficially like a drab New World warbler (pp. 424–443). Dusky olive-brown overall, with stout bill. White eyebrow prominent; thin white wing bar inconspicuous.

■ **VOCALIZATIONS** Song a slow trill of 8–20 unmusical notes on the same pitch. Call a shrill, short, buzzy *bzzeet*.

■ **BEHAVIORS** Nests in low-stature spruce-birch woods and dense willow thickets. Sings from low perches in the open, but forages close to cover.

■ **POPULATIONS** Widespread breeder across Old World taiga, penetrating well inland in Alas. It is one of *the* birds along the Denali Highway. Casual in fall to lower 48, mostly coastally.

BULBULS | PYCNONOTIDAE

Numbering well over 150 species worldwide, the bulbuls are a mostly Afro-Asian assemblage. They are related to the Wrentit, white-eyes (p. 320), and laughingthrushes (pp. 320–323). Like waxwings (p. 324), bulbuls are arboreal frugivores, sleek and crested, with bright splashes of color.

Established on Oahu

Red-vented Bulbul
Pycnonotus cafer | RVBU | L 8½" (22 cm)

- **APPEARANCE** Slim and dark, with crest variably raised and lowered. Both bulbuls in the West have red vents and white-tipped tails. Red-vented is larger and darker than Red-whiskered, with a uniformly dark brown head and breast.
- **VOCALIZATIONS** Low and rich: Calls include a sharp *aaawrp*, twangy and ascending; song a short gargle, *rawrp reer-a-a-rip*.
- **BEHAVIORS** A bird of woodland edges and forest fragments, including urban settings; common in downtown Honolulu.
- **POPULATIONS** Indigenous to Indian subcontinent; well established on Oahu. Recent sightings on the other main islands of Hawaii are worrisome, suggesting range expansion by this aggressive invader.

Established on Oahu

Red-whiskered Bulbul
Pycnonotus jocosus | RWBU | L 8" (20 cm)

- **APPEARANCE** Smaller and paler than Red-vented Bulbul, with long, pointed crest. Brown above, whitish below; the red "whisker" (more like eyeliner) is hard to see, but red undertail coverts are prominent.
- **VOCALIZATIONS** Higher and sweeter than Red-vented's: call a descending *ch'whee-ur;* song a short, whistled jumble.
- **BEHAVIORS** Habits and habitats in West like those of Red-vented; favors residential districts with trees.
- **POPULATIONS** Indigenous across a large swath of South Asia. Like Red-vented, restricted in Hawaii to Oahu, slowly expanding from Honolulu area; established and expanding in and around Los Angeles.

SYLVIID WARBLERS | SYLVIIDAE

This taxonomically vexed family, restricted almost entirely to the Old World, contains species that have been all over the checklist in recent decades. Many are slim and long-tailed, and most have excellent songs. Placement of the strange Wrentit within the Sylviidae is not universally agreed upon.

Wrentit
Chamaea fasciata | WREN | L 6½" (17 cm)

- **APPEARANCE** Long-tailed and drab, with a small bill; eye yellow. Has blurry streaks below. Female Bushtit (p. 316), with yellow eye, much smaller.
- **VOCALIZATIONS** Song of male, heard year-round, is the "bouncing ball" of coastal chaparral: a series of bright chirps, accelerating. Song of female does not accelerate. Both sexes give scratchy call notes.
- **BEHAVIORS** Utterly sedentary; furtive and hard to see. Occurs year-round in brushlands of the Pacific Slope. Most famously associated with fog-shrouded coastal scrub, where it sings within earshot of the surf, but also occurs in sun-kissed chaparral well upslope on west flank of Sierra.
- **POPULATIONS** Northern birds warmer buff; southern birds plainer, cooler gray. Placed by some authorities in the family Paradoxornithidae, a name befitting the puzzling status of this species.

BULBULS | SYLVIID WARBLERS
PYCNONOTIDAE | SYLVIIDAE

Red-vented Bulbul

very dark overall

juvenile

dark head with short crest

scaly back

adult

white rump and tail tips

Red-whiskered Bulbul

long crest

juvenile

adult

red undertail coverts

Wrentit

northern
phaea

long tail

typically seen in pairs

compare with smaller Bushtit

southern
henshawi

319

WHITE-EYES | ZOSTEROPIDAE

An Old World assemblage, the sociable and chattery white-eyes are concentrated heavily in Africa, South Asia, and Australasia. The white "eye" refers to feathering encircling the eye. Most are small and round, running in greens, yellows, and browns.

Swinhoe's White-eye
Zosterops simplex | SWWE | L 4" (10 cm)

Indigenous mostly to China; recently established and rapidly increasing in urban Southern Calif. Both of our white-eyes are bright yellow-green, with white eye ring interrupted by small black patch in front of eye (lores). Supraloral and forehead of Swinhoe's brighter than Warbling. Gives down-slurred short whistles (*slwee*, *slweet*), singly or in short, fast jumbles. Favors residential districts with ample plantings; often forages in canopy.

Warbling White-eye
Zosterops japonicus | WAWE | L 4" (10 cm)

■ **APPEARANCE** Basically identical to Swinhoe's White-eye, but lacks brighter yellow above bill (forehead) and in front of eyes (supraloral).

■ **VOCALIZATIONS** Like Swinhoe's, gives down-slurred whistles, but more complex and more typically in elaborate jumbles. Both species also give short chatters and whines.

Well established on main islands

■ **BEHAVIORS** All over in Hawaii, from resorts at sea level to high-elevation forest interiors; a flock of very small birds in the treetops is often this species.

■ **POPULATIONS** Indigenous to Pacific Rim from Japan to Indonesia; probably the most abundant bird species in Hawaii.

LAUGHINGTHRUSHES | LEIOTHRICHIDAE

This family hails from tropical Asia, with many of its representatives admired for their great songs. Shades of brown run strong in the Leiothrichidae, but some species are adorned with bright splashes of color. They are weak fliers and generally nonmigratory; our species are introduced.

Greater Necklaced Laughingthrush
Garrulax pectoralis | GNLA | L 13" (33 cm)

■ **APPEARANCE** Sturdy legs, ample tail, and elongate body suggest a scrub-jay (p. 296) or thrasher (pp. 336–341). Warm brown above and mostly pale below, with ornate face pattern; tail has buffy corners.

Established on Kauai

■ **VOCALIZATIONS** Song comprises whistled or chirping notes in series, often on same pitch. Call a short, fast trill: *brrrrrt*.

■ **BEHAVIORS** Furtive frugivore of shady woods, at sea level but also upslope.

■ **POPULATIONS** Indigenous mostly to China. Introduced to Kauai, where long established but usually outnumbered by Hwamei.

Hwamei
Garrulax canorus | HWAM | L 10" (25 cm)

■ **APPEARANCE** Smaller relative of Greater Necklaced Laughingthrush. Mostly warm brown with well-marked face: Baby blue eye ring has thin "bridle" extending rearward.

Established on main islands

■ **VOCALIZATIONS** Varied songs low, rich, and whistled; some are repetitious like Northern Mockingbird's (p. 340), others more caroling like *Pheucticus* grosbeaks' (p. 448).

■ **BEHAVIORS** Stays in well-wooded cover; a "mystery voice" in lowland groves of nonindigenous trees often proves to be this species.

■ **POPULATIONS** Indigenous to China; present on main islands of Hawaii.

WHITE-EYES | LAUGHINGTHRUSHES
ZOSTEROPIDAE | LEIOTHRICHIDAE

Swinhoe's White-eye
- bold white eye ring
- lemon yellow throat
- acts like a small warbler (pp. 424–443)

Warbling White-eye
- olive upperparts

Greater Necklaced Laughingthrush
- black-and-white streaked cheek
- reddish-brown sides
- large sturdy legs

Hwamei
- adult
- streaked head
- pale eye ring with "bridle"
- juvenile

Established on main islands

Red-billed Leiothrix
Leiothrix lutea | RBLE | L 5½" (14 cm)

- **APPEARANCE** Much smaller than confamilial laughingthrushes, genus *Garrulax* (p. 320). Gray overall, but with dashes of brilliant orange and yellow; note red bill, bright orange-yellow on throat and breast, and deep orange-yellow on wings.
- **VOCALIZATIONS** Song herky-jerky, rising and falling, suggesting a mainland tanager's (pp. 444–447), but sweeter.
- **BEHAVIORS** Fidgety, but stays in woods; accepting of diverse forest types.
- **POPULATIONS** Indigenous mostly to China, also west into Himalaya. Established on main islands of Hawaii, but long-term numerical fluctuations complex.

KINGLETS | REGULIDAE
What they lack in size, they more than make up for in personality. The tiny kinglets are hyperactive fussbudgets with remarkable songs. The family numbers six species, restricted mostly to the north temperate zone, especially where there are conifer forests.

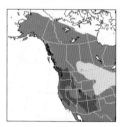

Ruby-crowned Kinglet
Corthylio calendula | RCKI | L 4¼" (11 cm)

- **APPEARANCE** A little green ball of a bird; stub-tailed. Male's ruby crown may be dazzlingly displayed, barely visible, or, more often than not, entirely hidden; female lacks ruby on crown. Wing bars thick and white. Bold eye ring broken above and below; face lacks black stripes of Golden-crowned. Similar Hutton's Vireo (p. 284) has thicker bill, blue-gray legs, and different song.
- **VOCALIZATIONS** Song remarkable for so small a bird. Starts with high notes, inaudible to some, followed by lower elements: *see see see see cheer cheer cheer cheer ch'whooda ch'whooda ch'whooda*. Call a rough *digit* or *fidget*, sometimes in series.
- **BEHAVIORS** Always fidgety. On breeding grounds, where often quite common, sticks to treetops, especially firs and spruces. Widespread in winter, typically singly, in thickets; visits feeders.
- **POPULATIONS** Migrates fairly early in spring, fairly late in fall. Occurs in small numbers well north of core winter range.

Golden-crowned Kinglet
Regulus satrapa | GCKI | L 4" (10 cm)

- **APPEARANCE** A bit smaller than Ruby-crowned Kinglet, with different face pattern: Golden-crowned has white eyebrow bordered black above and below; lacks eye ring of Ruby-crowned. Both male and female show yellow crown, variably suffused with orange on male.
- **VOCALIZATIONS** Very high call a series of three quick hisses: *sssss sssss sssss*. Song like that of Ruby-crowned, but higher, with less elaborate ending: *see see see si si si si si sick-sick-sick-suck-suck*.
- **BEHAVIORS** Occurs singly or in twos and threes, often joining mixed-species flocks. A metabolic marvel; even in the dead of winter, subsists mostly on animal matter. Golden-crowneds huddle together at night to preserve warmth; especially in winter, individuals lose substantial body weight overnight, only to regain it while feeding continually the next day.
- **POPULATIONS** Breeds widely in diverse conifer forests from Pacific coast to timberline in the high Rockies. Postbreeding movements complex: Some migrate far from breeding grounds; others stay put, including where winter conditions are severe.

WAXWINGS | BOMBYCILLIDAE

With their exquisite markings and "silky" plumages, the waxwings are among our most striking birds. They are famously nomadic, wandering widely in search of fruiting trees. Waxwings trill and twitter constantly as they fly and forage, but they apparently do not sing in the biological sense of the term.

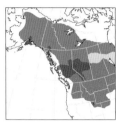

Accidental to Hawaii

Bohemian Waxwing
Bombycilla garrulus | BOWA | L 8¼" (21 cm)

■ APPEARANCE Recognized as a waxwing by crest, black mask, and waxy tips of flight feathers. Larger and grayer than Cedar, with chestnut undertail coverts. Both species have yellow-tipped tail, but wing of adult Bohemian more complexly marked with red, yellow, and white. Young grayer than adult, variably splotched below, but show telltale chestnut under tail.

■ VOCALIZATIONS Flocks call constantly: a loose trill, dropping in pitch—lower, louder, and more jangling than Cedar's.

■ BEHAVIORS Nests widely in conifer forests, especially fir, of Alas. and western Canada. Gregarious year-round, especially winter; flocks bunch tight in treetops and fly in compact formations.

■ POPULATIONS Winter wanderings notoriously erratic. Some winters, can be nearly the most abundant species (Christmas Bird Count data) in a region, then absent for multiple years following. Accidental to Hawaii.

Cedar Waxwing
Bombycilla cedrorum | CEDW | L 7¼" (18 cm)

■ APPEARANCE Smaller and warmer than Bohemian Waxwing; undertail coverts whitish. All ages, including juvenile, have waxy yellow tail tip. Red tips to secondaries acquired with age; male usually has more than female.

■ VOCALIZATIONS Calls, given mostly in flight, are higher, thinner, and more monotone than Bohemian's.

■ BEHAVIORS Where Bohemians and Cedars co-occur, they usually do not mix. Aggregations of either species, perched or flying, suggest starling (p. 340) flocks.

■ POPULATIONS Complex movements include broad-front but ill-defined migration in late spring, frequent irruptions in fall, and unpredictable midwinter wanderings. Birds with red (as opposed to yellow) tail tips occasionally seen; this aberration is diet-related.

SILKY-FLYCATCHERS | PTILIOGONATIDAE

"Flycatcher" may be a bit of a stretch, but "silky" works well for these sleek, soft-plumaged relatives of waxwings. The Phainopepla is the only silky-flycatcher that occurs regularly outside Middle America.

Phainopepla
Phainopepla nitens | PHAI | L 7¾" (20 cm)

■ APPEARANCE Slim and trim with prominent crest. Adult male, also called "Black Cardinal," shining black with red eye; wings flash white in flight. Adult female grayer; juvenile male mottled gray and black.

■ VOCALIZATIONS Primary call a rich, whistled, up-slurred *wurp?* Song, infrequently heard, a simple, soft, rather low warble.

■ BEHAVIORS Nomadic, wandering about open woodlands and arid shrublands; occurs singly or in loose assemblages of a few individuals. Fluttering flight erratic and twisting, recalling Townsend's Solitaire's (p. 344).

■ POPULATIONS Movements complex: breeds in mistletoe-infested mesquite woodlands late winter to early summer, then again in semiarid woodlands in foothills, late spring to summer.

WAXWINGS | SILKY-FLYCATCHERS
BOMBYCILLIDAE | PTILIOGONATIDAE

Bohemian Waxwing

reddish face

gray belly

cinnamon undertail coverts

juvenile

white tips to primary coverts

Bohemian

Cedar

yellow belly

Phainopepla

reddish eye in adults; juveniles brownish

♂

♀

Cedar Waxwing

juvenile

white undertail coverts

♂

prominent white wing patches; grayer in females

NUTHATCHES | SITTIDAE

These woodland birds are distinctively shaped: ovoid, stub-tailed, and spike-billed. Even more distinctive is their behavior: They cling to trunks and boughs, performing a "random walk" up, down, and sideways.

Red-breasted Nuthatch
Sitta canadensis | RBNU | L 5" (13 cm)
- **APPEARANCE** A midsize nuthatch. Bluish above with black cap, white eyebrow, black eyeline; washed orangey below.
- **VOCALIZATIONS** Call a petulant *aaank* or *ehhnk,* often repeated slowly.
- **BEHAVIORS** Most common in conifers. Easily irritated; responds instantly to pishing or whistling.
- **POPULATIONS** Annual movements complex: In some years, most stay on breeding grounds; in other years, fall dispersal heavy.

White-breasted Nuthatch
Sitta carolinensis | WBNU | L 5¾" (15 cm)
- **APPEARANCE** Our largest nuthatch. Blue-gray above; mostly white below, with undertail coverts orangey. Face white with beady black eye; broad dark crown expands to all of nape.
- **VOCALIZATIONS** Mellow song, steady and whistled, suggests a distant flicker (p. 256). Calls vary from harsh *yeern* on Pacific Slope to stuttering *st'st'st'st'* in interior to powerful *yarnk* of eastern bird ranging west into Canadian Rockies.
- **BEHAVIORS** Found in montane conifer forests with other nuthatches, but also at low elevations in old broadleaves.
- **POPULATIONS** Three groupings, best distinguished by calls, may be different species: Pacific Slope *aculeata,* interior *lagunae,* and eastern *carolinensis.*

Pygmy Nuthatch
Sitta pygmaea | PYNU | L 4¼" (11 cm)
- **APPEARANCE** Our smallest nuthatch, with plain, gray-brown cap and gray-buff underparts.
- **VOCALIZATIONS** Sharp *peep!* and *pleep!* notes, like Red Crossbill's (p. 376).
- **BEHAVIORS** In most of range, restricted to ponderosa pine and other yellow pine forests, where it travels in noisy, sociable flocks.
- **POPULATIONS** Pinewoods habitat forecast to dwindle with climate change, with worrisome prospects for this habitat-specialist nuthatch.

TREECREEPERS | CERTHIIDAE

Even more than nuthatches, treecreepers are birds of trunks and large boughs. With their tails pressed against the bark, they jerk their way up a tree, then flutter weakly to the base of another tree and ascend again.

Brown Creeper
Certhia americana | BRCR | L 5¼" (13 cm)
- **APPEARANCE** Body flattened, bill thin and decurved. Upperparts patterned like tree bark; gleaming white below reflects light in crevices, aiding the search for food.
- **VOCALIZATIONS** Call a high-pitched, wavering trill, *ssseee.* Song a short jumble of high, thin whistles.
- **BEHAVIORS** Occurs singly wherever there are old trees with rugose trunks. Cryptic and unobtrusive; easily overlooked.
- **POPULATIONS** Darker Mexican subspecies *albescens,* ranging north to Ariz. and N. Mex., may be a separate species.

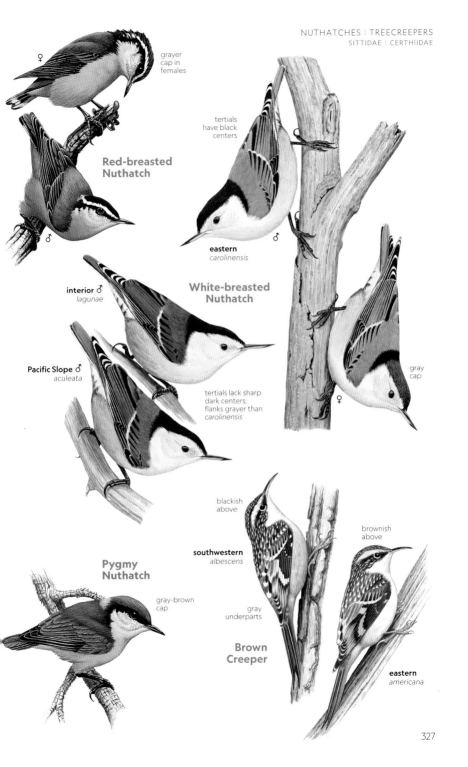

GNATCATCHERS | POLIOPTILIDAE

These bluish sprites occur in thickets, chaparral, and open woodlands. They are active and fidgety, with their long tails often cocked or flipped about.

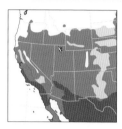

Blue-gray Gnatcatcher
Polioptila caerulea | BGGN | L 4¼" (11 cm)

- **APPEARANCE** Like other gnatcatchers, slight of build with a thin bill and long, thin tail. Pale blue overall, deeper above than below. Folded tail appears black above but mostly white below. Breeding male has thin black eyeline and complete white eye ring; slightly browner female has eye ring but lacks eyeline.
- **VOCALIZATIONS** Call a tinny *bzzzeeee*, nasal and falling in pitch. Song a giddy jumble of thin notes.
- **BEHAVIORS** Nests in woods and scrublands in canyons, along streams, and into foothills; Black-tailed nests at lower elevations, but some overlap.
- **POPULATIONS** Most are migratory, reaching midlatitudes of interior by May, departing in Sept. Breeding range expanding north.

Black-tailed Gnatcatcher
Polioptila melanura | BTGN | L 4" (10 cm)

- **APPEARANCE** Similar in build to Blue-gray Gnatcatcher, a bit duskier overall. Tail feathers black below with broad white tips; tail of Blue-gray appears mostly white below. Breeding male Black-tailed has black cap, reduced in winter, and white eye ring; female similar to female Blue-gray, but note differences in undertail pattern.
- **VOCALIZATIONS** Calls and songs harsher than corresponding repertoire of Blue-gray; some are similar to calls of Bewick's Wren (p. 332).
- **BEHAVIORS** A desert gnatcatcher, frequenting washes and pond edges. Blue-gray on migration and in winter routinely gets into such habitat.
- **POPULATIONS** Western subspecies *lucida* paler than more easterly nominate *melanura*; breakpoint around N. Mex. Bootheel.

California Gnatcatcher
Polioptila californica | CAGN | L 4¼" (11 cm)

- **APPEARANCE** Darker and dingier than Black-tailed. Tail mostly black; from below, shows just narrow white edging to rectrix tips. Like Black-capped, male California has black cap, reduced in winter; eye ring of California less distinct than on Black-tailed. Female California especially dark and dingy.
- **VOCALIZATIONS** Call a thin, nasal *wheeeeen*, like a distant Gray Catbird's (p. 336).
- **BEHAVIORS** Found in West only in chaparral near coast, where conflict with real estate development is intense.
- **POPULATIONS** Closely related to Black-tailed; almost no range overlap in West, but some contact in Riverside Co., Calif.

Black-capped Gnatcatcher
Polioptila nigriceps | BCGN | L 4¼" (11 cm)

Mexican gnatcatcher that barely reaches Ariz. and N. Mex. Bill long and thin. Breeding male has black cap like Black-tailed, but folded tail appears mostly white beneath like Blue-gray. Female like female Blue-gray; in good view or photos, note how rectrix tips of Black-capped are stacked, or graduated, below. Calls intermediate in timbre between those of Blue-gray and Black-tailed. Most sightings in West in recent years in Santa Cruz Co., Ariz.; has bred.

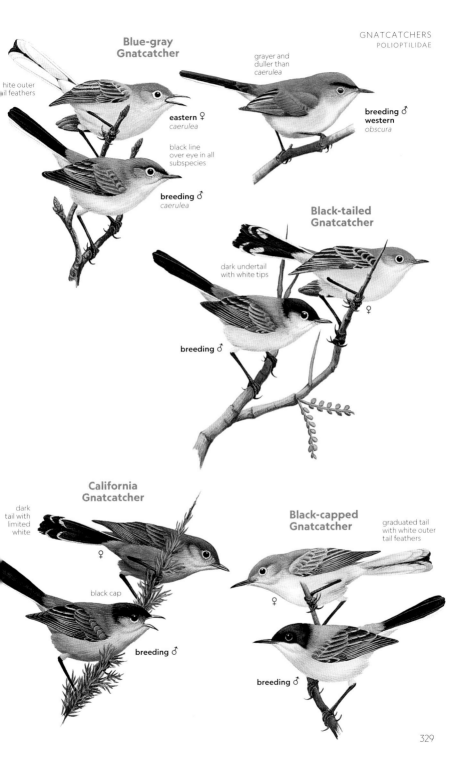

WRENS | TROGLODYTIDAE

They're small, brown, and often hidden from view, but wrens are also fidgety and vocal and usually can be seen with a bit of patience. Their tails, rather long to quite short, are often cocked. The wrens' center of diversity is Middle America, and their closest relatives are the gnatcatchers (p. 328).

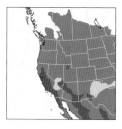

House Wren
Troglodytes aedon | HOWR | L 4¾" (12 cm)
- **APPEARANCE** Small and plump, with medium-length tail. Similar Pacific and Winter Wrens even smaller and notably shorter-tailed. Plain overall: eyebrow faint, barring on flanks indistinct.
- **VOCALIZATIONS** Song a vigorous and exuberant chatter, fading or falling off. Calls variable, but always harsh, *chet* or *chich*, suggesting Marsh Wren's (p. 332); often run together in short series.
- **BEHAVIORS** Common throughout breeding range in thickets, clearings, and forest edges; readily accepts nest boxes. Fall to winter, many make their way to wetlands, where confusion with Sedge (p. 332) and Marsh Wrens possible.
- **POPULATIONS** "Brown-throated Wren," formerly treated as distinct species, has warmer tones and bolder eyebrow; reaches mountains of Southeast Ariz. from Mexico.

Pacific Wren
Troglodytes pacificus | PAWR | L 3½–4½" (9–11 cm)
- **APPEARANCE** In most of range, stub-tailed, slender-billed, and very small. Underparts, especially throat and breast, warm buff; flanks densely barred.
- **VOCALIZATIONS** Amazing song a very fast, long-duration outpouring of soft chirps and thin trills. Call a dry, smacking, sharply descending *tsip* or *chip*, like Lincoln's Sparrow's (p. 404) or Broad-tailed Hummingbird's (p. 92).
- **BEHAVIORS** Most in West nest in moist, old-growth conifer forests; in treeless Bering Sea region, in dense grass. At all seasons, occurs right at ground level, even disappearing beneath the surface in roots and streambanks; scurries about mouselike, always in motion.
- **POPULATIONS** Bering Sea residents, represented by eight (!) subspecies, larger, paler, longer-tailed, and longer-billed. Away from Bering Sea region, including southern and southeastern Alas., partially or fully migratory.

Winter Wren
Troglodytes hiemalis | WIWR | L 3½" (9 cm)
- **APPEARANCE** Size and shape essentially identical to Pacific Wren (discounting the notably larger Pacific Wrens in Bering Sea region). Winter Wren is colder and grayer overall, especially on throat and breast. Both Pacific and Winter are told from larger House Wren by more densely barred flanks, shorter tail, and darker hues overall.
- **VOCALIZATIONS** Song like Pacific's, but more muscular, more musical. Call a shrill, nasal *chimp*, typically doubled: *chimp-chimp ... chimp-chimp ....* Call of Song Sparrow (p. 404) similar, but not usually delivered in incessant doublets like Winter Wren's.
- **BEHAVIORS** Much like Pacific Wren. Breeds in moist, often old-growth conifer forests; winters widely, often winding up in well-wooded floodplains.
- **POPULATIONS** Although it is the "eastern" counterpart of Pacific Wren, nests in Canada west to B.C. and Yukon. Closely related to Pacific Wren and split from that species in 2010; their breeding ranges overlap in the eastern foothills of the Canadian Rockies.

WRENS
TROGLODYTIDAE

House Wren

juvenile

medium-length tail

western *parkmanii*

brown overall with dull face pattern

"Brown-throated Wren"

Pacific Wren

dark above; warm brown below

Pacific Slope *pacificus*

large overall and long-billed

central and eastern Aleutians *kiskensis*

paler overall and grayer below than Pacific Wren, but variable; vocalizations diagnostic

Winter Wren

Sedge Wren
Cistothorus stellaris | SEWR | L 4½" (11 cm)
Rather nomadic close relative of Marsh Wren; at western limit of its range in our area, but gets well into Alta. and probably eastern Mont. in breeding season. Similar to Marsh Wren but smaller and slighter; wings boldly barred and checkered. In fleeting glimpse, note bright tan tones overall; flushing, the bird appears ruddy-tailed. Song comprises bright *chap* notes, accelerating. Call, like a single element from its song, cleaner and clearer than harsh call of Marsh Wren. Nests in wet meadows with sedges and other grasses. Migrants secretive and likely overlooked; annual to Rocky Mountain states, casual to West Coast.

Marsh Wren
Cistothorus palustris | MAWR | L 5" (13 cm)
- **APPEARANCE** A wren of average proportions. Dark cap contrasts with broad white eyebrow. Back dark with white streaks; wings dark rufous. Juvenile duskier and lower-contrast overall, inviting comparison with House Wren (p. 330).
- **VOCALIZATIONS** Gurgling song, with run-on quality, starts with a few stutters, then rambles on; sings at night. Call harsh and scratchy, *chit*.
- **BEHAVIORS** Found year-round in cattail marshes; locally abundant in summer at favored sites. Males build multiple "dummy nests" in addition to those actually used for rearing young; these ovoid nests can be conspicuous. Except when singing, hard to see, but pishing will often bring one out into the open.
- **POPULATIONS** Opportunistic; even within breeding season, may move around a lot, raising successive broods in quite different places. Eastern nominate subspecies group *palustris*, barely reaching eastern Mont., has much smaller song repertoire than *paludicola* group, widespread in West.

Carolina Wren
Thryothorus ludovicianus | CARW | L 5½" (14 cm)
Eastern species with small permanent population in and around Big Bend N.P., Tex. About the same length as Bewick's Wren, but shorter-tailed and more heavyset. Has prominent eyebrow like Bewick's, but more warmly colored: rich chestnut above, bright buffy below. Repetitious song bright and whistled: *cheerily cheerily cheerily* .... Varied calls include a descending trill, a froglike stutter, and a slow buzz. Very vocal; even vagrants routinely sing. Although largely resident range-wide, wanders annually west to foothills of Rockies in Colo. and at least to Rio Grande in N. Mex.

Bewick's Wren
Thryomanes bewickii | BEWR | L 5¼" (13 cm)
- **APPEARANCE** Slender and long-tailed. Gray-brown above, pale gray below. Face marked with prominent white supercilium; tail edged white.
- **VOCALIZATIONS** Many songs include an inhaled-sounding buzz followed by a jangling trill, but variations are endless; a "mystery bird" in its range often proves to be this species. Varied calls include a harsh *jrrrt* and a growling, nasal, slightly rising *grree*.
- **BEHAVIORS** Found in thickets and brush, especially in dry pinyon-juniper woods but also in broadleaf groves along streams and in suburbia. Flips tail about in swishing, sideways motion.
- **POPULATIONS** Interior birds, particularly in Desert Southwest, palest and grayest; Pacific Slope birds darker and more brown-chestnut, especially Ore. northward. Range slowly expanding north, mostly in interior.

WRENS
TROGLODYTIDAE

Sedge Wren

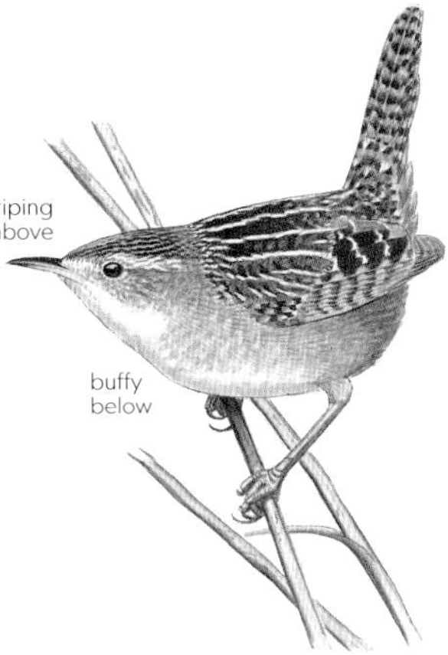

eak striping above

buffy below

Marsh Wren

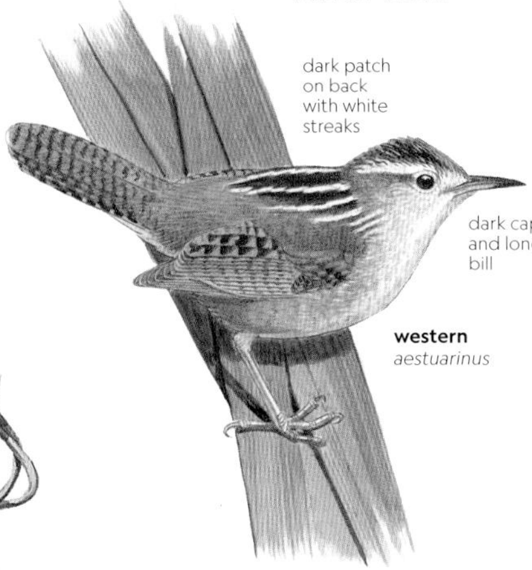

dark patch on back with white streaks

dark cap and long bill

western *aestuarinus*

Carolina Wren

bold white supercilium

richly colored underparts

Bewick's Wren

Pacific Northwest *calophonus*

richer brown than *eremophilus*

pale supercilium

long tail

western interior *eremophilus*

pale gray underparts

Rock Wren
Salpinctes obsoletus | ROWR | L 6" (15 cm)

■ **APPEARANCE** Large, rotund wren; short tail and long bill give it a top-heavy look. Pale and plain, gray overall, with low-contrast plumage; however, flying from rock to rock, splays out tail, revealing broad but broken band of orange-buff at tip.

■ **VOCALIZATIONS** Song incredibly varied, but always involving repeated elements, like a slowed-down Northern Mockingbird (p. 340): *bee ... bee ... bee ... b'dee ... b'dee ... b'dee ... bzheer ... bzheer ... bhzeer ....* Far-carrying call an explosive *ch'PEE!*

■ **BEHAVIORS** Usually seen singly around buttes and mesas with loose rock; also around quarries, rocky catchment basins, even outdoor masonry in suburbia. Engages in short flights among scree or boulders, then perches on a ledge or outcropping, where it bops up and down while surveying its domain.

■ **POPULATIONS** Partially migratory: arrives midlatitudes in interior in Apr., heads back south by late Sept. Northern limit of regular winter range poorly known and possibly shifting northward.

Canyon Wren
Catherpes mexicanus | CANW | L 5¾" (15 cm)

■ **APPEARANCE** Slimmer than Rock Wren, with even longer bill. Rusty overall, but with crown blotchy gray and throat gleaming white.

■ **VOCALIZATIONS** Song as invariant as the Rock Wren's is varied: a cascade of falling whistles, rated by many as one of the most haunting birdsongs on the continent. Call a shrill *beet!*

■ **BEHAVIORS** Magnificently adapted to life on sheer cliffs, often sandstone, in canyon country. The bill, proportionately longer than that of any other songbird in the West, is used for extracting spiders and centipedes from rock crevices; like Brown Creeper (p. 326), reflects light into crevices with white throat. Mouselike, scurries straight *up* rock faces, its orangey plumage perfectly matching the sandstone substrate.

■ **POPULATIONS** Unlike Rock Wren, largely nonmigratory; even in harsh winters, able to subsist on live arthropods. Some move downslope, especially to sunnier microhabitats, in cold weather.

Cactus Wren
Campylorhynchus brunneicapillus | CACW | L 8½" (22 cm)

■ **APPEARANCE** Huge wren, about the same length and heft as Sage Thrasher (p. 340). Large and spotted, with sturdy legs and bill and a broad tail; eyebrow prominent; no other wren in the West has dense spots on the breast like this species. Adult mostly brown, black, and white; juvenile washed buff below. Flying away, with white-cornered tail splayed out, suggests Sage Thrasher.

■ **VOCALIZATIONS** Song a rapid series of harsh *chuh* or *chugga* notes, increasing in loudness; sounds like an old car starting up in the distance. Calls varied, mostly rough and monosyllabic.

■ **BEHAVIORS** Equally at home in subdivisions and remote desert, but thorny plants always a requirement. The nest is strikingly conspicuous—and almost impossible to get close to, ingeniously protected by thorns and cactus spines.

■ **POPULATIONS** Nonmigratory; never wanders from core range. "Coastal Cactus Wren," occurring north in Calif. to at least Orange Co., has paler belly and more white in tail; it is threatened by development.

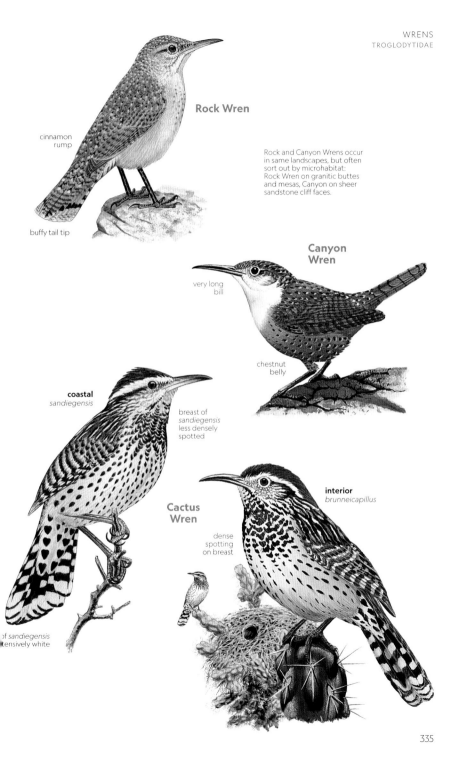

THRASHERS AND MOCKINGBIRDS | MIMIDAE

These midsize to large songbirds are slim with long tails, and several have notably curved bills. They may be hard to see in chaparral and streamside thickets, but they often give the game away with their superb songs; several are accomplished mimics.

Gray Catbird
Dumetella carolinensis | GRCA | L 8½" (22 cm)

■ **APPEARANCE** Slim and long-tailed; often perches erect, with tail pointing straight down. Mostly uniform gray, but with black cap and blackish tail; undertail coverts deep rufous.

■ **VOCALIZATIONS** Famous call a descending, nasal *rrwaaaa*, like a ticked-off cat. Song brilliant and insouciant: a jumble of squeaks, squawks, and whistles, with the "cat's meow" call thrown in for good measure. Other calls include a muffled *chutch* and a harsh cackle.

■ **BEHAVIORS** Skulks in shady thickets, where often hard to see. Breeders in West closely tied to broadleaf riparian habitats, from forests along streams and rivers to wooded canyons with just a trickle of water. Winterers invariably find their way to shrubs with berries.

■ **POPULATIONS** Within West, migration is more latitudinal (east-west) than longitudinal (north-south). Very rare but annual in winter, especially Colo. and N. Mex.

Curve-billed Thrasher
Toxostoma curvirostre | CBTH | L 11" (28 cm)

■ **APPEARANCE** Large and relatively long-billed, but bill length variable: Bill of juvenile almost as short as that of Sage Thrasher (p. 340). Dirty olive-brown all over at all ages, with breast diffusely spotted; eyes fearsome yellow-orange.

■ **VOCALIZATIONS** Song a disorganized, halting warble, a bit like Gray Catbird's song, but not as wild and squeaky. Call an explosive *WEET! WEET!* or *WEET! WEET! WEET!*

■ **BEHAVIORS** Occurs year-round in arid lowlands and foothills, from grasslands to canyons to towns and cities; cholla microhabitat favored, especially for nest placement.

■ **POPULATIONS** Two groups are well differentiated genetically: Sonoran Desert *palmeri*, found in West in Ariz., has indistinct breast spots; Chihuahuan Desert *celsum*, from N. Mex. eastward, has more prominent breast spots. Nonmigratory, but range of *celsum* spreading slowly northward.

Brown Thrasher
Toxostoma rufum | BRTH | L 11½" (29 cm)

■ **APPEARANCE** About the same length as Curve-billed Thrasher, but slimmer and shorter-billed. Warm rufous above from head to tail, unlike any other thrasher in West. Dark streaks below contrast with white background.

■ **VOCALIZATIONS** Run-on song suggests Northern Mockingbird's (p. 340), but Brown Thrasher sings elements twice; mockingbird repeats elements three to six times. Call a smacking, sharply down-slurred *tsack*, like a loud Lincoln's Sparrow's (p. 404).

■ **BEHAVIORS** Occurs in "eastern" habitats like woodland edges and broadleaf plantings around homes. Often feeds on ground, probing and kicking at leaf litter.

■ **POPULATIONS** Fairly common within limited range in West. Expanded in 20th century with changing land-use practices; now breeds west to foothills of Canadian Rockies in Alta. Migrates fairly early in spring.

THRASHERS AND MOCKINGBIRDS
MIMIDAE

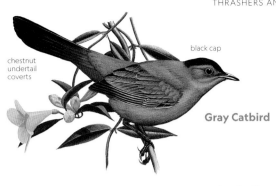

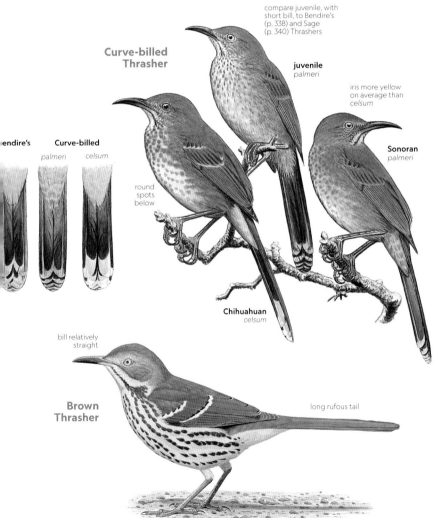

Bendire's Thrasher
Toxostoma bendirei | BETH | L 9¾" (25 cm)

- **APPEARANCE** Intermediate in size and structure between larger Curve-billed (p. 336) and smaller Sage (p. 340) Thrashers. Bill shorter and less decurved than that of Curve-billed; bill of Bendire's pale at base (dark on Curve-billed). Bendire's is warmer gray-brown overall than Curve-billed, with triangular marks on breast; worn Sage Thrasher (summer), with buffy tones and muted breast streaking, similar to Bendire's.
- **VOCALIZATIONS** Song fast and run-on; individual elements, strained and squeaky, delivered three or four times, with no pause between groupings. Calls include an abrupt *wut!* and short, muffled rattle.
- **BEHAVIORS** Scarce; occurs singly or in family groups. Found year-round in arid grasslands with yuccas and Joshua trees.
- **POPULATIONS** Migratory in much of range; returns to breeding grounds by late winter, starts heading back south again by midsummer.

California Thrasher
Toxostoma redivivum | CATH | L 12" (30 cm)

- **APPEARANCE** Large thrasher with strongly decurved bill. Dark-eyed and darker overall than Crissal Thrasher; LeConte's even paler.
- **VOCALIZATIONS** Song not as fast as that of Bendire's; individual elements lower and rougher. Calls include a nasal *pleent* and slurred *chrooey*.
- **BEHAVIORS** Occurs year-round in dense vegetation: in sagebrush-dominated chaparral coastally; in shrub layer of open oak woods in foothills; and in suburbs with adequate plantings.
- **POPULATIONS** Only thrasher in much of range; barely overlaps with Crissal and LeConte's. Nonmigratory; annual cycle influenced by onset of winter rains, with courtship underway by late autumn and nesting soon thereafter.

LeConte's Thrasher
Toxostoma lecontei | LCTH | L 11" (28 cm)

- **APPEARANCE** Similar in size and shape to Crissal Thrasher, but even paler, with dark eye; Crissal has pale eye. Undertail coverts of LeConte's apricot; darker rufous on Crissal.
- **VOCALIZATIONS** Song has relatively slow phrasing of California Thrasher's, but individual elements higher and purer, approaching that of Bendire's in timbre. Call distinct, a rising, whistled *hooweee?*
- **BEHAVIORS** Most extreme of the desert thrashers, often well out in flatlands with only saltbush and bursage. Runs on bare ground, stops with tail cocked; digs for arthropods.
- **POPULATIONS** Nonmigratory; like California Thrasher, courts and nests very early. In long-term decline in San Joaquin Valley, Calif.

Crissal Thrasher
Toxostoma crissale | CRTH | L 11½" (29 cm)

- **APPEARANCE** Similar in build to LeConte's Thrasher, but not as pale; however, eye is pale (dark on LeConte's). Undertail coverts of Crissal deep rusty; pale apricot on LeConte's.
- **VOCALIZATIONS** Song long and halting, with individual elements sweeter than those of California Thrasher; call an airy *cheery cheer*.
- **BEHAVIORS** Occurs in dense shrubbery in washes and foothills; less inclined than LeConte's to get out into open desert.
- **POPULATIONS** Nonmigratory, but some movement downslope and southward after nesting.

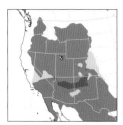

Sage Thrasher
Oreoscoptes montanus | SATH | L 8½" (22 cm)

- **APPEARANCE** Small and stout for a thrasher, with slight bill. Adult in fresh fall plumage has sharp arrow marks, fading by the following summer; worn birds suggest a small, short-billed Bendire's Thrasher (p. 338). Juvenile grayer above than adult, with more diffuse markings below.
- **VOCALIZATIONS** Song a long, run-on warble, with little variation in pitch. Calls include a muffled *chuck* and whistled *churr*.
- **BEHAVIORS** In nesting season occurs mostly in sagebrush; migrants and winterers more generalist, even getting into pinyon-juniper woodlands.
- **POPULATIONS** Migratory; arrives midlatitudes of interior by early Mar., is on the move again by late July.

Introduced to Hawaii

Northern Mockingbird
Mimus polyglottos | NOMO | L 10" (25 cm)

- **APPEARANCE** Long and slender. Gray overall, with thin black eyeline and white flashes in wings and tail; compare with shrikes (p. 292). Adult plain below with golden eye; juvenile, with darker olive eye, spotted below, suggesting Sage Thrasher.
- **VOCALIZATIONS** Amazing song, running on for many minutes, given day or night; individual song elements, including skillful imitations of other species' songs, delivered three to six times. Calls include a powerful *tsuck*, a descending *tssseeep*, and a harsh *rrrwwuuh*.
- **BEHAVIORS** Animated; prances around, flashing wings and tail. Defends winter feeding territories.
- **POPULATIONS** Withdraws from northern portion of range in winter. Introduced to Hawaii; well established on all major islands.

STARLINGS AND MYNAS | STURNIDAE
Indigenous to the Old World, sturnids are often dark and glossy with short tails. Many are superb vocalists, and several have become widely established far from their original ranges.

European Starling
Sturnus vulgaris | EUST | L 8½" (22 cm)

- **APPEARANCE** Compact with short tail, sturdy legs, and long, wedge-shaped bill; outspread wings triangular in flight. Adult in fresh fall plumage black with dense pale spots; glossy purple and green by spring. Bill dark fall to winter, yellow spring to summer. Juvenile dull dusky gray-brown all over.
- **VOCALIZATIONS** Fantastically varied; mimics other species. Song high-pitched, with clacks, rattles, and whistles. Calls include a squeaky *cheek* and rougher *chuuh* and, in flight, a purring *zzzzzzrrr*. Juvenile begs with down-slurred *jeeeuurrr*.
- **BEHAVIORS** Aggressive; nests in cavities, often displacing other species. Especially in nonbreeding season, gathers in immense roosts.
- **POPULATIONS** Introduced from Europe. Most withdraw in winter from northernmost portions of range.

Common Myna
Acridotheres tristis | COMY | L 10" (25 cm)

Indigenous to South Asia; introduced to Hawaii, now well established on all main islands. Clad in browns and blacks, with white wing patch and tail tip; bare parts, including exposed skin around eye, yellow. Calls sharp and whistled, often in short series, with strained quality; sometimes imitates other birds. Common; seen around parks and resorts, also roadsides and pastures.

DIPPERS | CINCLIDAE

Numbering only five species, this family is nevertheless widely distributed, being found in Eurasia, northern Africa, and the Americas. Dippers are rotund, bobtailed, and long-legged; supremely adapted to aquatic environments, they are always found along—and often submerged in—swift streams and rivers.

American Dipper
Cinclus mexicanus | AMDI | L 7½" (19 cm)

- **APPEARANCE** Adult uniform slaty with dark bill. Young paler below, with yellowish bill that darkens with age.
- **VOCALIZATIONS** Song, given year-round, is wild, long-duration, and thrasher-like. Flushes with shrill *dzeet*, often doubled or trebled.
- **BEHAVIORS** Mesmerizing: plunges headfirst into icy rapids; swims powerfully upstream; bobs ("dips") on rocks, fallen logs, or edge of ice.
- **POPULATIONS** In winter, pushed down from highest-elevation streams when completely iced over.

THRUSHES | TURDIDAE

With close to 200 species worldwide, the familiar thrushes haunt terrestrial, often wooded landscapes. Their diets trend omnivorous, and many are gifted songsters. They are generally strong-legged and small-billed, with plump and portly builds.

Eastern Bluebird
Sialia sialis | EABL | L 7" (18 cm)

Despite being "eastern," reaches southern Ariz. by way of Mexico; also reaches periphery of our region Mont. to N. Mex. Throat of male rich ruddy, paler on female; compare with Western Bluebird. Upperparts of both sexes lack rufescent intrusions shown by Western. Song a simple but joyous gurgle; flight call twangy and musical, *cheer-lee* or *chur-lee*.

Western Bluebird
Sialia mexicana | WEBL | L 7" (18 cm)

- **APPEARANCE** Darker than Mountain Bluebird, usually with obvious rufous; both sexes gray-bellied. Male has indigo hood; back and wings also indigo, with rufous intrusions. Female gray-hooded with paler throat and pale eye ring. Many female Mountains in winter also are washed rusty below.
- **VOCALIZATIONS** Year-round, gives muffled *pew* and *pyoof* notes; simple song consists of repeated *pew* notes.
- **BEHAVIORS** Like other bluebirds, a cavity nester; accepts both tree holes and nest boxes. More of a woodland species than Mountain Bluebird.
- **POPULATIONS** Negatively impacted by snag removal, fire suppression, and other efforts to "beautify" woodlands.

Mountain Bluebird
Sialia currucoides | MOBL | L 7¼" (18 cm)

- **APPEARANCE** Long-winged and thin-billed. Adult male intensely sky blue; female much grayer, but nevertheless with blue highlights in wings and tail. Many in winter suffused rusty below, suggesting Western.
- **VOCALIZATIONS** Calls, weak and muffled, midway in timbre between ringing Eastern's and gruff Western's.
- **BEHAVIORS** More of an open-country species than the other bluebirds; often seen in shrubs in desert and high mountain meadows.
- **POPULATIONS** Early spring migrant. Winter range shifting north as climate warms.

Townsend's Solitaire

Myadestes townsendi | TOSO | L 8½" (22 cm)

- **APPEARANCE** Slender; long-tailed and small-headed. Buff-orange in wings inconspicuous on perched bird but prominent in flight. Gray tones, eye ring, and wing markings may invite confusion with *Contopus* (p. 274) and *Empidonax* flycatchers (pp. 276–281). Juvenile boldly scalloped above and below.
- **VOCALIZATIONS** Song finchlike, a herky-jerky warble. Call a strange, whistled *eee*, in long, slow sequence, given while perched; also a thin, drawn-out buzz, given in flight.
- **BEHAVIORS** Found in conifer forests. Watches from exposed perch for flying insects; catches prey in high, twisting flight. Also feeds on fruit; hover-gleans awkwardly. Both sexes defend winter feeding territories with much fighting and singing.
- **POPULATIONS** Complexly migratory: Some stay behind in winter; others disperse widely in search of food, especially juniper berries.

Omao

Myadestes obscurus | OMAO | L 7" (18 cm)

Endemic to Hawaii I.

- **APPEARANCE** Plain, short-tailed solitaire, smaller and plumper than Townsend's, but no range overlap; Omao is the only thrush on Hawaii I. Adult cold brown above, smooth slate gray below; juvenile spotted below.
- **VOCALIZATIONS** Song a short outburst of sharply rising and falling whistles. Varied calls include a shrill, buzzy, slightly rising *bzhrrrr*, slowly repeated.
- **BEHAVIORS** Occurs widely in forests, but with decided affinity for wet, high-elevation tracts; seen singly and in small family groups, both in treetops and closer to ground. Flutters in front of fruiting trees; also spends long periods of time perched motionless.
- **POPULATIONS** Found only on south and east slopes of Hawaii I. Has declined from historic levels because of habitat loss, but still one of the more common indigenous passerines in the region. Nonmigratory, but a few wander downslope.

Puaiohi CR

Myadestes palmeri | PUAI | L 6½" (17 cm)

Endemic to Kauai

- **APPEARANCE** Even smaller than Omao; an alternative name, Small Kauai Thrush, is morphologically, geographically, and taxonomically efficient. Adult similar in plumage to adult Omao (no range overlap) but has pale eye ring and dark "whisker"; legs pale (dark on Omao). Juvenile heavily spotted below.
- **VOCALIZATIONS** Name is onomatopoetic; song short with whistled elements, *pu-a-i-o-hi*, with notes mostly on same pitch (song of Omao spans greater frequency range). Calls include a sharp, annoyed buzz, *chzzzt*, wrenlike.
- **BEHAVIORS** Restricted to extremely wet, high-elevation forests where streams course through steep ravines. More active than Omao; often engages in flycatching; perched, frequently cocks its tail.
- **POPULATIONS** Critically endangered; total number unknown, but surely fewer than 1,000. Occurs only in 14-square-mile (36 km^2) Alaka'i Wilderness Preserve of Kauai. Has likely always been rare, but range has contracted. At least five species of *Myadestes* solitaires have occurred in Hawaii, but only Omao and Puaiohi remain. Although not as famous as the Hawaiian honeycreepers (pp. 362–369), they are nevertheless an impressive case study for adaptive radiation.

THRUSHES
TURDIDAE

Veery
Catharus fuscescens | VEER | L 7" (18 cm)
- **APPEARANCE** Uniform rusty buff above; plain face has faint, broken eye ring. Below, soft butterscotch wash across lightly spotted, pale gray breast; belly white.
- **VOCALIZATIONS** Song starts with gruff rising note *(rrwaa?)*, then gushes downward with slurred *reeur* notes: *rrwaa? ree-ur ree-ur ree-ur ree-ur ree-ur ree-ur.* Low-pitched flight call rough and burry: *vrruurrr.*
- **BEHAVIORS** Nests near ground in wet clearings in broadleaf groves, especially around bogs and beaver dams.
- **POPULATIONS** Veeries in West have heavier spotting below and are less warmly colored than those in the East; compare especially with Swainson's Thrush on Pacific Slope. Rarely detected on migration in West.

Gray-cheeked Thrush
Catharus minimus | GCTH | L 7¼" (18 cm)
- **APPEARANCE** Cold gray-brown above, with vague, broken eye ring; gray breast densely spotted. Swainson's, with "spectacles" (including complete eye ring), has brighter, buffier breast. Hermit, with weak but complete eye ring and fairly dense spots below, quite similar to Gray-cheeked; reddish tail of Hermit contrasts with colder gray back and relatively short wings.
- **VOCALIZATIONS** Song, descending weakly, comprises one or two rough notes followed by slurred phrases, thin and nasal: *uh uh ssweew sweew sweeah sweeah sweeeurp.* Buzzy flight call an urgent *queeerrr!*
- **BEHAVIORS** Nests in spruce-fir forests, often near tundra or timberline. Rarely seen away from nesting grounds.
- **POPULATIONS** Incredible migrant, with some crossing Bering Strait to nest in Russia.

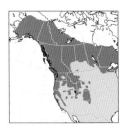

Swainson's Thrush
Catharus ustulatus | SWTH | L 7" (18 cm)
- **APPEARANCE** All have buffy "spectacles" (complete eye ring and lores) and buffy infusion on throat. Gray-cheeked has plain gray face with only vague, broken eye ring; breast of Gray-cheeked washed cold gray. Compare especially with Veery in Pacific Northwest.
- **VOCALIZATIONS** Song spirals upward, intensifying: *eewee eweee eweeah eeweewee eeeweewee! aurelia!* Call note a resonant *pilp;* flight call, often heard at night, a mellow, rising *pweee.*
- **BEHAVIORS** Nests in habitats like Veery, from mountains down to sea level. Migrants stop anywhere, including cities and desert oases.
- **POPULATIONS** "Russet-backed" subspecies (Pacific Slope), warmer overall and less boldly marked, similar to Veery.

Hermit Thrush
Catharus guttatus | HETH | L 6¾" (17 cm)
- **APPEARANCE** Variable, but all have rufous tail and thin but complete eye ring; breast and throat with small blackish spots.
- **VOCALIZATIONS** Song opens with achingly pure whistle, then goes into an airy, tinkling jumble like wind chimes. Calls include a hollow *tuck,* a rising, whining *rrweeeen,* and, in flight, a simple monotone whistle.
- **BEHAVIORS** Nests in conifer forests, from pines in foothills to spruce-fir near timberline. Perched, slowly raises and then quickly lowers its tail; Gray-cheeked never does this.
- **POPULATIONS** Interior West breeders grayish above and fairly large; Pacific Slope breeders smaller and darker. Only *Catharus* thrush in winter in West.

THRUSHES
TURDIDAE

Veery

compare *salicicola* Veery with *ustulatus* Swainson's Thrush

western
salicicola

eastern
scescens

faintly spotted breast

most immature (first-year) *Catharus* thrushes have buff tips to wing coverts

Gray-cheeked Thrush

gray lores

1st fall

grayish-brown flanks

Swainson's Thrush

buffy "spectacles"

upperparts with russet cast

"Russet-backed" Pacific Slope
ustulatus

Alaska
incanus

olive-gray upperparts

"Olive-backed" interior
swainsoni

all juvenile *Catharus* thrushes have spots below

juvenile
faxoni

eastern
faxoni

Hermit Thrush

brownish flanks

all subspecies have rufous tail and thin eye ring

Interior West
auduboni

Pacific
guttatus

auduboni large and pale

guttatus small and dark

Eyebrowed Thrush
Turdus obscurus | EYTH | L 8½" (22 cm)
Asian relative of American Robin; regular in spring to western Aleutians, sometimes in numbers (exceptional one-day count of 180 in 1998). Both sexes mostly olive-brown with white belly and white eyebrow; head of male grayish, head of female olive-gray. Vagrants to West usually silent, but some give a high, wavering, buzzy flight call. Very rare at best in Alas. in fall and away from western Aleutians; accidental to Calif. and Hawaiian Is.

Dusky Thrush
Turdus eunomus | DUTH | L 9½" (24 cm)
Like Eyebrowed Thrush, an Asian species that regularly reaches our area in Bering Sea region. More boldly marked than Eyebrowed, with thicker eyebrow, bold blackish and white on face, extensive rufous in wings, and blackish spotting below. Call a sharp *pleck*. Fewer reach the West than Eyebrowed, but Dusky vagrates more broadly, with records for Ore., Wash., B.C., Yukon, and widely scattered locales in Alas. away from Bering Sea.

Rufous-backed Robin
Turdus rufopalliatus | RBRO | L 9¼" (23 cm)
West Mexican robin that wanders annually in winter to southern Ariz. Differs from American Robin by trademark rufous on back (also wing coverts) and prominent black streaking on throat and upper breast. Varied calls include a high buzz and lower whistle. Has strayed to Calif., Utah, Colo., N. Mex., and West Tex.

American Robin
Turdus migratorius | AMRO | L 10" (25 cm)
- **APPEARANCE** Large thrush with "robin redbreast" rufous below, dark on males, orangey on females; red or orange color below extends to under-wing coverts, obvious in flight. Bill yellow; throat and vent mostly white. Juvenile heavily spotted below.
- **VOCALIZATIONS** Song loud and exuberant, a series of four to eight slurred whistles. Many calls include a descending *WHEE WHEE whew whew whew whew*, a scolding *tut tut tut*, an explosive *PLEEP*, and, in flight, a buzzy *dzeeeet*.
- **BEHAVIORS** Found anywhere, from remote wilderness to our largest cities: hunts worms on lawns; gorges on berries in fruiting trees; nests on homes and other structures. Occurs singly, in family groups, and in large flocks.
- **POPULATIONS** Common to abundant in much of range. Highly migratory; even in regions where present year-round, seasonal movements pronounced.

Varied Thrush
Ixoreus naevius | VATH | L 9½" (24 cm)
- **APPEARANCE** Stockier and shorter-tailed than American Robin, with high-contrast orange-and-blackish plumage. Adult male has black breastband and face mask; breastband and face mask duskier on adult female. Both sexes have extensive orange in wings, prominent in flight.
- **VOCALIZATIONS** Song a long-duration, very buzzy whistle with odd mechanical or electronic quality. Call a hollow, down-slurred *tsooh* or *chook*.
- **BEHAVIORS** Nests in deep conifer woods, seeming to favor the darkest, dankest haunts.
- **POPULATIONS** Migratory, with most birds simply dropping straight south; however, a few spread out to interior and even east of Rockies.

OLD WORLD FLYCATCHERS | MUSCICAPIDAE

This huge family, numbering 350+ species globally, is, as its name implies, of rather peripheral occurrence in the Americas. Our muscicapids occur on or near the ground, supported by stout legs. Despite the name, they are not closely related to tyrant flycatchers (pp. 266–283).

Established Kauai, Oahu, and Maui

White-rumped Shama
Copsychus malabaricus | WRSH | L 10" (25 cm)

Indigenous to South Asia; introduced to main islands of Hawaii, where widely established except, oddly, on Hawaii I. Terrestrial; common and confiding around yards, gardens, and tourist traps. Tail very long, often cocked; white rump prominent in all plumages. Orange-and-blue male unmistakable. Female and immature, much browner, recognized by long tail, gleaming white rump, terrestrial haunts, and, relative to indigenous honeycreepers (pp. 362–369), large size. Song relatively low-pitched, comprising rich whistles, gurgles, and imitations of other birds. May be a reservoir for avian malaria, devastating indigenous Hawaiian bird species.

Bluethroat
Cyanecula svecica | BLUE | L 5½" (14 cm)

■ **APPEARANCE** Small, stout, and bobtailed. All plumages show white eyebrow and rufous in tail, prominent in flight. Breeding male dazzling; other plumages more muted.

■ **VOCALIZATIONS** Song an exuberant jumble of whistles and buzzes, with halting quality overall. Call a wrenlike *clack*, often given in quick succession.

■ **BEHAVIORS** Active; runs around, bobs and spreads tail, but stays close to cover. Found amid shrubs and tussocks at edge of tundra, in denser vegetation than Northern Wheatear.

■ **POPULATIONS** Most breeders in West funnel narrowly across Bering Strait on migration to and from Southeast Asia.

Siberian Rubythroat
Calliope calliope | SIRU | L 6" (15 cm)

Asian species that vagrates annually to Bering Sea region. Build similar to slightly smaller Bluethroat; Bluethroat paler below. Cocks tail. Extent of ruby on throat variable; all plumages show white moustache and eyebrow. Some females are quite plain. Vagrants typically silent but sometimes give a thin, slurred whistle or a gruff *cluck*. More sightings in Alas. spring to early summer than fall.

Northern Wheatear
Oenanthe oenanthe | NOWH | L 5¾" (15 cm)

■ **APPEARANCE** Small; portly and short-tailed, with upright posture. Adult male pale gray above, pale buffy below; wings and face mask blackish. Female similarly patterned, but duskier overall with more extensive buff; wings and face mask not as dark. In all plumages, tail distinctive in flight: broad and splayed out, with coverts and basal (proximal) half of outer rectrices white but central rectrices and distal half of outer rectrices black.

■ **VOCALIZATIONS** Song a short, stuttering jumble; calls short, sharp whistles and clucks.

■ **BEHAVIORS** Restless denizen of fairly dry, open country, especially stony tundra. Flits among loose rocks, calling and sometimes singing in flight.

■ **POPULATIONS** Breeders in West are nominate subspecies *oenanthe*, astonishing migrants that winter in Africa.

OLD WORLD FLYCATCHERS
MUSCICAPIDAE

OLIVE WARBLER | PEUCEDRAMIDAE

This family comprises a single species found in pine forests of Middle America and the southwestern U.S. It is not a wood-warbler, family Parulidae (pp. 424–443), and it is not even particularly closely related to the parulids.

Olive Warbler
Peucedramus taeniatus | OLWA | L 5¼" (13 cm)

■ **APPEARANCE** Similar in size, build, and color to wood-warblers. All plumages have white wing bars and tail edges: Adult male gray overall with orange-rust hood and black mask; female has yellower hood with fainter mask; immature paler, rather gray-white, but with overall suggestion of adult pattern.

■ **VOCALIZATIONS** Song a bright, rich *peedo peedo peedo* or *cheery cheery cheery*. Call a down-slurred whistle, *tyee-eeo*.

■ **BEHAVIORS** Forages actively in treetops, typically longleaf pines, in a manner suggesting Grace's Warbler (p. 438). Another name for the species, "Ocotero" ("of the pines"), nicely indicates the biology of *P. taeniatus*, neither a warbler nor olive-colored.

■ **POPULATIONS** Restricted in U.S. to mountains of N. Mex. and Ariz.; found in winter in our region only close to border with Mexico. Vagrant to West Tex.

WEAVERS | PLOCEIDAE

Numbering more than 100 species, the weavers hail from the Old World, mostly sub-Saharan Africa. Most are vocal and sociable, and many build elaborate nests bunched close together in trees or large shrubs. Several species have become established well outside their indigenous ranges.

Northern Red Bishop
Euplectes franciscanus | NRBI | L 4" (10 cm)

Indigenous to equatorial Africa; introduced and established in Los Angeles area, where found singly or in small flocks, in weedy places. Small and rotund; short-tailed and thick-billed. Breeding male brilliant scarlet and black; other plumages, pale and buffy with blurry streaks, oddly suggestive of Grasshopper Sparrow (p. 388). Calls high and short, buzzy and scratchy; often in quick series of three to five notes. Taxonomy and nomenclature for this species are confused; often called Orange Bishop. Escapes of other bishop species—for example, Yellow-crowned, *E. afer*—sometimes seen in the wild in our area.

WHYDAHS | VIDUIDAE

Closely related to the weavers, this family is indigenous to Africa. Like the Brown-headed Cowbird (p. 420), whydahs and other viduids are obligate brood parasites, meaning they lay their eggs in other species' nests. Male whydahs are astonishingly long-tailed when breeding.

Pin-tailed Whydah
Vidua macroura | PTWH | L 4–12" (10–30 cm)

Indigenous to sub-Saharan Africa; introduced and established in Los Angeles area, with escapes sometimes noted elsewhere in West. Breeding male distinctive but occasionally misreported as Fork-tailed Flycatcher (p. 461). Other plumages relatively short-tailed and pale buffy overall, with black stripes on head. Call a rapid, scratchy chatter. Lays eggs in nests of Scaly-breasted Munias (p. 354), in a novel host-parasite relationship involving nonindigenous species with nonoverlapping indigenous ranges.

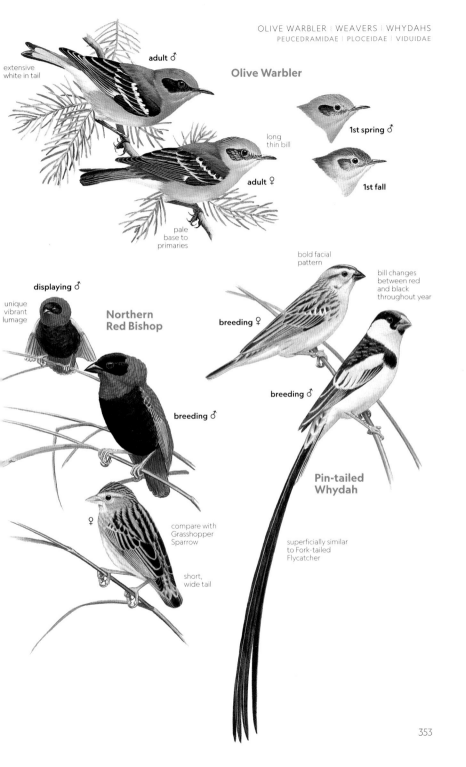

WAXBILLS | ESTRILDIDAE
These very small songbirds are indigenous to the Old World, especially at lower latitudes. Numbering well over 100 species, they are popular cage birds; many have been released from captivity in the Americas. Successful establishments here tend to be in brushy and weedy habitats, often near humans.

African Silverbill
Euodice cantans | AFSI | L 4" (10 cm)
Indigenous to equatorial Africa and Arabian Peninsula; introduced and established in Hawaii (all main islands). Adult soft brown with weak vermiculations; juvenile plainer. Bill large and blue-gray; thin eye ring also blue-gray. Call a high, descending *chik,* often run together in rapid chatter. Roams about in energetic flocks, often in degraded, weedy habitats.

Java Sparrow 🆎
Padda oryzivora | JASP | L 6" (15 cm)
Endangered in indigenous range (Indonesia), but widely introduced, including Hawaii (all main islands). Our largest estrildid. Adult gray with huge reddish bill and large white oval on face; juvenile pale gray-brown with dusky bill. Calls include a squeaky *squenk* and a fast twitter. Widespread, including parks, resorts, even restaurants.

Scaly-breasted Munia
Lonchura punctulata | SBMU | L 4½" (11 cm)
Indigenous to South and Southeast Asia. Introduced Hawaii, established all main islands; also Calif., where well established along southern coast, has reached San Francisco, and is still expanding. Bill large and dark. Adult warm brown above, scaly black and white below; juvenile pale gray-brown below. Calls high and squeaky: *chee, t'cheet,* etc. Common in brushy places; loves the Los Angeles R.

Chestnut Munia
Lonchura atricapilla | CHMU | L 4¼–4¾" (11–12 cm)
Indigenous to Southeast Asia. Established Hawaii; present on main islands, although scarce on Hawaii I. Size and shape like Scaly-breasted Munia; large bill is blue-gray. Body mostly uniform deep chestnut; head and breast darker. Calls, squeaky and rising, like Scaly-breasted's. Flocky; found in brushy places and along roadsides.

Red Avadavat
Amandava amandava | REAV | L 4½" (11 cm)
Indigenous to South Asia. Introduced to Hawaii; present on main islands, but scarce on Maui. Adult male red with white spots, like a strawberry; adult female and nonbreeding male mostly gray, but with red bill, eyes, and uppertail coverts. Call a high, piercing, descending *tseeeoo.* Singles and small flocks found in tallgrass pastures and along roadsides.

Common Waxbill
Estrilda astrild | COWA | L 4" (10 cm)
Indigenous to southern Africa. Introduced to Hawaii; all main islands, abundant on Oahu. Both sexes have red bill and face mask. Adult male vermiculated gray and cold brown; adult female also vermiculated, but with warmer tones. Call down-slurred, squeaky and harsh, typically in fast series by many birds at once. Like other estrildids in Hawaii, found mostly in pastures and along roadsides, but also gets into woodland edges.

OLD WORLD SPARROWS | PASSERIDAE

Small, stout, and brown, the passerids are exemplary sparrows. Their name even derives from the Latin word for sparrow, *passer*. However, they are not particularly closely related to the New World sparrows, family Passerellidae (pp. 386–409).

Introduced to Hawaii

House Sparrow
Passer domesticus | HOSP | L 6¼" (16 cm)
- **APPEARANCE** Compact; wings and tail short. Adult male has whitish cheeks, gray crown, chestnut nape, and black bib; paradoxically, fresh feathers in fall wear down to spiffy-looking breeding plumage in spring. Sandy brown female has broad buffy eyebrow, white wing bar, and plain breast.
- **VOCALIZATIONS** Song sounds like monotonous chirps to humans, but each utterance is caroled; our brains can't hear "fast enough" to process the variation and musicality in the song. Calls include a muffled *chiff*, a nasal *laah* or *laah-laah*, and a dry rattle.
- **BEHAVIORS** Cheery and sociable; almost always around humans, from farm country and roadside rest stops to major cities.
- **POPULATIONS** Indigenous to Eurasia and North Africa. Mainland birds spread after introduction to New York in 1840s; Hawaiian birds were brought in 1870s from a population previously introduced to New Zealand. House Sparrows in the Americas have diverged morphologically from those in Eurasia, an important result for evolutionary biology.

WAGTAILS AND PIPITS | MOTACILLIDAE

Slender and terrestrial, the motacillids are often seen strutting across shores, pastures, and tundra. They are thin-billed and long-tailed, and their flight calls can be very useful in field identification.

Accidental to Hawaiian Is.

Eastern Yellow Wagtail
Motacilla tschutschensis | EYWA | L 6½" (17 cm)
- **APPEARANCE** Slim and delicate. Outer tail feathers mostly white, prominent in flight. Adult olive-gray above, extensively yellow below; dark auriculars ("ears") bordered by thin white eyebrow and variably whitish throat. Young are less colorful but nevertheless show white eyebrow and wing bars; most are pale below with just a hint of dusky yellow.
- **VOCALIZATIONS** Song a fast, short series of shrill, descending chirps. Flight call a piercing, buzzy *tk'ZZZt!*
- **BEHAVIORS** At all seasons, found in open habitats; nests on low-elevation tundra. Like other motacillids, walks purposefully, pumping its tail; even the drabbest birds recognized by their movements.
- **POPULATIONS** Most winter in Asia, but a few make their way down the West Coast; fewer records in recent years, though. Accidental to Hawaiian Is.

White Wagtail
Motacilla alba | WHWA | L 7¼" (18 cm)
Asian species with breeding range that barely, and irregularly, reaches western Alas. Even longer-tailed and a bit larger than Eastern Yellow Wagtail. Plumage variation complex, but all are gray, black, and white; face mostly white, bib mostly black, and wings with broad white panel. Alas. breeders are gray-backed subspecies *ocularis;* black-backed subspecies *lugens* ("Black-backed Wagtail") annual vagrant to Bering Sea region. Flight call a sharp *pip* or *peep*, often doubled or trebled; not buzzy. Annual fall to winter on West Coast (both *ocularis* and *lugens*), with more records than Eastern Yellow. Accidental to Hawaiian Is.

OLD WORLD SPARROWS | WAGTAILS AND PIPITS
PASSERIDAE | MOTACILLIDAE

Accidental to Hawaiian Is.

Red-throated Pipit
Anthus cervinus | RTPI | L 6" (15 cm)

- **APPEARANCE** A bit more compactly built than American Pipit. Legs and bill pale; tail edged white. Olive back has several white stripes. Head and throat of breeding male washed in orange-red, reduced in breeding female. Nonbreeding adult and immature resemble American Pipit, but note pale legs and white stripes above; most show at least a hint of a pale orangey wash on breast.
- **VOCALIZATIONS** Song fast and repetitious: one element repeated many times, then another element repeated many times, then another, with no pause at all between phrases; runs on for 15 seconds. Flight call piercing, rises very sharply then slurs downward, slightly wavering: *t'sseee*.
- **BEHAVIORS** Nests in peat meadows with sedges and dwarf willow. Migrants and winterers often on shortgrass near coast. Pumps tail as it walks.
- **POPULATIONS** Rare but regular fall to winter on West Coast, especially Calif. (also to Baja peninsula); usually seen singly. Accidental to Hawaiian Is.

Accidental to Hawaiian Is.

American Pipit
Anthus rubescens | AMPI | L 6½" (17 cm)

- **APPEARANCE** Slim, long-tailed, and slender-billed. Legs and bill dark on most; dark tail flashes white edges in flight. Back plain olive-brown, unstreaked. Freshly molted in fall, shows bold streaking below; by following spring to summer, plainer below with pumpkin-orange wash. All pipits have long tertials, covering primary tips on bird at rest.
- **VOCALIZATIONS** Flight call, often the first cue to the bird's presence, a sharp, down-slurred *psleet* or *psleeit*, often doubled (*pipit*). Flight call of Horned Lark (p. 308) sweeter and thinner. Underrated song a powerful, pulsing series given in flight low above the ground.
- **BEHAVIORS** Nests on tundra, often dry; away from the breeding grounds, favors beaches, muddy pastures, and lake edges. Pumps tail as it walks.
- **POPULATIONS** Breeders in Rockies and the Sierra Nevada (*alticola*) brighter below with almost no streaking in summer. Asian subspecies (*japonicus*), with paler bill and legs, similar to Red-throated Pipit; *japonicus* rare but annual to Bering Sea region, accidental to West Coast, Hawaiian Is.

Sprague's Pipit
Anthus spragueii | SPPI | L 6½" (17 cm)

- **APPEARANCE** Plump; stockier and shorter-tailed than American Pipit. Legs and bill pale; tail edged white. Beady black eye stares from plain face. Upperparts and breast streaked. Juvenile Horned Lark (p. 308), common where Sprague's occurs, is similarly streaked and scaly, but has dark legs, longer tail, and different calls.
- **VOCALIZATIONS** Incredible song, given in towering flight, gushes downward; suggests Veery's (p. 346), but longer duration. Sings at such altitude that the bird can be difficult to see. Flight call a sharp *pleet*, often doubled.
- **BEHAVIORS** Secretive; difficult to detect except when singing. Migrants and winterers hide in tall grass; are best detected by their flight calls. Does not pump tail when walking.
- **POPULATIONS** Declining. Migratory routes in West poorly known. Wintering grounds in West neatly coincident with those of Baird's Sparrow (p. 402), another secretive species.

FINCHES | FRINGILLIDAE

Most are small, many are attired in reds and yellows, and several are famously nomadic. Bill structure and flight calls are often important in identification. The Hawaiian honeycreepers (pp. 362–369), a lineage deep within the finch family, are perhaps the most celebrated example of adaptive radiation in all of biology.

Brambling
Fringilla montifringilla | BRAM | L 6¼" (16 cm)

Widespread in forests of northern Eurasia; annual in small numbers to Bering Sea region, especially outer Aleutians. Bill stout and conical. Breeding male has extensive black above. Other plumages grayer, but all have orangey wash on breast, bright orange wing patch, and black-spotted flanks. Flight call a rising *wink*, often doubled or trebled; also gives a muffled *wup* or *woop*. One of the most frequent Asiatic vagrants to Canada and the lower 48, with notable accumulation of records from B.C. to Ore. Accidental also to Hawaii; even shows up on container ships at sea.

Evening Grosbeak
Coccothraustes vespertinus | EVGR | L 8" (20 cm)

- **APPEARANCE** Large finch; stocky and huge-billed. All have golden hues and dark wings with white patches. Male relatively dark, with yellow brow; white in wing confined to secondaries. Female paler overall, with gray on head and back, and apple-green bill; has broad white patch on inner secondaries like male, but also white in primaries.
- **VOCALIZATIONS** Far-carrying, descending *CLEEP*, often given in flight. Also a burry *zrrrrt*, especially when feeding.
- **BEHAVIORS** Nests in mixed broadleaf-conifer forests in mountains and foothills. Where available, box elder, *Acer negundo*, seeds (samaras) are a favored foodstuff. Pairs and small flocks feed quietly and unobtrusively spring to early fall. Completely the opposite in winter: Nicknamed "Flying Pigs" or "the Motorcycle Gang," they clean out bird feeders in a riot of sound, color, and voraciousness.
- **POPULATIONS** Many stay in montane forests in winter, but others wander widely; a few of our winterers likely come from the East. Geographic variation complex, with multiple "types" possibly indicating different species.

Pine Grosbeak
Pinicola enucleator | PIGR | L 9" (23 cm)

- **APPEARANCE** Our longest finch; long-tailed, with a stubby, rounded bill. All have white wing bars. Older males extensively red-pink with gray highlights; younger males grayer. Females variably clad in yellows, olives, and even greens; some adult females and young are warmer russet.
- **VOCALIZATIONS** Call a short, desultory warble: *t'leedle* or *t'lweeooli*. In flight, a short, slurred whistle.
- **BEHAVIORS** Occurs year-round in boreal forests; despite name, more strongly associated with spruce-fir than pine in large swaths of range. Amazingly tame; along trails in woods and at feeders, singles and small flocks are easily approached. Actions sluggish and deliberate; some greenish females, with their exaggeratedly slow movements and rounded bills, are oddly parrotlike.
- **POPULATIONS** Only weakly irruptive. Most stay well within conifer forests in winter, but a few wander to lowlands.

Akikiki 🆑

Oreomystis bairdi | AKIK | L 5" (13 cm)

Imperiled Kauai endemic; only a few remain in Alaka'i Wilderness Preserve, and the species' survival prospects are poor. Compactly built, with short tail, stout, decurved, orange-pink bill, and sturdy, orange-pink legs; plumage plain gray. Active: Feeding behavior and general body plan recall nuthatches (p. 326). Calls include a short, rising whistle and huskier, rising *schwiit*. Occurred in small flocks until recently; so few left that only singles and pairs are seen.

Maui Alauahio 🆔

Paroreomyza montana | MAAL | L 4½" (11 cm)

Endemic to Maui

■ **APPEARANCE** Built like Akikiki; compact, with short tail and stout, decurved bill. Adult male olive above extending to crown and yellow below extending to face; recalls female Wilson's Warbler (p. 442). Female grayer with yellow highlights; immature almost wholly gray like Akikiki, but no range overlap.
■ **VOCALIZATIONS** Song, short but vigorous, a series of bright chirps. Call a monosyllabic scratchy note.
■ **BEHAVIORS** Like Akikiki, feeds in the manner of a nuthatch (Maui Creeper is another name). Often at eye level in shrub layer of forests; active and fairly tame.
■ **POPULATIONS** Lanai subspecies went extinct in 1930s; although endangered, still fairly easy to find on upper slopes of Maui.

Palila 🆑

Loxioides bailleui | PALI | L 7½" (19 cm)

Endemic to Hawaii I.

■ **APPEARANCE** Fairly large honeycreeper, with thick bulbous bill. Head and breast yellow; otherwise mostly gray.
■ **VOCALIZATIONS** Beautiful song comprises slurred whistles, recalling that of Pine Grosbeak (p. 360) of mainland.
■ **BEHAVIORS** Restricted to dry forest, where it feeds mostly on mamane, *Sophora chrysophylla*, seeds of a certain ripeness. Feeds slowly in trees, but also flies well above canopy. In its movements, vocalizations, body structure, and plumage, oddly reminiscent of adult female Pine Grosbeak.
■ **POPULATIONS** Endemic to dry slopes of Mauna Kea, Hawaii I.; Palila Discovery Trail is the best bet to see one. Fewer than a thousand remain.

Laysan Finch

Telespiza cantans | LAFI | L 6½" (17 cm)

Endemic to tiny Laysan I., less than two square miles (5 km^2) in total area, where it is the only Hawaiian honeycreeper. Large and stocky, with thick, stout bill; mostly yellow. Easily found on Laysan, especially in vegetation, but also out in the open; ridiculously tame. Song like Palila's, but higher, squeakier, and longer. Still common on Laysan, but avian malaria and sea-level rise are threats. Has been introduced to Pearl and Hermes Atoll, where established.

Nihoa Finch 🆑

Telespiza ultima | NIFI | L 6" (15 cm)

Nihoa I. counterpart of Laysan Finch; Palila, Laysan, and Nihoa Finches together are the "finch-billed" Hawaiian honeycreepers. Similar in bill structure and color to Laysan Finch, but no range overlap; like Laysan, very tame. Song similar in timbre and phrasing to Laysan's, but delivery slower. Widespread and fairly common on Nihoa I., less than a quarter of a square mile (.65 km^2) in total area.

FINCHES
FRINGILLIDAE

Endemic; found on all main islands

Apapane
Himatione sanguinea | APAP | L 5" (13 cm)

- **APPEARANCE** Stoutly built, with decurved bill and short tail. Crimson adult roughly similar in structure and appearance to Iiwi, but smaller, with less impressive bill and duskier hues overall; undertail coverts white, bare parts black. Juvenile olive-gray, transitioning through olive greens and patchy reds on its way to adult plumage.
- **VOCALIZATIONS** Songs incredibly variable, often with wing-clicking thrown into the mix. A fast, bright "mystery song" with sharp, short whistles often proves to be this species.
- **BEHAVIORS** Feeds on nectar of ohia, genus *Metrosideros*, often in treetops. Active; cocks tail. Dusky juvenile may suggest fidgety elepaios (p. 290), but bill structure very different.
- **POPULATIONS** Remains fairly common, even locally abundant, on main islands, although declining on Oahu. Susceptible to avian malaria, but also may be evolving resistance. Hard to miss at popular Hawai'i Volcanoes N.P.

Endemic; found on main islands, but declining

Iiwi
Drepanis coccinea | IIWI | L 6" (15 cm)

- **APPEARANCE** Spectacular; larger than Apapane, with longer and more decurved bill. Intensely scarlet, with jet-black wings and tail; bill and feet also scarlet. Juvenile variable; in transition to adult plumage, splotchy red and chartreuse, recalling some Summer Tanagers (p. 444).
- **VOCALIZATIONS** Calls not as impressive as Apapane's: sudden squeaks and whistles, just a note or two, then a pause, then another utterance.
- **BEHAVIORS** Like Apapane, an ohia specialist. Active; small, disorganized family groups bound through the treetops.
- **POPULATIONS** Highly susceptible to avian malaria and declining range-wide; very local now on Kauai, and apparently gone from Oahu. Still fairly common, but getting scarcer, on Maui (Haleakalā N.P. is especially good) and central highlands of Hawaii I.

Maui Parrotbill CR
Pseudonestor xanthophrys | MAPA | L 5½" (14 cm)

Declining, extremely rare resident of wet forests around Haleakalā N.P., central Maui. Upper mandible strongly decurved. Adult dirty olive above, dusky yellow below; yellow eyebrow bordered by olive crown and olive auriculars (ear patches). Juvenile more monochrome, but told by bill structure. Song a short jumble that trips down the scale. Seeing a "Kiwikiu," as it also is known, is currently possible only on permitted guided tours.

Endemic to Hawaii I.

Akiapolaau EN
Hemignathus wilsoni | AKIA | L 5½" (14 cm)

- **APPEARANCE** Midsize honeycreeper with unique bill: upper mandible very thin and sharply decurved; straight, sharp, shorter lower mandible like a chisel. Compact: head big and blocky, tail very short. Male mostly yellow, with broad black lores; female dusky olive-yellow with gray lores.
- **VOCALIZATIONS** Song like Maui Parrotbill's played backward: a short jumble that rises in pitch. Call a sharp, descending chirp.
- **BEHAVIORS** Uses lower mandible to hammer into bark, then yanks out food with upper mandible (maxilla). Small family groups move quickly through treetops in wet forest.
- **POPULATIONS** Endangered and declining; restricted to central Hawaii I. highlands, where maybe 2,000 remain.

FINCHES
FRINGILLIDAE

Endemic to Kauai

Anianiau
Magumma parva | ANIA | L 4" (10 cm)
- **APPEARANCE** Smallest extant Hawaiian honeycreeper. Bill pinkish, decurved, and fine-tipped. Male entirely yellow, luminous even in the darkest forest interior; female entirely olive-yellow. On both sexes, no darkening or other contrast around lores or on wings.
- **VOCALIZATIONS** Song a fast, bright, repetitious warble, with each phrase the same; no change in pitch from beginning to end.
- **BEHAVIORS** Very active; like many other Hawaiian honeycreepers (pp. 362–369), feeds on nectar and blossoms. Occurs in forests, from dry mid-elevation slopes to extremely wet high mountains.
- **POPULATIONS** Kauai only, and recent sightings only in northwestern quadrant of the island, where still fairly common.

Endemic to Maui, Molokai, and Hawaii I.

Hawaii Amakihi
Chlorodrepanis virens | HAAM | L 4¼" (11 cm)
- **APPEARANCE** Similar in structure and color scheme to smaller Anianiau, but no range overlap. All amakihis have decurved bills, stout legs, and mostly yellow plumage with dark lores; bare parts gray-black on all. Hawaii Amakihi is the brightest yellow of the three species. No range overlap with Oahu and Kauai Amakihis, but compare with Hawaii Creeper (p. 368).
- **VOCALIZATIONS** Song bright and ringing, 5–10 repeated elements, recalling the rollicking delivery of *Geothlypis* warblers (p. 430) of mainland. Call a scratchy, rising, catlike *rrriihn?*
- **BEHAVIORS** More generalist than most other Hawaiian honeycreepers in both habits and habitats. Diet broad and feeding methods diverse; occurs from sunny, mid-elevation slopes to interior of wet forests.
- **POPULATIONS** Widespread and still fairly common on islands of Hawaii and Maui; also occurs on Molokai I., where very scarce. Evolving malaria resistance at low elevations, and the hope is that this trait will be transmitted upslope.

Oahu Amakihi
Chlorodrepanis flava | OAAM | L 4¼" (11 cm)

Restricted to Oahu, where it is the only small, mostly yellow honeycreeper with a fairly short, decurved bill. On all the islands, however, poorly glimpsed white-eyes (p. 320) can be mistaken for small honeycreepers. Like Hawaii Amakihi (no range overlap), mostly yellow (male) with dark lores; but Oahu Amakihi is not as vibrantly yellow and some have wing bars. Song a simple trill that changes pitch or timbre toward the end; recalls that of Orange-crowned Warbler (p. 426) of mainland. Call similar to Hawaii Amakihi's, but tinnier. Like Hawaii Amakihi, evolving resistance to malaria; in fact, may even be increasing in lowlands.

Kauai Amakihi
Chlorodrepanis stejnegeri | KAAM | L 4½" (11 cm)

Found only on northwestern Kauai, where it co-occurs with similar Anianiau. Kauai Amakihi is duller yellow overall and larger than Anianiau, with dark lores and longer, darker, stout-based bill. Compare also with rare, range-restricted Akekee (p. 368). Song, clear and ringing, a fast series of short whistles, with the first note often different from all the others; call more whistled than other amakihis'. Declining: Unlike other amakihis, restricted mostly to wet forest, and also unlike other amakihis, no evidence for recent evolution of resistance to malaria.

FINCHES
FRINGILLIDAE

Anianiau

small pinkish bill

♂

juvenile

♀

Hawaii Amakihi

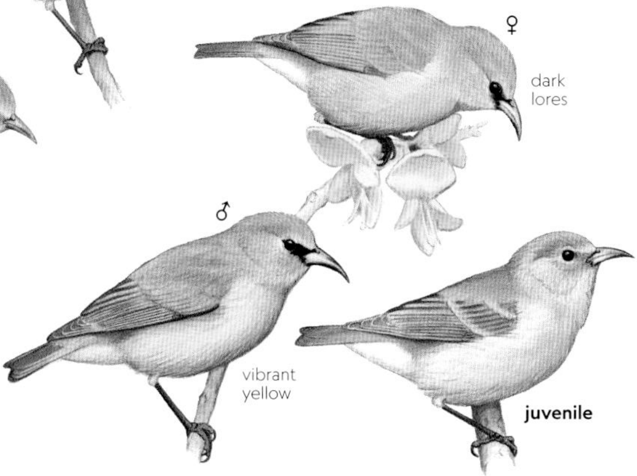

♀

dark lores

♂

vibrant yellow

juvenile

dusky lores

♂

faint wing bars

Oahu Amakihi

♀

dark lores

Kauai Amakihi

♂

♀

Endemic to Hawaii I.

Hawaii Creeper
Loxops mana | HCRE | L 4½" (11 cm)

■ **APPEARANCE** Small and plump; short-tailed. Like Hawaii Amakihi (p. 366), but bill of Hawaii Creeper shorter and straighter. Hawaii Creeper is olive-yellow overall (Hawaii Amakihi more yellowish), darker above than below, with throat especially pale; Hawaii Creeper has thick black face patch (more of a black stripe on Hawaii Amakihi). Keep in mind that nonindigenous white-eyes (p. 320)—small, yellow-green, active, and flocky—can be common anywhere indigenous specialty species are sought.
■ **VOCALIZATIONS** Song a loose, dry trill that drops in pitch. Call an ascending, emphatic *wheet!*
■ **BEHAVIORS** Active; picks at bark on large boughs, but also flits among flowers in ohia groves. Found in extremely wet ohia-koa forests; unlike Hawaii Amakihi, does not usually range into drier habitats.
■ **POPULATIONS** Occurs only on Hawaii I. Recent sightings almost exclusively on windward (wet) slopes of Mauna Kea, mostly Hakalau Forest N.W.R. and, less so, Kona Forest N.W.R.

Endemic to Hawaii I.

Hawaii Akepa 🇪🇳
Loxops coccineus | HAAK | L 4" (10 cm)

■ **APPEARANCE** Small, short-tailed honeycreeper; bill, slightly decurved, is short and pale. Adult male blaze orange with darker wings and tail; superficially similar to larger Apapane and Iiwi (both on p. 364), but bill smaller. Many females have orange wash on breast, but others are almost entirely devoid of reddish hues; they are told from Hawaii Amakihi and Hawaii Creeper by their plain olive-gray face and smaller, paler bill. In close view of either sex, offset bill tips (both sexes) are diagnostic.
■ **VOCALIZATIONS** Song, bright and cheery, a series of slurred notes descending in pitch. Varied calls include a short, slurred *chirrup.*
■ **BEHAVIORS** Often in or near treetops, where it pries insects from flowers and leaves with its unusual bill.
■ **POPULATIONS** Range closely matches that of Hawaii Creeper; currently restricted mostly to east slopes of Mauna Kea. Stable or possibly even increasing in core range.

Akekee 🇨🇷
Loxops caeruleirostris | AKEK | L 4½" (11 cm)

Restricted to Kauai, with recent detections only in and around Alaka'i Wilderness Preserve. Yet another in the suite of morphologically similar Hawaiian fringillines that might usefully be grouped as the guild of small yellow honeycreepers, but notably long-tailed and usually black-masked; tips of mandibles offset, like Hawaii Akepa's (no range overlap). Song a series of sharp notes that trip down the scale; call a sharp, ascending *shriii!* Found in very wet forests, often in or near treetops; pries insects from flowers with unusual bill. Declining; population probably fewer than 1,000.

Akohekohe 🇨🇷
Palmeria dolei | AKOH | L 7" (18 cm)

Restricted to Maui, with recent sightings only in and around Haleakalā N.P. A large honeycreeper of stocky build, but with extraordinary plumage, quite dark overall with bold white and orange spotting; the head and face, with brilliant shocks of papaya and a striking silvery pompadour, would be the envy of any punk rocker. Noisy and active; bounds among blossoms in the manner of Iiwi. Calls raucous: throaty croaks and abrupt chirps. Declining; barely 1,000 left.

FINCHES
FRINGILLIDAE

Gray-crowned Rosy-Finch
Leucosticte tephrocotis | GCRF | L 5½–8¼" (14–21 cm)

- **APPEARANCE** Like all "rosies" (rosy-finches), dark with pink highlights; field ID to genus straightforward, but ID at species level often challenging. Fresh adults of all rosy-finches, in fall and early winter, have fine, gray scaling; their plumage wears down over the winter, becoming deeper and darker. Bill color changes seasonally too: in all rosies, yellowish fall to early winter, blackish by spring. Gray-crowned always has much gray on cap; body mostly plum-brown.
- **VOCALIZATIONS** All rosy-finches call in flight, giving muffled *chfff* and brighter *cheer* notes. Calls of Gray-crowned seem to average sweeter than the other species, but much overlap; more study needed.
- **BEHAVIORS** Nests on tundra; feeds around snowbanks. Tends feeders, often at high elevations, in winter.
- **POPULATIONS** Geographic variation extensive. "Hepburn's" nests in far western mountains, winters throughout interior; entirely gray face instantly separates it from other species. Nonmigratory Bering Sea breeders resemble "Hepburn's," but larger and body blacker. Trickiest are those breeding in central and northern Rockies but wintering widely, with head pattern like Black Rosy-Finch's and body color like Brown-capped Rosy-Finch's.

Black Rosy-Finch 🇪🇳
Leucosticte atrata | BLRF | L 6" (15 cm)

- **APPEARANCE** Size and shape same as Brown-capped and non–Bering Sea Gray-crowned Rosy-Finches. Adult male unmistakable by midwinter: body deep black with glorious pink highlights; face black with broad silver swatch. Female, freshly molted male (fall), and especially young grayer, often weirdly pale; tones always colder than those of other rosies, especially Brown-capped.
- **VOCALIZATIONS** Flight calls perhaps intermediate in timbre between sweeter Gray-crowned's and rougher Brown-capped's, but much overlap.
- **BEHAVIORS** Habits like Gray-crowned's. In winter, mixes freely with the other rosies if present; often the scarcest rosy in mixed-species flocks.
- **POPULATIONS** Migrates farther than Brown-capped but not nearly as far as many Gray-crowneds; Black wanders throughout interior, sometimes to valley floors, in fall and winter. The restricted ranges of both this species and the closely related Brown-capped are predicted to diminish under most climate change forecasts.

Brown-capped Rosy-Finch 🇪🇳
Leucosticte australis | BCRF | L 6" (15 cm)

- **APPEARANCE** Size and shape same as Black Rosy-Finch. Adult male mostly warm brown by midwinter, but cap actually gray or black (not really brown); often lacks gray swatch on face. Reddish hues of Brown-capped also warmer than on other species. Female and young Brown-capped and Black Rosy-Finches, with gray of face greatly reduced and general gray-brown hue overall, closely resemble one another.
- **VOCALIZATIONS** Like other rosy-finches', but flight calls a bit rougher.
- **BEHAVIORS** Like other rosy-finches'. Even in ideal habitat on alpine tundra, scarce as a breeder; flocks to feeders in winter, usually in mountain villages.
- **POPULATIONS** Restricted to southern Rockies, mostly Colo.; doesn't wander much in winter. The N. Amer. rosy-finches have been treated as a single species; Brown-capped and Black are suspected to hybridize near Utah-Colo. border. As with Black, the already limited range of this species is forecast by most climate change models to become even smaller.

Introduced to Hawaii

House Finch
Haemorhous mexicanus | HOFI | L 6" (15 cm)
- **APPEARANCE** Similar to Purple and Cassin's Finches, but bill stubbier and more rounded; tail longer, not as deeply notched. Plumage variation extreme. Typical adult male combines broad reddish eyebrow with concolor "breastplate." Extent and hue of red variable; some individuals orange and, more rarely, yellow. Younger males and most females dirty brown all over with blurry streaking below.
- **VOCALIZATIONS** Song a bright, twangy warble, ending with rising, nasal *chweeEER*. Ascending call, frequently given in flight, a rough *wriitt* or brighter *wreeet*.
- **BEHAVIORS** Nesting habitats exceedingly varied: on buildings in large cities, but also far from human habitation in forests and deserts. Often in small flocks.
- **POPULATIONS** Indigenous to western N. Amer., but an unknown percentage in West, especially east of Continental Divide, are descendants of birds introduced to New York around 1940. Introduced to Hawaii in 19th century; well established on all main islands.

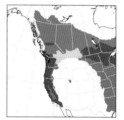

Purple Finch
Haemorhous purpureus | PUFI | L 6" (15 cm)
- **APPEARANCE** Similar in build to House Finch, but has shorter, strongly notched tail, slightly longer wings, and straighter culmen (ridge of upper mandible). Older males suffused in raspberry; belly streaking of House Finch, and undertail coverts unmarked white. Females and younger males have stronger face pattern than House Finch. Compare also with Cassin's Finch, largely nonoverlapping in range.
- **VOCALIZATIONS** Song a rich warble like House Finch's, but faster; lacks rising terminal note. Calls include a vireo-like *sl'wooeee?* and, in flight, a hard *plick*.
- **BEHAVIORS** Less inclined to human-dominated landscapes than House Finch. Favors moist conifer forests for nesting; more generalist in winter.
- **POPULATIONS** Two distinctive subspecies in West. Nominate eastern *purpureus* breeds west to Canadian Rockies; female has bolder face markings. Western *californicus* breeds from Salish Sea region southward; female has blurrier face markings, blurrier streaking below, and variable olive tinge.

Cassin's Finch
Haemorhous cassinii | CAFI | L 6¼" (16 cm)
- **APPEARANCE** Similar to Purple Finch, but wings slightly longer and culmen perfectly straight; often appears weakly crested. Markings crisper than on Purple and House Finches, with sharp streaking below extending to undertail coverts. Red crown of adult male well set off from rest of head. Female and younger male like corresponding plumages of Purple, but streaking below, reaching undertail coverts, darker and crisper.
- **VOCALIZATIONS** Song, often lengthy, a fast warble; not as rich and twangy as House Finch's, can be confused with that of Townsend's Solitaire (p. 344). Calls include a short stutter, a bit like Western Tanager's (p. 444), and a wheezy whistle.
- **BEHAVIORS** Found year-round in conifer forests, often in treetops, but also feeds on ground and along roads.
- **POPULATIONS** Numbers vary annually: can be scarce across large swaths of breeding grounds some years; irruptions to foothills and, more rarely, valleys and plains, likewise variable.

Yellow-fronted Canary
Crithagra mozambica | YFCA | L 4½" (11 cm)

Indigenous to sub-Saharan Africa. Introduced and well established, islands of Oahu and Hawaii; occasional reports also from Kauai and Maui. Adult male olive-backed with bright yellow underparts; head gray and yellow with prominent black "whisker." Female and especially immature have more muted colors and indistinct markings on face. An unexpected point of confusion can be House Finch (p. 372); yellow-variant male House Finches are relatively common in Hawaii. Song bright and sweet, short or long, a fast jumble of slurred whistles. Common around parks and resorts at sea level, but also in wilder settings at higher altitudes. Has increased in recent years in Los Angeles area and may become established there.

Island Canary
Serinus canaria | ISCA | L 5–5½" (13–14 cm)

This is *the* canary, indigenous to the Canary Is. (also other islands in the eastern Atlantic). Introduced and well established on tiny Midway Atoll; several thousand occur there. The Midway birds are a pale, sickly yellow, unlike their darker, streakier Atlantic Ocean antecedents; some are mostly white. The song is an exuberant outpouring of chirps and whistles. Island Canary and Common Myna (p. 340) are the only regularly occurring passerines on Midway.

Accidental to Hawaii

Common Redpoll
Acanthis flammea | CORE | L 5¼" (13 cm)

■ **APPEARANCE** Smaller and slightly larger than House Finch, with small bill. Pale overall and streaky with well-defined red cap (the poll) and small black patch on throat; bill yellow. Adult male has extensive pink on breast, reduced on female; juvenile, plain and streaky, differs from juvenile Pine Siskin (p. 378) in bill structure. Compare with Hoary Redpoll.

■ **VOCALIZATIONS** Dry *chit* notes, strongly descending, as birds flush or fly over. Feeding, they murmur among themselves with twangy, siskin-like *eeeawee* and *eeyay* notes.

■ **BEHAVIORS** Active and flocky; easily approached. Nests on taiga with spruce and birch; tends feeders, especially in irruption years, but, even then, finds its way to catkin-bearing birches if available.

■ **POPULATIONS** Famously irruptive; movements south roughly biennial. In strong irruption years, winters in sometimes sizable flocks to midlatitudes of U.S.; vagrant even to Hawaii.

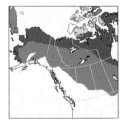

Hoary Redpoll
Acanthis hornemanni | HORE | L 5½" (14 cm)

■ **APPEARANCE** Morphology and plumage reflect its high-latitude biology: larger-bodied than Common (to retain body heat) but smaller-billed (to prevent heat loss), with frostier plumage overall. Streaking on rump and undertail coverts reduced compared to Common.

■ **VOCALIZATIONS** Like Common Redpoll's, but flight call, sharply descending, seems to be buzzier, rougher: *chzzt*.

■ **BEHAVIORS** Nests on ground on treeless tundra. Incredibly hardy; in many winters, stays on high Arctic breeding grounds, flourishing in perpetual darkness where food is mostly under deep snow cover.

■ **POPULATIONS** Irruptive like Common Redpoll, but doesn't disperse nearly as far south, ordinarily reaching only far northern tier of lower 48. Closely related to Common and treated by some as conspecific.

Red Crossbill
Loxia curvirostra | RECR | L 5½–7¾" (14–20 cm)

■ **APPEARANCE** Block-headed and short-tailed, with crossed bill tips, an adaptation for extracting conifer seeds. All plumages normally plain-winged, lacking bold white wing bars of White-winged Crossbill. Adult male brick red, adult female olive-yellow, juvenile streaked brown. See Cassia Crossbill.

■ **VOCALIZATIONS** Far-carrying flight call monosyllabic, often doubled: for example, *kyew! kyew!* or *tyip tyip*. Also gives a softer, muffled whistle, called the "toop note," when feeding. Song an amorphous jumble of slurred whistles and rich chirps.

■ **BEHAVIORS** Exceedingly nomadic. Hard to detect when feeding in tops of conifers, but flocks and pairs call loudly in flight high above canopy. Sometimes drops to ground level to pick up grit.

■ **POPULATIONS** Around a dozen "types" occur in N. Amer. They may be valid species, and most occur in the West; these types differ consistently in flight call, feeding ecology, bill size, and even the morphology of the *insides* of their bills. Most frequent in the West are widespread, large-billed Type 2, specializing on ponderosa pine; small-billed Type 3 (Pacific Northwest, hemlock); small-billed Type 4 (Pacific Northwest, Douglas fir); large-billed Type 5 (southern Rockies, lodgepole pine); large-billed Type 6 (Mexico north to Ariz., Apache pine); and small-billed Type 10 (Pacific coast, Sitka spruce). Spectrograms of cellphone audio, reviewed by eBird experts, establish ID.

Cassia Crossbill
Loxia sinesciuris | CACR | L 7–7½" (18–19 cm)

Discovered in 1990s and elevated to full-species status by the AOS in 2017. Until recently thought to occur only in lodgepole pine forests without squirrels in Cassia Co. and Twin Falls Co., Idaho. External morphology essentially identical to that of large-billed Type 2 and Type 5 Red Crossbills, but flight call different: hollow and clucking, quite unlike Type 2's (clear and piercing) but more similar to Type 5's (drier and flatter than Type 2's). Spectrograms probably required for ID. Beginning in 2021, crossbills with spectrograms perfectly matching those of Cassia have been widely documented in squirrel-infested lodgepole pine forests of central Colo., complicating our understanding of this problematic "species."

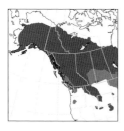

White-winged Crossbill
Loxia leucoptera | WWCR | L 6½" (17 cm)

■ **APPEARANCE** Crossed mandibles identify it to genus; head smaller and rounder than Red Crossbill's, tail not as short. All have black wings with thick white wing bars; some Red Crossbills show vague wing bars, but never as thick as on White-winged. Adult male extensively pink, adult female yellow-gray, juvenile streaked.

■ **VOCALIZATIONS** Flight call a muffled *chut*, down-slurred; not as bright and ringing as most Red Crossbills' flight calls. Trilled song simpler and much faster than Red's.

■ **BEHAVIORS** Conifer specialist, with decided preference for spruce. Can be quite tame.

■ **POPULATIONS** Irruptions driven by availability of spruce seeds; not as nomadic as Red Crossbill. Limits of breeding range in southern Rockies not well known.

Pine Siskin
Spinus pinus | PISI | L 5" (13 cm)

■ **APPEARANCE** A bit slimmer than American Goldfinch. Tail notched; bill thin, straight, and pointed. Streaked dusky all over, like female or young *Haemorhous* finch (p. 372). All have yellow in wings and tail, extensive in older males, more muted in other plumages.

■ **VOCALIZATIONS** Flight call a rising, nasal *zraaay*; perched, a rising, wavering, very buzzy *zzzhhrrRRR*. Song twittery and twangy.

■ **BEHAVIORS** Nests mostly in conifer forests; most detections on breeding grounds are of birds calling in flight. Tends feeders with "seed socks," filled with costly thistle seed.

■ **POPULATIONS** Widespread and often common, but distribution and abundance vary considerably from year to year.

Lesser Goldfinch
Spinus psaltria | LEGO | L 4½" (11 cm)

■ **APPEARANCE** Smaller than American Goldfinch. Bill dark; white flash at base of primaries prominent in flight. Breeding male has black cap, olive upperparts, and yellow below; other plumages, less boldly marked, show dusky yellow above and below, extending to undertail coverts.

■ **VOCALIZATIONS** In flight, alternates pure-tone whistles, rising or falling, with rough chatter: *chleee… ch'ch'ch'ch… plweee….* Song, rapid and tinkling, includes imitations of other birds' songs.

■ **BEHAVIORS** Nests in foothills and forest clearings. Often in small mixed-species flocks; tends feeders.

■ **POPULATIONS** Winter range expanding north. Wanders annually but erratically to valleys and plains by midsummer.

Lawrence's Goldfinch
Spinus lawrencei | LAGO | L 4¾" (12 cm)

■ **APPEARANCE** A bit daintier in build than American Goldfinch. All are pale-billed with extensive yellow in wings. Male has gray head with black "goatee"; female, especially in winter, told by wing and bill color.

■ **VOCALIZATIONS** Flight call a sharp, clearly two-noted *see-tyoop*. Song a rapid twitter, jumbling and tinkling, with imitations of other birds' calls.

■ **BEHAVIORS** Nests in sunny oak savannas, especially in foothills. Migrants and winterers occur in small flocks in weedy pastures, meadows, and desert washes.

■ **POPULATIONS** Most depart breeding grounds by early fall, but extent of movements varies annually; reaches well into Ariz. and even N. Mex. some years.

American Goldfinch
Spinus tristis | AGOL | L 5" (13 cm)

■ **APPEARANCE** Largest goldfinch in West. Breeding male bright yellow above and below, with black cap and wings. Dull birds in winter distinguished by whitish undertail coverts (yellow on Lesser), pale bill (dark on Lesser), and dirty-white wing bars (wings of Lawrence's more extensively yellow).

■ **VOCALIZATIONS** Flight call a rapid stutter of three to five falling notes, *chih-chih-chih-chit (potato chip)*; rising *tsuWEEE* given perched or in flight. Run-on song, bright and fast, comprises sharply rising and falling notes.

■ **BEHAVIORS** Habitat generalist; visits gardens with sunflowers and other composites. Nests in midsummer, later than most other songbirds.

■ **POPULATIONS** Like Lesser, wintering farther north in recent decades.

LONGSPURS AND *PLECTROPHENAX* BUNTINGS | CALCARIIDAE

These lovers of open country were long classified as New World sparrows (pp. 386–409) because of their general similarity in size, shape, and plumage. But they differ importantly in everything from gait to flocking behavior, from molt to flight calls. The calcariids have a long claw on the hind toe, an adaptation for ambling over uneven terrain.

Lapland Longspur
Calcarius lapponicus | LALO | L 6¼" (16 cm)

■ **APPEARANCE** Sparrowlike in overall size, but structure subtly different: longer-winged and shorter-legged, giving it a crouched or prostrate look when on the ground. All longspurs follow this basic model. Outermost rectrix (tail feather) of Lapland mostly white; other longspurs show two or more rectrices white. For most birders in the West, encounters with Lapland are of drab birds in winter, told by their long primaries, reddish panel on wings, and face with black markings. Males acquire spiffy breeding plumage in early spring, just before heading north to breed.

■ **VOCALIZATIONS** Calls include a descending *pyoo* and a short rattle, slowly alternated in flight. Jangling song heard mostly on breeding grounds, although some start singing while migrating north in spring.

■ **BEHAVIORS** Nests on tundra, wet or dry, but not usually as rocky as microhabitat favored by Snow Bunting (p. 384). Winters wherever there is extensive, sparse, open country, especially fallow cornfields, overgrazed range, and roadsides. Walks like Horned Lark (p. 308), with which it often consorts in winter.

■ **POPULATIONS** Abundant on breeding grounds; in many places the most numerous nesting passerine species, and in some places the *only* nesting passerine. Arrives mid-autumn in lower 48 states, heads back north in late winter. Bulk of wintering range a bit to our east, but immense flocks winter west to High Plains; winters widely in much smaller numbers to northern Great Basin and coastally. Midwinter movements induced by heavy snow cover.

Smith's Longspur
Calcarius pictus | SMLO | L 6¼" (16 cm)

■ **APPEARANCE** Like Lapland, a long-winged and relatively large longspur. Smith's is the most warmly colored longspur; even in its drab winter plumage, Smith's is washed buff below. Tail mostly dark, but two outermost rectrices mostly white; on Lapland, only the outermost rectrix appears white. Broad white upper wing bar, often prominent in flight, forms crescent on bird at rest. Like Lapland, begins to acquire breeding plumage—male has harlequin face pattern and bright butternut breast—by end of winter.

■ **VOCALIZATIONS** Rattling flight call not as rapid as other longspurs'. Song, heard on spring migration and on breeding grounds, tinkling and rising, recalls Horned Lark's (p. 308).

■ **BEHAVIORS** Nests in wetter, shrubbier habitats than Lapland Longspur, from taiga-tundra ecotone out onto open tundra if sufficiently wet. In winter and on migration favors disturbed habitats: corn stubble, grazed pastures, airfields, etc. Breeding biology unusual among passerines: Females have multiple mates, a system known as polyandry; phalaropes (p. 132) too are polyandrous.

■ **POPULATIONS** Winter range well to our east; practically unrecorded as vagrant in Interior West, but scattered records fall to winter to West Coast and western High Plains.

LONGSPURS AND *PLECTROPHENAX* BUNTINGS
CALCARIIDAE

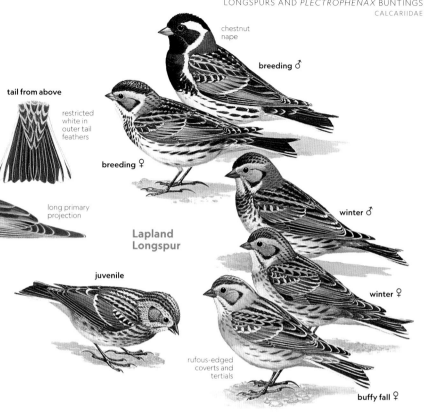

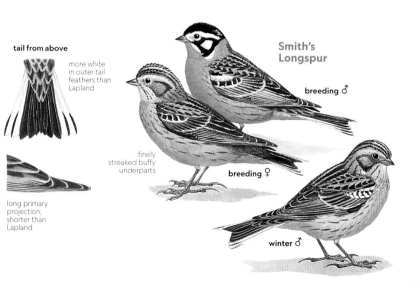

Chestnut-collared Longspur
Calcarius ornatus | CCLO | L 6" (15 cm)

- **APPEARANCE** The smallest longspur; slimly built and small-billed. Breeding male, with chestnut collar and extensive black below, suggests breeding male Lapland Longspur (p. 380), but no range overlap on breeding grounds. Outside breeding season, told from Smith's (p. 380) and Lapland Longspurs by pale gray plumage, short primaries, and, in flight, mostly white tail with broad black tips on inner rectrices; Thick-billed in winter is even paler than winter Chestnut-collared, with larger bill and gray (male) or buff (female) breast. Male Chestnut-collared in winter retains hint of black breast; all plumages relatively dark-billed in winter.
- **VOCALIZATIONS** Rattling flight call averages shorter (two or three syllables) than in other longspurs, but also can run longer. Song rich and gurgling; shorter than song of Thick-billed.
- **BEHAVIORS** Along with Thick-billed, one of the two "prairie longspurs"; however, Chestnut-collared is more closely related to Lapland and Smith's. Where Chestnut-collared and Thick-billed co-occur as nesters, they sort out by microhabitat: Chestnut-collared prefers taller grasses, often in lower and wetter settings. On migration and in winter too, Chestnut-collared goes for taller grass than Thick-billed and other longspurs.
- **POPULATIONS** In long-term and worrisome decline, but still fairly common in lush grasslands west almost to foothills of central and Canadian Rockies. Migration routes not well understood; casual at best in much of Great Basin and southern Rockies, yet routinely gets to coastal states, especially Calif.

Thick-billed Longspur
Rhynchophanes mccownii | TBLO | L 6" (15 cm)

- **APPEARANCE** Pale and gray overall, with odd, front-heavy build: big-billed and large-headed, short-winged and short-tailed; the "spur" (hind toe) is shorter than on other longspurs. Tail mostly white, but gets black on all but outermost rectrix; similar to tail of Chestnut-collared, but with even more white. Breeding adult, unstreaked below, is mostly plain gray-black (male) or gray-brown (female), with chestnut patch on gray wings. Differs in winter from Chestnut-collared by plain gray (male) or gray-buff (female) breast and large, pale bill; even drab birds in winter usually show chestnut patch on wings; larger, stockier body a good mark year-round. Compare also with female House Sparrow (p. 356), plain and pale with short wings and short tail.
- **VOCALIZATIONS** Flight call a short rattle like other longspurs', but also gives a mellow *plip*, unlike other longspurs. Song an amorphous, run-on twittering, given in flight.
- **BEHAVIORS** Compare with Chestnut-collared, the other "prairie longspur." Where the two species co-occur as breeders, Thick-billed is inclined to dry, often disturbed, shortgrass prairie. Tolerates light grazing, but range has withdrawn considerably due in part to fire suppression.
- **POPULATIONS** Distribution in West outside breeding season mirrors that of Chestnut-collared: largely unrecorded in Great Basin and southern Rockies, but annual farther west, especially Calif. Despite its name, Thick-billed Longspur is more closely related to *Plectrophenax* buntings (p. 384) than *Calcarius* longspurs (pp. 380–383).

LONGSPURS AND *PLECTROPHENAX* BUNTINGS
CALCARIIDAE

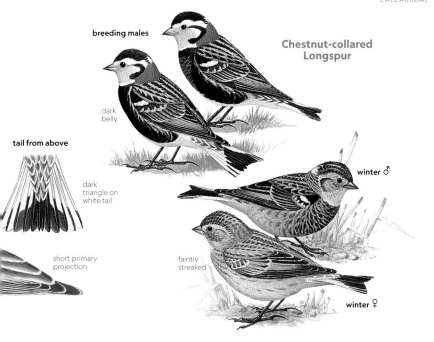

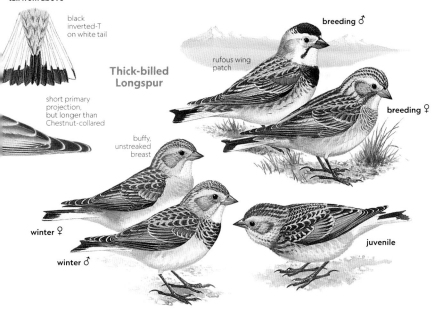

Vagrant to Hawaiian Is.

Snow Bunting
Plectrophenax nivalis | SNBU | L 6¾" (17 cm)

- **APPEARANCE** Slightly larger than longspurs; small-billed. In winter, when most are seen, mottled black and white or black and buff above and mostly white below; in breeding plumage, starting to show by late winter, whiter overall, with males black above, females grayer. All species in family Calcariidae acquire their striking breeding garb largely via feather wear during the winter, rather than a complete spring molt. Year-round, both sexes flash huge white patches on wings in flight. Aberrantly whitish (leucistic) juncos (p. 396) and other songbirds can suggest Snow Bunting. Compare also with McKay's Bunting, rare and range-restricted.
- **VOCALIZATIONS** Like Lapland Longspur (p. 380), alternates short rattle with descending whistle when flying over. Song a run-on twitter, tinkling and high-pitched.
- **BEHAVIORS** Nests in rock crevices on tundra, where well-protected from predators. Winters in barren habitats: beaches, lakeshores, fallow agricultural fields. Flocky; often with Horned Larks (p. 308) and Lapland Longspurs.
- **POPULATIONS** Although most in winter in our area don't get south of northern border states, several have vagrated to Hawaii. Recent declines conjectured to result from high-latitude climate change.

McKay's Bunting
Plectrophenax hyperboreus | MKBU | L 6¾" (17 cm)

Breeds mostly on St. Matthew and Hall Is., well out in the Bering Sea; winters on west coast of Alas., primarily from Seward Peninsula to Alaska Peninsula. Same size and shape as Snow Bunting, with vocalizations and most details of behavioral ecology apparently very similar. All plumages whiter than corresponding plumages of Snow Bunting. Breeding male mostly white, with just a bit of black on wings and tail; both sexes browner in winter like Snow Bunting, but nevertheless with extensive white in wings and tail. Field ID complicated by apparent interbreeding with Snow Bunting on Bering Sea islands other than St. Matthew and Hall; hybrids especially challenging in winter on coast where both species occur.

OLD WORLD BUNTINGS | EMBERIZIDAE

The terms "bunting" and "emberizid" have been confusingly applied to various groups of songbirds, but the current scheme is straightforward: The family Emberizidae comprises the single genus *Emberiza*, an Old World assemblage with jangling songs, stripy heads, and yellow and rusty highlights. They are most closely related to the longspurs (pp. 380–383) and *Plectrophenax* buntings.

Rustic Bunting
Emberiza rustica | RUBU | L 5¾" (15 cm)

Widespread across northern Eurasia in wet taiga. Regular in small numbers in spring to Bering Sea islands. Casual much farther south, to Calif., mostly coastally and mostly fall to winter; accidental to Hawaii. Sparrowlike in general plumage, streaked brown and rusty, but body structure a bit different, with fine bill, long tail, and short crest. Breeding male has high-contrast black-and-white face pattern, more muted on breeding female; both sexes have pale patch at rear of crown, a diagnostic field mark but hard to see. Lovely song not usually heard in N. Amer., but vagrants sometimes give a light, high *tyip*, repeated slowly. Appears to be in recent and widespread decline.

LONGSPURS AND *PLECTROPHENAX* BUNTINGS | OLD WORLD BUNTINGS
CALCARIIDAE | EMBERIZIDAE

NEW WORLD SPARROWS | PASSERELLIDAE

"Small" and "brown" are recurring themes in this family, but so are "biodiverse" and "melodious." Most have short, conical bills, and many are terrestrial. Field ID depends on study of face pattern, streaking below, and, even more so, vocalizations and body structure.

Rufous-winged Sparrow
Peucaea carpalis | RWSP | L 5¾" (15 cm)

■ **APPEARANCE** Midsize sparrow with rounded tail, large bill, and weak crest. Adult, with plain gray underparts, has reddish cap, vague eye ring, dark line through eye, and two black "whiskers" on either side of face. Chipping Sparrow (p. 392) similar in plumage, but smaller and slimmer, with longer, narrower tail; lacks whiskers. Rufous-crowned Sparrow (p. 406) also similar, but has one moustache, thick and black.

■ **VOCALIZATIONS** Song an accelerating series of sharply descending chips. Call a weak, very high *zyee*, descending.

■ **BEHAVIORS** Occurs year-round in washes and lower foothills with mesquite and paloverde. Nests in spiny shrubs out in the open.

■ **POPULATIONS** Onset of nesting triggered by monsoon rains in midsummer; nonmigratory, but wanders opportunistically in wetter years.

Botteri's Sparrow
Peucaea botterii | BOSP | L 6" (15 cm)

■ **APPEARANCE** Quite similar to Cassin's Sparrow: Both are large-billed (bill of Botteri's slightly larger) and flat-headed, with a rounded, fairly long tail. Botteri's more reddish-hued overall. Tail of Botteri's lacks white edges and transverse barring of Cassin's; wings plainer than Cassin's. Streaky juvenile, with plain plumage overall, very similar to Cassin's; note slightly larger bill and warmer tones of juvenile Botteri's.

■ **VOCALIZATIONS** Song, sputtering and disorganized, never seems to end; consists of chips and trills, randomly uttered. Call a high, light, rising *tsi?*

■ **BEHAVIORS** Found in grasslands with thorny shrubs and scattered oaks; locally common where *Sporobolus* grass abounds.

■ **POPULATIONS** Intrinsically variable; numbers vary naturally from year to year, especially in response to rainfall. Previously thought to occur in our area only in summer, but many records in recent winters.

Cassin's Sparrow
Peucaea cassinii | CASP | L 6" (15 cm)

■ **APPEARANCE** Very similar in build to Botteri's Sparrow. Fresh Cassin's are scaly-backed with white edgings on wing; tail barred and white-edged, and most show thin white eye ring. Cassin's is most similar to Botteri's, but compare also with Grasshopper Sparrow (p. 388), warmer and somewhat smaller, and with Brewer's Sparrow (p. 392), colder and notably smaller.

■ **VOCALIZATIONS** "Minor key" song, high and sweet, comprises one or two pure whistles, then a high trill, then two lower whistles; often performed in flight. Call a high, weak *phit* often in slow, stuttering series.

■ **BEHAVIORS** Nests on prairie with scattered yucca and sand sage. Where Cassin's and Botteri's overlap, Cassin's tends to go for mixed grass-shrub microhabitats, Botteri's more for grass monocultures.

■ **POPULATIONS** Highly variable from year to year, and even within a single breeding season; birds move around in response to rainfall or lack thereof.

NEW WORLD SPARROWS
PASSERELLIDAE

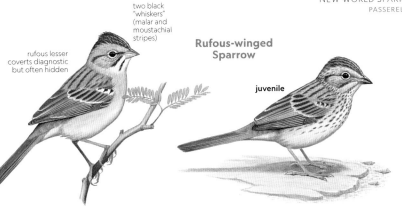

Grasshopper Sparrow
Ammodramus savannarum | GRSP | L 5" (13 cm)

- **APPEARANCE** Rotund; short-tailed, flat-headed, and large-billed. Adult buffy overall, brightest on unstreaked breast. Note also white crown stripe and thin white eye ring of adult; most have ochre patch in front of eye. Juvenile, which can be seen on fall migration, has streaks on buff-brown breast; compare with Cassin's (p. 386) and LeConte's (p. 402) Sparrows and, unexpectedly, Northern Red Bishop (p. 352).
- **VOCALIZATIONS** Song comprises two to four high chips followed by a thin buzz like a grasshopper: *p't'tk'ZZZZZZZZZZ*. Also an ill-formed jumble of short, high buzzes. Flight call a high, shrill, thin buzz.
- **BEHAVIORS** Nests in dry tallgrass where hard to see except when teed up and singing; Savannah Sparrow, with similar song, in wetter settings. Habitually sings at night.
- **POPULATIONS** Most in West are relatively pale subspecies *perpallidus*; subspecies *ammolegus*, from Southeast Ariz. and Bootheel of N. Mex., darker and more rufescent.

Savannah Sparrow
Passerculus sandwichensis | SAVS | L 5½" (14 cm)

- **APPEARANCE** Most in West are small, plump, and short-tailed, with a slight bill. Variable, but all are streaked below, most rather crisply so; most also show broad yellow eyebrow and thin white central crown stripe.
- **VOCALIZATIONS** Song begins with a few clipped notes, very high, followed by long buzz then short buzz; song of Grasshopper Sparrow has just one buzz following introductory notes. Call note a light *tsip*; flight call a descending *tsyee*.

Accidental to Hawaiian Is.

- **BEHAVIORS** Nests in wet meadows; winters widely in open habitats. More prone than other, often rarer, grassland and marshland sparrows to venture into the open.
- **POPULATIONS** "Belding's Sparrow" (subspecies *beldingi*), resident in coastal salt marshes of Southern Calif., dark and small with proportionately long bill. Duller "Large-billed Sparrow" *(rostratus)*, large overall and large-billed, with blurrier streaking, breeds to our south, with some wintering north to Salton Sea and San Diego; buzzy elements in song much harsher. "Large-billed" suspected by many to be a separate species; "Belding's" may be more closely related to "Large-billed" than to other Savannah Sparrows. Vagrant (multiple records) to Hawaiian Is.

Five-striped Sparrow
Amphispizopsis quinquestriata | FSSP | L 6" (15 cm)

Range-restricted; occurs only in northwestern Mexico and a bit of Ariz., especially Santa Cruz Co. Built like Botteri's and Cassin's Sparrows (both on p. 386), large-billed and flat-headed; like those two, hard to see except when singing. Dark brown above, with five white markings on head; gray and white below, with dark breast spot. Black-throated Sparrow (p. 390) superficially similar, but slimmer and grayer. Ecology like that of Rufous-winged Sparrow (p. 386): Nests in shrubs on brushy, rocky hillsides when monsoon rains begin in midsummer. Song, high and sputtering, alternates sharp *spit* notes with short buzzes; delivery "random" like that of Botteri's. Probably resident in limited range in West, but winter detections difficult because of species' secretive behavior when not singing.

Black-throated Sparrow
Amphispiza bilineata | BTSP | L 5½" (14 cm)

- **APPEARANCE** Slim and trim. Head and breast of adult boldly marked black and white. Dark tail has white corners; rest of body plain grayish. Juvenile, less boldly marked, nevertheless shows bold white eyebrow and bold white malar, with smudgy wash on breast and white corners on tail.
- **VOCALIZATIONS** Song a few short chirps and a loose trill: *chit chit ch' twe-e-e-e-e-e-e*. Calls clipped and high-pitched: *tee* and *tsee*.
- **BEHAVIORS** Occurs year-round in arid habitats; even in midsummer, flourishes in creosote bush flatlands.
- **POPULATIONS** Frequently wanders beyond mapped range, even offshore.

Lark Sparrow
Chondestes grammacus | LASP | L 6½" (17 cm)

- **APPEARANCE** Large overall, with ample tail. Harlequin face pattern of adult highlighted by extensive chestnut; more monochrome on immature and winter adult. Plain breast has single dark splotch; dark, rounded tail has white corners. Juvenile streaky below, with distinctive tail of adult and hint of adult head pattern.
- **VOCALIZATIONS** Song a hesitant jumble of short buzzes, whistles, and trills, with elements often doubled: *chit-chit zzzz syee-syee whit e-e-e-e-e seet-seet*…. Flight call an abrupt *tsip*.
- **BEHAVIORS** Year-round favors open, semiarid country around hedgerows, outbuildings, and woodland edges.
- **POPULATIONS** Large movements of adults by early summer are likely a molt migration.

Lark Bunting
Calamospiza melanocorys | LARB | L 7" (18 cm)

- **APPEARANCE** One of the largest New World sparrows; bill large, tail fairly short. White wing panel prominent in all plumages. Snazzy black-and-white breeding male unmistakable. Other plumages, especially young in winter, told by large size, chunky bill, and, in flight, white on wing.
- **VOCALIZATIONS** Pulsing song mixes series of buzzes with loose trills: *rzz rzz rzz jree-jree-jree-jree-jree dzzzzzzz chre-chre-chre-chre* …. Flight call a rich, rising *whooee?*
- **BEHAVIORS** As with Cassin's Sparrow (p. 386), numbers vary with precipitation. Breeds on shortgrass prairie, where it performs marvelous flight displays. Winters in large flocks on grasslands and desert.
- **POPULATIONS** Range-restricted, short-distance migrant, yet wanders in small numbers every year to coast.

Black-chinned Sparrow
Spizella atrogularis | BCSP | L 5¾" (15 cm)

- **APPEARANCE** Typical of *Spizella* sparrows (pp. 390–395), slender but pot-bellied, with long tail. Adult mostly gray, but with brown wings; black on face in spring and summer, especially prominent on male, contrasts with pink bill. Winter adult more gray-faced; juvenile like winter adult, but with blurry streaking below.
- **VOCALIZATIONS** Song an accelerated series of descending chips, not unlike that of Rufous-winged (p. 386). Call a simple, high *tik*, repeated frequently.
- **BEHAVIORS** Found year-round in rugged, arid country, especially canyon mouths and lower slopes at juniper-shrubland interface.
- **POPULATIONS** Local breeding density varies annually, responding in part to moisture and wildfire.

NEW WORLD SPARROWS
PASSERELLIDAE

Black-throated Sparrow

white throat

juvenile

Lark Sparrow

distinct harlequin head pattern

prominent white tail corners

juvenile

Lark Bunting

breeding ♂

pale wing patch in all plumages

early spring ♂

winter ♂

streaking darker than females

♀

Black-chinned Sparrow

streaked brown back

pink bill

breeding ♂

breeding ♀

juvenile

Chipping Sparrow
Spizella passerina | CHSP | L 5½" (14 cm)

- **APPEARANCE** Potbellied, long-tailed, and quite small. Adult brownish above, but with plain gray rump; underparts plain gray. In breeding plumage, has rusty cap, white eyebrow, and black line through eye. Winter adult and immature resemble Clay-colored and Brewer's Sparrows, but black line extends through eye on Chipping. Clay-colored buffier overall, with brown rump and gray nape; Brewer's paler and grayer, with thin white eye ring. Juvenile Chipping, streaked below, shows dark line through eye.
- **VOCALIZATIONS** Song a simple trill of monotone *chip* notes with dry, unmusical quality; some songs of Dark-eyed Junco (p. 396) very similar. Flight call a high, rising *seen?*
- **BEHAVIORS** Nests in forests, not too dense, from dry pinewoods in southern Rockies to taiga in Alas. Flocky on migration and in winter, often mixing with Clay-colored and Brewer's.
- **POPULATIONS** Interior West breeders are molt migrants that engage in midsummer nocturnal migration eastward.

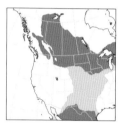

Clay-colored Sparrow
Spizella pallida | CCSP | L 5½" (14 cm)

- **APPEARANCE** Same size and shape as Chipping Sparrow. Breeding adult has intricate head pattern, with white central crown stripe, prominent auriculars (ear patches), and bold "whisker." Immature and nonbreeding adult buffier overall; distinguished from nonbreeding Chipping by brown rump (gray on Chipping), gray nape (also gray on Chipping, but not as cleanly set off), and dark line behind but not in front of eye (goes through eye on Chipping). Compare especially with Brewer's Sparrow.
- **VOCALIZATIONS** Song a few pulses of white noise, *ZZZZ ZZZZ ZZZZ*, utterly unmusical. Flight call a weak, rising *see?*
- **BEHAVIORS** Nests in tallgrass prairie with scattered trees. Like Chipping and Brewer's, joins mixed-species foraging flocks on migration and in winter.
- **POPULATIONS** Migrates mostly east of Rockies; huge pushes to foothills in stormy easterlies in spring. Uncommon in fall to West Coast.

Brewer's Sparrow
Spizella breweri | BRSP | L 5½" (14 cm)

- **APPEARANCE** Even slighter and slimmer than Chipping and Clay-colored Sparrows. Sandy gray-brown and finely marked, suggesting washed-out Clay-colored. Brown crown has thin white streaks; face has narrow white eye ring and thin dark "whisker." Similar to Clay-colored fall to winter, but Clay-colored brighter and buffier, more boldly marked on face, and usually with pale central crown stripe and prominent gray nape. Striped juvenile plumage, held briefly, begins to show eye ring and streaky crown.
- **VOCALIZATIONS** Long-duration song has buzzes like Clay-colored's and trills like Chipping's. Flight call like Clay-colored's.
- **BEHAVIORS** Widespread interior population nests in open expanses of sagebrush. Joins Chipping and Clay-colored (and other) Sparrows in mixed-species flocks on migration and in winter.
- **POPULATIONS** Largely isolated subspecies *taverneri* ("Timberline Sparrow") has higher-contrast plumage, more closely resembling Clay-colored; reaches Alas., where it nests alongside Wandering Tattlers (p. 128). Winter range of "Timberline" poorly known; may include Ariz.

Field Sparrow
Spizella pusilla | FISP | L 5¾" (15 cm)

■ **APPEARANCE** Like other *Spizella* sparrows (pp. 390–395), small overall and long-tailed. Adult has gray head with faint rusty highlights, bright pink bill, and white eye ring; breast gray. Streaky juvenile plumage shows reddish highlights and hint of eye ring.

■ **VOCALIZATIONS** Song an accelerating series of whistles, recalling Black-chinned Sparrow's (p. 390), but no range overlap in breeding season. Call note a smacking *tsik*; flight call a high, clear, descending buzz.

■ **BEHAVIORS** In most of range, nests in overgrown pastures transitioning to early successional woods; in limited breeding range in West, gets into broadleaf riparian habitats where available, but also brushy tallgrass and even sagebrush.

■ **POPULATIONS** At periphery of range in West; ours are subspecies *arenacea*, paler and grayer than birds farther east. Immature female Black-chinned Sparrow an unexpected point of confusion with *arenacea* Field Sparrow out of range.

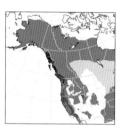

Fox Sparrow
Passerella iliaca | FOSP | L 7" (18 cm)

■ **APPEARANCE** Large; same size as Hermit Thrush (p. 346), which it superficially resembles. Thick-billed and sturdy-legged. Variable, but all have pale bill, dense spotting below, and rufous tail and wings contrasting with grayer back and head.

■ **VOCALIZATIONS** Songs variable, but all combine slurred whistles with short buzzes, with "wilder" quality than other sparrows. Call of most a smacking *tsook*, like that of Lincoln's Sparrow (p. 404).

■ **BEHAVIORS** In dense brush at all seasons. Away from nesting grounds, found singly in leafy understory, where it forages with exaggerated kicking.

■ **POPULATIONS** Four subspecies groups in West may constitute four separate species: "Red" (nominate *iliaca*), breeding Alas. and northern Canada, rich rusty all over with sweet song; "Slate-colored" (*schistacea*), breeding Interior West, gray-headed with song like Green-tailed Towhee's (p. 408); "Sooty" (*unalaschcensis*), breeding Pacific Slope, dark and muddy, with slow, halting song; and "Thick-billed" (*megarhyncha*), breeding Sierra Nevada and southern Cascades, with huge bill and call note a squeaky, ascending *squee*, very different from other Fox Sparrows'.

American Tree Sparrow
Spizelloides arborea | ATSP | L 6¼" (16 cm)

■ **APPEARANCE** Slender overall, with relatively long wings and tail. Bill bicolored, dark above and yellowish below; tail edges frosted white. Pale gray overall, with variable rusty highlights; plain breast has smudgy black spot at center. Juvenile plumage, rarely seen away from breeding grounds, streaky; note bicolored bill of juvenile.

■ **VOCALIZATIONS** All winter, gives whistled *tleet* and *tlee-eetle* calls. Song, rarely heard away from breeding grounds, comprises slurred whistles. Flight call a high, clipped *sseee*.

■ **BEHAVIORS** Nests at taiga-tundra transition. Winters in marshes, meadows, and thickets; flocks are flighty, flushing along hedgerows while calling.

■ **POPULATIONS** Migrates late in fall, heads back north early in spring. Midwinter movements extensive, related to food availability.

Dark-eyed Junco
Junco hyemalis | DEJU | L 6¼" (16 cm)

■ **APPEARANCE** A very average sparrow in size and build, the distinctive color scheme and funny name notwithstanding. All are pale-billed and dark-eyed, with outermost rectrices (tail feathers) mostly white. Juvenile mostly dark, with smudgy streaking above and below; compare with paler Vesper Sparrow (p. 402), also with white outermost rectrices and about the same size.

■ **VOCALIZATIONS** Song a loose trill, similar to Chipping Sparrow's (p. 392), but more jangling. Varied calls include a smacking *tsik* and a rapid series of *zing* notes.

■ **BEHAVIORS** Nests in forests. In winter, feeds in mixed-subspecies flocks in brushy habitats, especially foothills and canyon mouths; roosts in conifers if available.

■ **POPULATIONS** No other bird in the U.S. and Canada exhibits as much geographic variation as this species; it has been treated as five or more species, but current thinking is that it is just one. "Slate-colored Junco" (*hyemalis* subspecies group) uniform gray-black (male) or gray-brown (female) above, with contrasting white belly; breeds in West in Canada and Alas., winters widely in West. "Oregon Junco" (*oreganus* group) told by "executioner's hood" (black on male, gray on female) contrasting with brownish upperparts and white belly; breeds central Rockies and Pacific coast states, winters widely in West. "Cassiar Junco," comprising "Slate-colored" x "Oregon" intergrades, intermediate; breeds northern Rockies, winters southern Rockies, West Tex., and Great Plains. "Pink-sided Junco," a distinctive subspecies *(mearnsi)* within "Oregon" group, has pale blue-gray head with dark lores and buff-orange bulges on sides that nearly meet at center of breast; breeds central Rockies and Cypress Hills of Sask., winters central and southern Rockies. Range-restricted "White-winged Junco" (monotypic subspecies group *aikeni*) like a large "Slate-colored," but paler, with dark lores, white wing bars, and more white in tail; breeds mostly Black Hills of S. Dak., winters mostly on east flank of southern Rockies. "Gray-headed Junco" (*caniceps* group, including southern subspecies *dorsalis*, with some characters of Yellow-eyed Junco) pale gray overall, with darker lores and bright russet inverted triangle on back; breeds and winters southern Rockies. Despite striking differences in adult (especially male) plumages, introgression is extensive where ranges meet.

Yellow-eyed Junco
Junco phaeonotus | YEJU | L 6¼" (16 cm)

■ **APPEARANCE** Identical in build to Dark-eyed Junco, but instantly told from that species by staring yellow eye; note also bicolored bill and rufous on wings. Except for eye color, a close match for *dorsalis* subspecies of Dark-eyed Junco; *dorsalis* Dark-eyed, like Yellow-eyed, is pale gray all over, with dark lores and russet back, along with bicolored bill and some intrusion of rufous onto wings.

■ **VOCALIZATIONS** Song trilled as in Dark-eyed Junco, but considerably more complex, with three different trilled phrases in each song, suggesting that of Lincoln's Sparrow (p. 404); some songs of *dorsalis* Dark-eyed similar. Call smacking like Dark-eyed Junco's, but higher and clankier.

■ **BEHAVIORS** Common year-round in coniferous tracts in high country, often foraging amid fallen pine needles on forest floor.

■ **POPULATIONS** Mostly permanent resident throughout range, but some descend in winter to foothills.

White-crowned Sparrow
Zonotrichia leucophrys | WCSP | L 7" (18 cm)

- **APPEARANCE** Large, long-tailed sparrow. Adult, plain gray below, has black and white head stripes and colorful bill. Immature plumage, held until early spring of second calendar year, told by rust and tan head stripes. Heavily streaked juvenile plumage rarely seen away from breeding grounds.
- **VOCALIZATIONS** Song begins with one to three thin whistles, then goes into jumble of short trills or buzzes. Call a sharp *chink;* flight call a high, up-slurred *tsweee.*
- **BEHAVIORS** Nests in shrubby habitats, from timberline in Rockies to river valleys in Arctic tundra to fog-shrouded headlands on West Coast. Winters in flocks in hedgerows, brush piles, and thickets; like all *Zonotrichia* sparrows (pp. 398–401), habitually sings throughout winter.
- **POPULATIONS** Subspecies vary in migratory strategy, head pattern, and song. "Gambel's" *(gambelii),* breeding in the north and wintering widely, has pale lores and yellow-orange bill; simple song ends with discrete buzzes. "Mountain" *(oriantha),* breeding interior mountains and wintering mostly Mexico, has dark lores and pink bill; song ends with complex twitter. Coastal *nuttalli* (breeding central Calif., largely resident) and *pugetensis* (breeds north of *nuttalli,* but somewhat migratory), have pale lores, yellow bill, and dirty brownish wash below; variable song often ends in powerful trill.

Golden-crowned Sparrow
Zonotrichia atricapilla | GCSP | L 7" (18 cm)

- **APPEARANCE** Similar in build to White-crowned Sparrow; duskier and browner overall, with bill vaguely two-toned (darker above than below). Breeding adult has broad black eyebrow and bright yellow crown; same pattern in winter, but colors more muted. Winter immature has just a hint of yellowish on forecrown, resembling immature White-crowned.
- **VOCALIZATIONS** Song a descending series of three to seven achingly pure whistles. Call a descending *chick,* more like that of Lincoln's (p. 404) than White-crowned Sparrow; flight call higher than White-crowned's.
- **BEHAVIORS** Found in denser microhabitats than White-crowned: nests in shrubby tundra and stunted forest; winters in woodlots, gardens, and parks.
- **POPULATIONS** Very rare but annual in interior in winter from western Great Basin eastward, where typically in larger flocks of White-crowned Sparrows.

Harris's Sparrow
Zonotrichia querula | HASP | L 7½" (19 cm)

- **APPEARANCE** Even larger than White-crowned and Golden-crowned Sparrows. All have pink bill and at least some black on head and breast. Adult, especially in breeding plumage, has deep black on face and breast. First-winter, with less extensive black, told by white belly, rich brown flanks, and telltale pink bill and large size.
- **VOCALIZATIONS** Song a series of two or three long whistles, usually on same pitch. Call note a hard *chink;* flight call a steady, breathy *sseeeh.*
- **BEHAVIORS** Like Golden-crowned, nests mostly in shrubby tundra and forest; on migration and in winter, found in thickets and woodland edges.
- **POPULATIONS** The only bird species whose breeding range is restricted entirely to Canada. Migrates and winters mostly to our east, but wanders annually to West Coast.

NEW WORLD SPARROWS
PASSERELLIDAE

White-throated Sparrow
Zonotrichia albicollis | WTSP | L 6¾" (17 cm)

■ **APPEARANCE** Smaller than other *Zonotrichia* sparrows (pp. 398–401). Adults, with rufous wings and dusky gray below, have variably contrasting head and face patterns: Bright individuals (white morph) have bulging white throat patch, sharp black and white head stripes, and bright yellow lores; duller individuals (tan morph), with indistinct streaking below, have dirty white throat patch, duskier head stripes, and duller yellow on lores. Immature in winter, with dull face pattern, blurry streaks below, rufous in wings, and indistinct throat patch, similar to Swamp Sparrow (p. 404).

■ **VOCALIZATIONS** Until recently, sang two main songs, but in a striking example of avian cultural evolution, added a third in 2010s; all comprise wavering whistles on different pitches. Call note a hard *chink*; flight call ends with a stutter, *seeet't*.

■ **BEHAVIORS** Nests in boreal forest; in winter in West, often finds its way to feeders. White-morph individuals are socially dominant over tan-morph individuals.

■ **POPULATIONS** The two color morphs are not subspecies, but rather, a balanced polymorphism; white morphs mate only with tan morphs, and vice versa. Uncommon to West Coast, where often found with White-crowned Sparrows (p. 398).

Sagebrush Sparrow
Artemisiospiza nevadensis | SABS | L 6" (15 cm)

■ **APPEARANCE** Slim and long-tailed. Pale overall; sandy gray with neat streaking above, sparsely streaked below with blurry central spot. Head of adult mostly gray with white eyebrow and eye ring. Head of immature browner, with facial markings less distinct. Closely related Bell's Sparrow an obvious point of similarity, but compare also with Black-throated Sparrow (p. 390), especially immature.

■ **VOCALIZATIONS** Pulsing song, far-carrying but not loud, is strangely mechanical: *sixty-three, sixty-three, niiiiine-ty*. Call a high, sharp, insistent *teep!*

■ **BEHAVIORS** Occurs year-round in arid shrublands with no tree cover. Tees up nicely atop prominent shrubs, but quickly drops to ground level when approached. On ground, trots at an impressive clip with long tail cocked up like a tiny roadrunner (p. 76).

■ **POPULATIONS** Early spring migrant, back on territory in breeding range by first week of Mar. Locally common, but also absent from large swaths of seemingly appropriate habitat.

Bell's Sparrow
Artemisiospiza belli | BESP | L 6¼" (16 cm)

■ **APPEARANCE** Same build as Sagebrush Sparrow, but darker. Back of adult largely unstreaked (streaked on Sagebrush); "whisker" (malar) solid and dark (fainter, grayer on Sagebrush).

■ **VOCALIZATIONS** Song not as pulsing and mechanical as Sagebrush Sparrow's; more warbled, finchlike. Calls similar.

■ **BEHAVIORS** Like Sagebrush Sparrow, occurs year-round in shrublands, using sagebrush where available, but also bursage, saltbush, chamise, etc.

■ **POPULATIONS** Resident coastal subspecies (nominate *belli*) dark overall, with very little white in tail. Interior subspecies *canescens*, breeding east of the Sierra Nevada, intermediate in appearance between Sagebrush Sparrow and nominate *belli*; breeding ranges of Sagebrush and *canescens* nonoverlapping, but flocks co-occur in winter.

NEW WORLD SPARROWS
PASSERELLIDAE

Vesper Sparrow
Pooecetes gramineus | VESP | L 6¼" (16 cm)

- **APPEARANCE** A hefty sparrow. Gray-brown overall, with fine streaks below, small rusty patch on wing, and white outermost tail feather. Brownish face, bordered by white below, sports white eye ring.
- **VOCALIZATIONS** Song opens with two or three monotone whistles, followed by buzzes and trills. Call note a light *psip;* flight call a rising *sssee*.
- **BEHAVIORS** Nests in grasslands, sagebrush, and aspen parkland; winters widely in open habitats.
- **POPULATIONS** Migrates in broad front across West; migrants usually seen in small numbers, but large fallouts result from spring snowstorms in and near southern and central Rockies.

LeConte's Sparrow
Ammospiza leconteii | LCSP | L 5" (13 cm)

- **APPEARANCE** Slim; bill small, tail spiky. Brightly colored; flushing, shows orange rump and black-streaked back. Buffy-faced adult has white crown stripe, fine dark streaks on golden breast, and lavender streaking on nape. Juvenile, quite streaked below, has buffy crown stripe; like juvenile Grasshopper Sparrow (p. 388), can be seen on southbound migration in fall.
- **VOCALIZATIONS** Song a few weak clicks followed by shrill buzz, recalling Grasshopper Sparrow's. Flight call a fine, descending buzz.
- **BEHAVIORS** Nests in marshes and wet meadows. Secretive on migration and in winter; always in dense cover. Instead of flushing, often runs away.
- **POPULATIONS** Common on breeding grounds west to Canadian Rockies, but migrates to our east; vagrants to coast and interior mostly fall to winter.

Nelson's Sparrow
Ammospiza nelsoni | NESP | L 4¾" (12 cm)

- **APPEARANCE** Spiky-tailed like LeConte's Sparrow but more heavyset. Adult has gray crown stripe, dull orange face, and gray nape; orange wash across breast contrasts with white belly. Juvenile, bright pumpkin-orange, relatively unstreaked below.
- **VOCALIZATIONS** Song a short buzz with sharp ending: *k'jzzzz-k'*. Flight call like that of LeConte's, but more level.
- **BEHAVIORS** Nests in same Canadian prairie marshlands as LeConte's, but goes for microhabitats with deeper water.
- **POPULATIONS** All in West are nominate subspecies, similar in breeding ecology and appearance to LeConte's. Migrates to our east, yet annual to coastal Calif., especially San Francisco Bay.

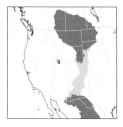

Baird's Sparrow
Centronyx bairdii | BAIS | L 5½" (14 cm)

- **APPEARANCE** A subtly stocky and short-tailed sparrow. Streaky overall, with pale ochraceous highlights. Head of adult lacks eyeline, but has two dark spots on face, a thin black "whisker," and an ochre crown stripe. Breast has fine black streaks; wing coverts and flanks flecked chestnut. Juvenile darker, with heavier streaking below and scalloped upperparts. Compare with shorter-tailed Savannah Sparrow (p. 388).
- **VOCALIZATIONS** Song, airy and tinkling, starts with one to three short whistles, then a loose trill, then a faster trill. Flight call a clipped, high *si*.
- **BEHAVIORS** Nests in tallgrass, especially undisturbed fescue, usually drier than microhabitat favored by Savannah Sparrow. Also winters in tallgrass.
- **POPULATIONS** In steady, long-term decline. Likely regular on migration east of Continental Divide, but detections very few.

NEW WORLD SPARROWS
PASSERELLIDAE

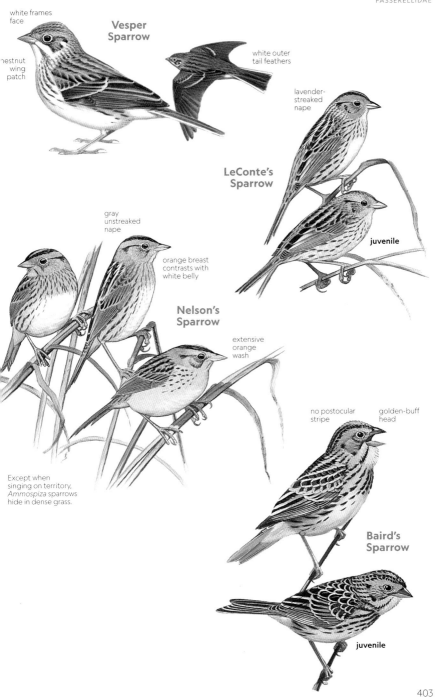

Song Sparrow
Melospiza melodia | SOSP | L 5½" (14 cm)

■ **APPEARANCE** Largest, longest-tailed *Melospiza* sparrow. Most are dark brown with coarse streaking; breast streaking coalesces in central splotch. Juvenile like adult, but buffier. Compare with Savannah (p. 388) and Vesper (p. 402) Sparrows.

■ **VOCALIZATIONS** Songs highly varied, but all follow basic pattern of a few pulsing chirps, followed by a trill and a medley of short whistles. Call note a nasal *chimp*. Flight call a high, thin *tseee*.

■ **BEHAVIORS** Occurs year-round at wooded edges of ponds, marshes, and streams. Often winters well within marshes, where it can co-occur with (and usually outnumbers) rarer sparrows. Flushing, flips tail in motions that seem uncontrolled.

■ **POPULATIONS** Exhibits tremendous geographic variation, ranging from bulky, muddy brown *maxima* of Aleutians to small, pale, rufous-brown *fallax* of Lower Colorado River Valley.

Lincoln's Sparrow
Melospiza lincolnii | LISP | L 5¾" (15 cm)

■ **APPEARANCE** Smaller-bodied, shorter-tailed, and smaller-billed than Song Sparrow. Adult finely streaked above and below; background color of breast yellow-buff. Thin white eye ring and crisply marked crown enhance sharp look overall—not as "pixelated" as a Song Sparrow. Juvenile Lincoln's browner overall; similar to juvenile Song, but shorter-tailed and slighter-billed. Compare also with juvenile Swamp Sparrow.

■ **VOCALIZATIONS** Song tripartite, three fast trills or stutters, each on a different pitch: *layda-layda-layda jurr-jurr-jurr-jurr-jurr plee-plee-plee-plee*. Call note a hard *tsik* like Dark-eyed Junco's (p. 396). Flight call a high, fine buzz.

■ **BEHAVIORS** Nests around boreal and taiga bogs and streams, especially in willow carrs, often in immediate vicinity of Wilson's Warbler (p. 442). Widespread on migration and in winter in brushy habitats, especially at edges of swampy woods. Although skulky, comes right out to pishing.

■ **POPULATIONS** Increased detections in winter in interior recent years likely the result in part of better effort by birders, but also thought to reflect milder winters. Breeding range appears to be expanding in Alta.

Swamp Sparrow
Melospiza georgiana | SWSP | L 5¾" (15 cm)

■ **APPEARANCE** Built like Lincoln's Sparrow. In all plumages, has white throat like larger White-throated Sparrow (p. 400). Breeding adult has rusty cap, browner in winter; wings have broad reddish-brown panel, flanks warm buff. First-winter dark and muddy, but with rusty wings, buffy flanks, and white throat; rusty wings already prominent on briefly held juvenile plumage.

■ **VOCALIZATIONS** Song a loud, loose trill; slower, more jangling than Dark-eyed Junco's and Chipping Sparrow's (p. 392). Call note a bright chip, sharply descending. Flight call like that of Lincoln's.

■ **BEHAVIORS** Occurs year-round in marshes (not swamps), especially in dense cattail beds. Like Lincoln's, readily responds to pishing.

■ **POPULATIONS** Away from Canadian breeding grounds, uncommon to rare throughout West as migrant and winterer, with concentration late fall to early winter, especially in marshes haunted by Song Sparrows.

Canyon Towhee
Melozone fusca | CANT | L 8" (20 cm)
- **APPEARANCE** Very large sparrow, the heft of a Red-winged Blackbird (p. 418). Adult mostly gray, but with rufous cap and apricot undertail coverts; throat pale buff. Breast has variable streaking and diffuse black splotch at center. Juvenile muddy brown, diffusely streaked below, with warm coloration on crown and undertail coverts starting to show.
- **VOCALIZATIONS** Call note an abrupt, nasal *schwrint!* Song a variable trill often preceded by the call note: *schwrint! le-le-le-le-le-le-le-le-le-le*.
- **BEHAVIORS** Occurs year-round in arid landscapes: dry washes, rocky slopes, and grasslands dotted with cholla and yucca.
- **POPULATIONS** Sedentary permanent resident throughout range; rarely wanders even short distances from core range.

Abert's Towhee
Melozone aberti | ABTO | L 9½" (24 cm)
- **APPEARANCE** More slender than Canyon Towhee. Mostly uniform brown-gray with apricot undertail coverts; black face contrasts with pale bill.
- **VOCALIZATIONS** Call a clanking *pleet.* Song bipartite, beginning with notes like call, followed by rougher series: *pleet pleet pleet ch'ch'ch'ch'ch'ch'.*
- **BEHAVIORS** Occupies lusher, lower-elevation sites than Canyon Towhee: cottonwood, willow, and mesquite along permanent streams.
- **POPULATIONS** Nonmigratory; range restricted almost entirely to U.S. portions of Sonoran Desert. Numbers increase following removal of cattle from riparian corridors.

California Towhee
Melozone crissalis | CALT | L 9" (23 cm)
- **APPEARANCE** Midway in structure and plumage between Canyon and Abert's Towhees. Darker, browner, and more uniformly colored than Canyon; rufous cap not as prominent as on Canyon, and lacks central breast spot.
- **VOCALIZATIONS** Call an explosive, descending *pink!* Song an accelerating series of notes like calls.
- **BEHAVIORS** Occurs year-round in shrubby habitats, from coastal chaparral to brushy understory of oak woodlands farther inland; readily takes to urban districts.
- **POPULATIONS** Essentially no range overlap with other *Melozone* towhees. Isolated but poorly defined subspecies *eremophila*, found only in arid Argus Mts. of Inyo Co., Calif., may be slightly smaller; numbers fewer than 1,000 individuals.

Rufous-crowned Sparrow
Aimophila ruficeps | RCSP | L 6" (15 cm)
- **APPEARANCE** Midsize sparrow with long tail. Breast plain gray; general color, including rusty cap and vague white eye ring, suggests Rufous-winged Sparrow (p. 386). Rufous-crowned has one moustache, thick and black; Rufous-winged has two thinner "whiskers," also black. Juvenile, streaked below, shows dark moustache and hint of eye ring.
- **VOCALIZATIONS** Bubbly song a rapid outpouring of chirps, suggesting House Wren's (p. 330). Call a twangy, descending *gwaay* or *gweer.*
- **BEHAVIORS** Found year-round on rock-strewn lower slopes of foothills and canyons.
- **POPULATIONS** Nonmigratory; rarely strays from core range. Despite the name "sparrow," is the closest relative of the *Melozone* towhees, with similar morphologies, vocal arrays, habitats, and dispersal patterns.

Green-tailed Towhee
Pipilo chlorurus | GTTO | L 7¼" (18 cm)

- **APPEARANCE** A small towhee, midway in size between Rufous-crowned Sparrow (p. 406) and Spotted Towhee. Our most colorful sparrow: wings, back, and tail of adult suffused with yellow-green; head, mostly gray, highlighted with bright rufous crown, gleaming white throat, and thick black "whisker" (malar); underparts also mostly gray, but with yellow-washed undertail coverts. Heavily streaked juvenile washed in dark olive-yellow.
- **VOCALIZATIONS** Song a striking outburst of buzzes and slurred whistles, not unlike song of *schistacea* ("Slate-colored") Fox Sparrow (p. 394). Distinctive call a mewing, nasal *mreeeay*. Like Spotted Towhee, gives a fine buzz, probably a flight call.
- **BEHAVIORS** Nests in shrubby habitats, including mountain mahogany–covered slopes, sagebrush flats in high valleys, and recently burned pinewoods; generally at higher elevations than Spotted Towhee. Winters widely in shrubby habitats, with mesquite-dominated desert washes favored.
- **POPULATIONS** The most migratory towhee. Is part of a suite, or guild, of western passerines that move upslope in appreciable numbers following nesting, routinely reaching alpine tundra; eventually migrates to main wintering grounds. Small breeding outposts in West Tex. (Davis Mts., Big Bend N.P.) are declining.

Spotted Towhee
Pipilo maculatus | SPTO | L 7½" (19 cm)

- **APPEARANCE** Midsize towhee. Adult recognized by combination of dark hood (black on male, browner on female), heavy white spotting on black mantle, and extensive rufous on flanks and undertail coverts; eyes red, tail with white corners. Juvenile, streaked and smudgy, already shows white tail corners of adult and, usually, some spotting on back and wings.
- **VOCALIZATIONS** Songs vary geographically, but almost all have one or more trilled elements with dry, raspy quality overall. Calls, also geographically variable, slurred and nasal. All give a fine, high, wavering buzz, probably a flight call.
- **BEHAVIORS** During nesting season is common and conspicuous in shrubby habitats, especially pinyon-juniper woods in foothills. More reclusive in winter, although a frequent visitor to feeders. Feeds on the ground with jerky "double kickbacks" that send leaf litter flying.
- **POPULATIONS** Where range overlaps with other towhees, usually nests at higher elevations than Canyon Towhee (p. 406) but lower elevations than Green-tailed Towhee. Exhibits much geographic variation, especially with regard to vocalizations. Southern Rockies *montanus*, with relatively limited white spotting above, sings a few dry chips followed by an unmusical trill; call note a harsh *wraaay*, level or slightly rising. Northern Great Plains *arcticus*, with more white above and on tail than other subspecies, sings a jangling trill with or without introductory notes; call note, harsh like *montanus*, an ascending *wraaynt?* Multiple Pacific coast subspecies, darker above in north than south, sing a simple trill, fast and buzzy; call note anguished and nasal, an ascending *wree-eent?* A final complication is that birds at limit of eastern range of *arcticus* hybridize with Eastern Towhee (p. 465); a few of these intergrades make it into our area, especially in winter.

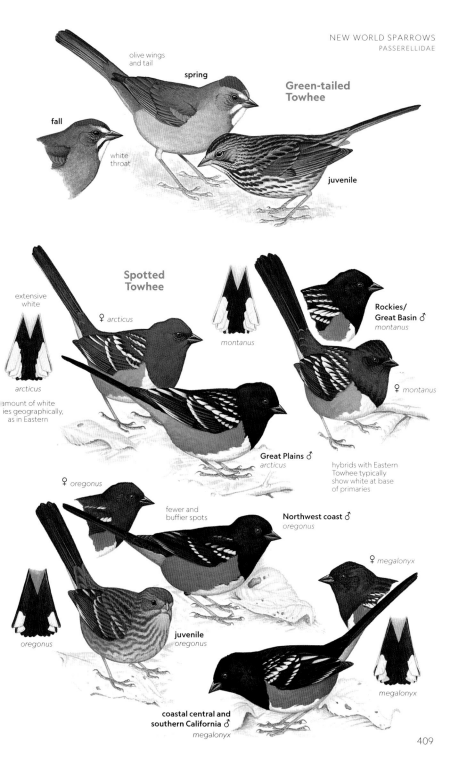

YELLOW-BREASTED CHAT | ICTERIIDAE

Many birds go by the name of "chat," but *Icteria virens* is sui generis, the sole representative of its family. The closest relatives of this bird are probably the blackbirds (pp. 410–423), an arrangement that makes sense in terms of the Yellow-breasted Chat's body structure and complex vocalizations.

Yellow-breasted Chat
Icteria virens | YBCH | L 7½" (19 cm)

- **APPEARANCE** Slender; long-tailed and thick-billed. Brown above with white "spectacles" and brilliant yellow below.
- **VOCALIZATIONS** Incredible run-on song a halting mix of laser-tag outbursts, white-noise shrieks, endearingly sad sighs, and soft chuckles. Mimics birds and other animals. Calls include a harsh *raaa* and knocking *plock*.
- **BEHAVIORS** Except in flight display, always in dense tangles. Impossible to miss when singing, but otherwise difficult to spot.
- **POPULATIONS** Timing of departure from breeding grounds and onset of fall migration not fully worked out, as chats become notoriously secretive after early July.

BLACKBIRDS | ICTERIDAE

Birds called "blackbird" are in this family, but so are birds called meadowlarks, orioles, and others. Most have strong bills and loud vocalizations, and many build unusual nests and engage in complex mating systems. Several species in the West form huge flocks during the nonbreeding season.

Yellow-headed Blackbird
Xanthocephalus xanthocephalus | YHBL | L 9½" (24 cm)

- **APPEARANCE** Large blackbird; stout-billed and peculiarly long-toed. Black-bodied adult male has saffron yellow head, white wing patch, and yellow feathers around cloaca. Adult female brown with dirty yellow face and bib. Variable subadult male told by extensive but muted yellow on head and breast, combined with large size and white in wing. Immature female can nearly lack yellow.
- **VOCALIZATIONS** Song of male, rated by many human hearers as decidedly unpleasant, a couple of mechanical clanks followed by a tormented scream. Diverse female calls include a rolling chatter. Both sexes give hollow, stuttering clucks.
- **BEHAVIORS** Nests in cattail marshes; mating system is polygynous. Joins mixed-species blackbird flocks, often at feedlots, following nesting.
- **POPULATIONS** Local abundance variable. Can be abundant in one marsh, absent in another nearby; sensitive to both habitat quality and tract size.

Bobolink
Dolichonyx oryzivorus | BOBO | L 7" (18 cm)

- **APPEARANCE** Small blackbird; tail short and spiky. Breeding male black below, with creamy nape and white on wings and rump; nonbreeding male buffy overall with streaking on back and flanks. Adult female year-round has plumage like nonbreeding adult male; note pink bill of all females and nonbreeding males.
- **VOCALIZATIONS** Song squeaky and metallic, with abrupt changes in pitch (the "R2-D2 bird"). Flight call a nasal, rising *weenk?*
- **BEHAVIORS** Nests in wet meadows and hayfields; after breeding, sexes separate.
- **POPULATIONS** Late spring migrant (May); heads back south again by Aug. Winters mostly Bolivia, Paraguay, and northern Argentina.

YELLOW-BREASTED CHAT | BLACKBIRDS
ICTERIIDAE | ICTERIDAE

white submoustachial stripe

Yellow-breasted Chat

Yellow-headed Blackbird

variable white in wing

juvenile ♂

variable white in wing

spring adult ♂

Bobolink

pink bill

heavily streaked

fall

long wings

breeding ♂

buffy edges when fresh

early spring ♂

breeding ♂

breeding ♀

Eastern Meadowlark
Sturnella magna | EAME | L 9½" (24 cm)

Barely enters our area in Colo., probably reaching to near Front Range foothills annually. Identical in structure to Western and Chihuahuan Meadowlarks: All three are rotund, short-tailed, and short-winged, with long bill and sturdy legs. Plumages also very similar: All three are brown above with fine striping and scalloping; brilliant yellow below with huge black breastband and coarsely streaked flanks. Eastern differs from Western by white malar, more white in tail, warmer upperparts, and especially in vocalizations: Song of Eastern comprises sweet, thin whistles; call a harsh, flatulent *dzzrnt*. Very similar Chihuahuan has more white in tail than Eastern; no range overlap on breeding grounds. Has strayed far from range (Alta., Wash., Calif.); documentation of vagrants requires photos and audio.

Western Meadowlark
Sturnella neglecta | WEME | L 9½" (24 cm)

Introduced to Hawaii

■ **APPEARANCE** Identical in size and shape to Chihuahuan, and similar in plumage. Malar region of Western yellow (white on Chihuahuan); Western has one or two mostly white outer rectrices (three or four on Chihuahuan). Head stripes of Western average blurrier than on Chihuahuan. Compare also with Eastern Meadowlark, but almost no range overlap in West.

■ **VOCALIZATIONS** Entire vocal array lower and richer than that of Chihuahuan. Song of Western a series of descending notes, loud and gurgling, with six to eight or more elements (four or five in Chihuahuan). Call a rich, musical *clurk* (corresponding call of Chihuahuan a harsh *dzert*); also gives a low, whistled *lurrr*. Flight call a rising whistle, *wheent*; flushing or landing, a low-pitched rattle.

■ **BEHAVIORS** Conspicuous in open country: farms, prairie, sagebrush, etc. In zone of overlap with Chihuahuan, Western occurs in wetter habitats: irrigation ditches, farm ponds, etc. All meadowlarks fly low to the ground in level flight, suggesting profile of European Starling (p. 340).

■ **POPULATIONS** Subspecies *confluenta*, resident from B.C. to Ore., darker above. Introduced to Kauai, where widespread in small numbers; the birds on Kauai are believed to be derived from *confluenta* stock.

Chihuahuan Meadowlark
Sturnella lilianae | CHME | L 9½" (24 cm)

■ **APPEARANCE** Body structure the same as other meadowlarks. Plumage differences slight, but Chihuahuan is, on average, the plainest, palest, and whitest meadowlark. Upperparts washed-out and face plain; flanks, streaked as on other meadowlarks, have off-white ground color. Malar of Chihuahuan white (also white on Eastern, but yellow on Western). Outer three or four (usually four) rectrices of Chihuahuan mostly white (two or three on Eastern, one or two on Western).

■ **VOCALIZATIONS** Vocal array closer to Eastern's than to Western's. Song a bit lower than Eastern's (more like Western's in this regard), but with simple, whistled, ungurgled elements (like Eastern's). Call repertoire basically matches that of Eastern.

■ **BEHAVIORS** Occurs year-round in desert grasslands. Western Meadowlark usually in less xeric habitats, especially in breeding season.

■ **POPULATIONS** Formerly treated as subspecies of Eastern Meadowlark, but recent research shows that Eastern and Western are more closely related to each other than either is to Chihuahuan.

Orchard Oriole
Icterus spurius | OROR | L 7¼" (18 cm)
- **APPEARANCE** Our smallest oriole; slim overall, with fairly long tail and slender, decurved bill. Adult male, black and rich chestnut, distinctive, but often appears all-dark in shade; mostly yellow female, with thin white wing bars, as likely to be mistaken for a warbler (pp. 424–443) as another oriole. Male in first spring to summer (second calendar year) like female but with narrow black bib and haphazard chestnut patches; this delayed plumage maturation is uncommon in passerines (pp. 266–453) but the norm for orioles. Compare female, especially immature, with female Hooded; normal ranges barely overlap, but vagrants out of range can be challenging.
- **VOCALIZATIONS** Song rapid and bubbly; Bullock's Oriole (p. 416) has slower, lower song. Calls include snappy *chack* and Bullock's-like chatter.
- **BEHAVIORS** Like other orioles, a mostly arboreal blackbird. In limited range in West, nests in broadleaf groves around ranches and along streams. Active in treetop foliage; its fidgetiness contributes to its warblerlike mien, especially in the case of the extensively yellow female.
- **POPULATIONS** Although at periphery of range in West, is locally common where it does occur in our area. An early fall migrant, with most gone from midlatitudes by late summer; migration generally to our east, but a few show up every year on coast and in Desert Southwest.

Hooded Oriole
Icterus cucullatus | HOOR | L 8" (20 cm)
- **APPEARANCE** Fairly small oriole with thin, decurved bill. Black face of adult male encircled by yellow-orange cowl; adult male Bullock's Oriole (p. 416), with roughly similar scheme of color and pattern, has black cap, black eyeline, and just a narrow patch of black on chin. Female and second-summer male Hooded can be quite similar to corresponding plumages of Orchard; Hooded is larger and more orange, with bill longer and thinner. Female and immature Bullock's also similar, but more yellowish than Hooded, with mostly white belly and straighter bill. Compare also with Streak-backed Oriole.
- **VOCALIZATIONS** Rambling song suggests Orchard Oriole's or even a thrasher's (pp. 336–341). Distinctive call a rising, nasal *veent?*
- **BEHAVIORS** Nests well up in trees, with palms especially favored; found around resorts, desert oases, and in palm-studded downtown districts of major metropolises.
- **POPULATIONS** Most in West are smaller-billed subspecies *nelsoni*, but nominate *cucullatus* reaches Big Bend region from Mexico; adult male *cucullatus* has more black on face, giving it more of a bibbed appearance, encircled by blaze orange.

Streak-backed Oriole
Icterus pustulatus | SBAO | L 8¼" (21 cm)
Mid. Amer. oriole; annual in recent years to Southwest, mostly fall to spring, with preponderance of records to Southeast Ariz., where breeding has occurred. Similar in all plumages to Hooded Oriole, but bulkier Streak-backed has thicker, straighter bill and broken black streaks on back. Adult male brighter orange than *nelsoni* Hooded. Adult male, as well as adult female and most subadults, shows limited black on face extending to breast in narrow vertical patch; compare with Bullock's Oriole (p. 416). Song a slow, low, simple warble; calls include a rising *tuwee*, not unlike House Finch's (p. 372), and a harsh, rapid clacking. Has strayed throughout Interior West, to West Tex., and to Calif. coast.

Bullock's Oriole
Icterus bullockii | BUOR | L 8¼" (21 cm)

■ **APPEARANCE** Midsize oriole with nearly straight bill. Adult male bright orange with white wing panel and black back, cap, eyeline, and bib. Most male Hooded Orioles in West not as orange, lack black cap and eyeline of male Bullock's. Adult male Baltimore Oriole has all-black head and orange in wing. Female and immature male Bullock's distinguished from corresponding plumages of Hooded by larger size, straighter bill, shorter tail, and pale belly; from Baltimore, with care, by pale yellowish face (dusky on Baltimore), gray rump (yellowish on Baltimore), pale belly (yellow-orange on Baltimore), and hint of dark eyeline (absent on Baltimore). Compare also with immature male Orchard Oriole (p. 414); immature male Bullock's is larger, straight-billed, and more orangey.

■ **VOCALIZATIONS** Song a few squeaks followed by short whistles. Call a dry chatter; flight call a rising, nasal *wreent?*

■ **BEHAVIORS** Nests in shade trees, especially cottonwoods in foothills, floodplains, and canyons. Spring migrants visit feeders provisioned with orange slices and grape jelly; molt migrants tend flowering yuccas in deserts.

■ **POPULATIONS** Large-scale molt migration underway by late June; many adults spend more time on molting grounds than breeding grounds.

Baltimore Oriole
Icterus galbula | BAOR | L 8¼" (21 cm)

■ **APPEARANCE** Same size and shape as Bullock's Oriole. All plumages of Baltimore darker-faced than corresponding plumages of Bullock's: Adult male Baltimore has black head and limited white in wing; adult female Baltimore has brown-black face and orange-yellow tones overall; young Baltimore dusky-faced, body with orangish wash. Hybridizes with Bullock's where ranges overlap; hybrid male has "messy" black-and-orange face; many hybrid females not possible to ID in field.

■ **VOCALIZATIONS** Song richer, slower, and lower than Bullock's. Chatter call and rising flight call like Bullock's.

■ **BEHAVIORS** Haunts and habits much as those of Bullock's; both species build pendant nests, exquisitely woven of plant matter and spider webs, prominent in treetops.

■ **POPULATIONS** Despite being the "eastern" oriole, nests west to foothills of Canadian Rockies. Very rare but annual, mostly fall, to coast and southern border states.

Scott's Oriole
Icterus parisorum | SCOR | L 9" (23 cm)

■ **APPEARANCE** Slightly larger than Bullock's Oriole, with long, straight, fine bill. Adult male sports black hood and bright lemon below; adult female similar, but black hood more mottled, lemon underparts suffused grayish. Trickier are first-fall birds, especially females; note broken black streaks on back, uniform lemon-gray underparts, and bill structure.

■ **VOCALIZATIONS** Gurgling song like Western Meadowlark's (p. 412), but steadier, not descending in pitch. Call a rough, scratchy, descending *chick*.

■ **BEHAVIORS** Nests in deserts and arid woodlands; spends much time at flowering yuccas.

■ **POPULATIONS** Northern limits of range something of a mystery; breeds well north, annually but erratically, of core range in Desert Southwest.

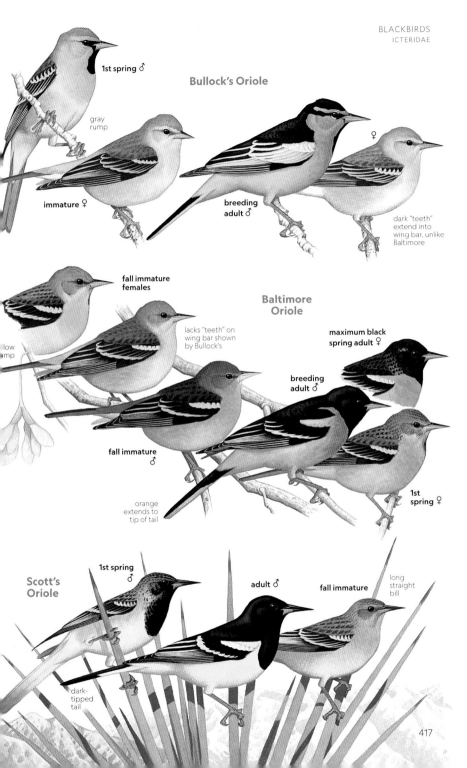

Red-winged Blackbird
Agelaius phoeniceus | RWBL | L 8¾" (22 cm)

■ **APPEARANCE** A fairly small blackbird; compactly built, with straight, thick-based bill and relatively short tail and wings. In most of range, breeding adult male unique, with completely black body except for wings with broad red-and-yellow "epaulets." Freshly molted male in fall has variable, sometimes extensive, bronzy scalloping. Female, smaller than male, one of the most frequently misidentified birds in the U.S. and Canada; rufous and black tones, with heavy streaking overall, suggestive of New World sparrows (pp. 386–409), but compare bill structure (almost all sparrows have shorter, thicker, more rounded bills).

■ **VOCALIZATIONS** Incredibly noisy. Song, loud and grating, a few clicks, then a harsh trill, then a trailing hiss: *ch'ch'LEEEEEEEsss;* reminiscent of Savannah Sparrow's (p. 388), but much louder. Calls, amazingly varied, include a musical *cluck*, an annoyed *pyit*, a sharply descending and ear-piercing whistle, a mechanical *k'dick* like Virginia Rail's (p. 96), and many others.

■ **BEHAVIORS** Nests in wetlands anywhere, from vast marshes in remote mountain valleys to tiny floodwater basins in our largest cities; sometimes also in meadows and weedy roadsides some distance from wetlands. Polygynous males display constantly, puffing out their epaulets while singing. Winters, sometimes in gargantuan flocks, around feedlots; many also stay behind in the marshes where they nested.

■ **POPULATIONS** Common to locally abundant; winter roosts persecuted for their supposed threat to agriculture. "Bicolored Blackbird," a complex of subspecies from Mexico and around Central Valley and central coast of Calif., distinctive: Epaulets of male have little if any yellow fringing; female mostly blackish on belly.

Tricolored Blackbird 🆔
Agelaius tricolor | TRBL | L 8¾" (22 cm)

■ **APPEARANCE** Slightly slimmer overall than similar Red-winged Blackbird, and wings of Tricolored less rounded and bill longer and thinner. Breeding adult male has red "epaulets" fringed gleaming white (yellow on Red-winged); however, freshly molted male in fall has buffier fringe, suggesting Red-winged. Female patterned like female Red-winged, but colder, grayer, and darker overall, except for paler gray-white throat; note also the long, thin bill and mostly black belly of female Tricolored. Keep in mind that core range of Tricolored overlaps extensively with that of "Bicolored Red-winged Blackbird," females of which are more Tricolored-like (grayer, darker, less rufescent) than are female Red-wingeds elsewhere in West.

■ **VOCALIZATIONS** Song less hissing, more mewing and growling, than Red-winged's. Calls, more nasal and generally less remonstrative than those of Red-winged, include an abrupt *skroonch,* a buzzy *fzznt,* and nasal, descending whines like Cliff Swallow's (p. 314).

■ **BEHAVIORS** Nests in large wetlands, typically in dense concentrations; usually not found in small, degraded, marginal wetlands of the sort readily accepted by Red-winged Blackbirds. After nesting, makes its way to agricultural districts, where it enlists in mixed-species flocks with other blackbirds.

■ **POPULATIONS** In steady and long-term decline in core range in central Calif., but also somewhat opportunistic and establishes colonies beyond periphery of normal range, for example, Gem Co. and Payette Co., Idaho, in 2020s. Partnerships with rice growers in Calif.'s Central Valley, whereby farmers are compensated for crops lost to voracious Tricolored Blackbirds, are beginning to yield conservation dividends.

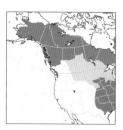

Rusty Blackbird
Euphagus carolinus | RUBL | L 9" (23 cm)

■ **APPEARANCE** The two *Euphagus* blackbirds are like scaled-down grackles (p. 422), shorter-tailed and smaller-billed. Bill of Rusty thinner than that of Brewer's, and slightly decurved. All plumages of Rusty pale-eyed. Freshly molted female Rusty in fall coppery overall, with buff supercilium and black face mask; freshly molted male Rusty darker, with copper-fringed feathers above. On both sexes, copper coloration wears down by spring, making the birds more Brewer's-like. Female Rusty, spring to summer, told by pale eyes; male Rusty, with care, by relatively dull plumage and bill structure. Compare also with rusty-scalloped Red-winged Blackbird (p. 418) in fall, thicker-billed and slightly smaller overall.

■ **VOCALIZATIONS** Song a rough gurgle followed by a rising whistle: *schwrlklwsh-WEEE*. Call a resonant *chenk*.

■ **BEHAVIORS** Entire annual cycle involves shady, swampy woods: breeding, wintering, and, even on migration, finds its way to well-wooded bogs, lakeshores, and streambanks. Does not tend feeders.

■ **POPULATIONS** In sharp decline, for reasons poorly understood. Rare in West away from breeding grounds, but sometimes winters in small flocks in western Great Plains.

Brewer's Blackbird
Euphagus cyanocephalus | BRBL | L 9" (23 cm)

■ **APPEARANCE** Size and shape much like Rusty Blackbird's but bill straighter. Male dull glossy black in fall, wearing to bright glossy blue-black by spring; some males in fall have dark rusty scalloping like Rusty, but wings of Brewer's wholly dark. Female Brewer's gray-black with dark eye; female Rusty has pale eye.

■ **VOCALIZATIONS** Song, weak and gasping, a few clacks or clanks, then a tinny buzz: *tsick-tsicka wzzzzzzzz*. Call a dull *chet*.

■ **BEHAVIORS** Nests in open places near water and woody vegetation, especially high-elevation bogs, but also on plains and even deserts if water is nearby. Away from breeding grounds, an assertive and conspicuous visitor to feedlots and parking lots.

■ **POPULATIONS** Has declined in urban habitats in western Great Plains, perhaps because of competition from invading Common Grackles (p. 422).

Brown-headed Cowbird
Molothrus ater | BHCO | L 7½" (19 cm)

■ **APPEARANCE** Small, short-tailed blackbird, with short, conical, sparrowlike bill. Male has brown hood and metallic blue-green body. Adult female, plain grayish brown, one of the most frequently sighted "mystery birds"; juvenile like adult female, but more streaked. Compare with larger-billed, larger-bodied Bronzed Cowbird (p. 422).

■ **VOCALIZATIONS** Song a liquid gurgle ending in a sharply rising whistle. Calls include a rattle and, in flight, a rising whistle.

■ **BEHAVIORS** Female lays eggs in other birds' nests; young raised by host parents, which are often much smaller. Year-round, tends livestock; fall to winter, also enlists in large, mixed-species blackbird flocks, especially around feedlots.

■ **POPULATIONS** In late 20th century, was enthusiastically and euphemistically "controlled" (killed) because of impacts on at-risk host species like Bell's Vireo (p. 284); evidence for effectiveness has been equivocal, however, and cowbird control is less common today.

BLACKBIRDS
ICTERIDAE

Rusty Blackbird

gray rump

♂

distinctive head pattern with rusty crown and pale supercilium

fall ♀

pale eye

breeding ♀

fall ♂

breeding ♂

Brewer's Blackbird

wings uniformly dark

immature ♂

♂

in good light, plumage more glossy than Rusty

♀

small conical bill

♂

immature ♂ in molt

♂

Brown-headed Cowbird

parasitic female Brown-headed Cowbirds lay eggs in nests of other species

heavily streaked above and below

juvenile

Yellow Warbler

♀

Bronzed Cowbird
Molothrus aeneus | BROC | L 8¾" (22 cm)

- **APPEARANCE** Longer-billed and bulkier than Brown-headed Cowbird (p. 420). Breeding male has red eye, puffed-out "ruff" on neck, and dark bronze cast overall. Variable female also red-eyed, but smaller, not as glossy, and with ruff somewhat reduced.
- **VOCALIZATIONS** Song high and wheezy, with less frequency sweep than Brown-headed's. Call a dull *chup*.
- **BEHAVIORS** Like Brown-headed Cowbird, an obligate brood parasite. Habitat generalist, but less inclined than Brown-headed to enter forests. Like Brown-headed, finds its way to feedlots after breeding season.
- **POPULATIONS** Female and young vary geographically: relatively pale in *loyei*, breeding west of Rio Grande; almost as dark as adult male in nominate *aeneus*, breeding in our area in southeastern N. Mex. and West Tex. More common in summer in West, although *loyei* increasingly noted in winter.

Common Grackle
Quiscalus quiscula | COGR | L 12½" (32 cm)

- **APPEARANCE** Longer and larger than Brewer's Blackbird (p. 420), with noticeably longer tail. Tail shape strange: corners angled in flight like a sharply cut wedge. Metallic blue hood of adult male contrasts strongly with lustrous bronzy cast of rest of body. Female and winter male less glossy; juvenile dull gray-brown, lacking any gloss.
- **VOCALIZATIONS** Song a loud *wshrishAAAANK*, like a louder, more clanging Brewer's Blackbird's. Call, often given in flight, a loud, toneless *chack*.
- **BEHAVIORS** Loosely colonial breeder, often nesting in planted conifers in yards and parks. Noisy and gregarious: Flocks strut on lawns, shriek from shrubs and trees, and fly over in disorganized assemblages.
- **POPULATIONS** Has expanded greatly in our area in past half-century, possibly to detriment of Brewer's Blackbird. Returns later in spring than at comparable latitudes farther east.

Accidental (ship-assisted) to Hawaii

Great-tailed Grackle
Quiscalus mexicanus | GTGR | L 15–18" (38–46 cm)

- **APPEARANCE** Huge and spectacular; tail massive (especially male's) and bill very long. Adult male, glossed deep purple, as long as or longer than American Crow (p. 300), but not nearly as bulky. Female variable, with some as small as Common Grackle but others larger, although never as large as male; pale eye and pale eyebrow of female contrast with dark face mask.
- **VOCALIZATIONS** Full song of male highly complex, with well-spaced chirps, explosive rattles, wild wailing, and wheezy whistles, but often gives just a few elements of song. Female call a rising, nasal *mraah*, like Black-billed Magpie's (p. 298); male call a resonant *cluck*.
- **BEHAVIORS** In southern portions of our area, nests widely wherever there is vegetation near water; farther north in our area, nests mostly in cattail marshes. In winter, gathers in dense flocks, both mixed-species and monospecific; sexes segregate somewhat in winter.
- **POPULATIONS** Birds from Ariz. westward smaller (especially females), but much variation. Frequently gets north of mapped range, typically singly spring to summer and sometimes forming large flocks in winter. A record for Hawaii (Honolulu area) pertains to a long-staying, ship-assisted male.

BLACKBIRDS
ICTERIDAE

"ruff" often raised

♂

Bronzed Cowbird
loyei

juvenile

thick conical bill

♀

short-tailed and big-headed

♂

Texas ♀
aeneus

♂

"Bronzed Grackle" ♂
versicolor

juvenile

bill shorter and thicker than Great-tailed

Common Grackle

Great-tailed Grackle
monsoni

juvenile

plumages other than breeding male extremely variable

♀

♂

♂

no contrast between head and body as in Common Grackle

long bill

western ♀
nelsoni

WOOD-WARBLERS | PARULIDAE

Active and mostly arboreal, the birds in this speciose assemblage are, for the most part, brightly colored and strongly patterned. They are small to very small songbirds, and all have fine bills that aid in the capture of their diminutive arthropod prey.

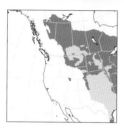

Ovenbird
Seiurus aurocapilla | OVEN | L 6" (15 cm)
- **APPEARANCE** A large and plump warbler. Olive-brown above; white below with bold streaks. Orange crown stripe has black borders.
- **VOCALIZATIONS** Rollicking song increases in amplitude: *er-teech er-teech er-TEECH er-TEECH!* Also gives soft, rambling flight song that ends with fragments of primary song. Call note a sharp *chip*; flight call a piercing *pseek!*
- **BEHAVIORS** The most un-warblerlike of the wood-warblers. Nests in lush broadleaf groves with dense understory, where it struts about the forest floor with tail cocked up. Heard far more often than seen; sings from ground or low shrub. Flight song typically given at night; secretive on migration.
- **POPULATIONS** Breeds west to foothills of Rockies, where it can be surprisingly common; apparent recent increases there due in part to improved effort by birders, but perhaps also to a real range shift.

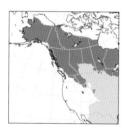

Northern Waterthrush
Parkesia noveboracensis | NOWA | L 5¾" (15 cm)
- **APPEARANCE** Superficially similar in structure and plumage to Ovenbird. Solid olive-brown above; pale below with dark streaks on faint creamy background. Mostly dark face marked with long, pale eyebrow. Compare with Louisiana Waterthrush (p. 465), which is very rare but annual to Ariz., late summer to winter.
- **VOCALIZATIONS** Explosive song a rapid series of notes dropping in pitch: *CHEE CHEE chip chip chup chup.* Call note a sharp, clanking *plink*, recalling White-crowned Sparrow's (p. 398); flight call a fine, rising buzz.
- **BEHAVIORS** Terrestrial and aquatic. Nests near ground in spruce bogs with willow and alder admixtures. Migrants invariably on or near ground at edges of well-wooded ponds, seeps, and draws. With tail twitching, constantly bobs body.
- **POPULATIONS** Probably the most frequently encountered "eastern" warbler on migration in West—although "eastern" is a misnomer, given that the species breeds west to the Bering Sea.

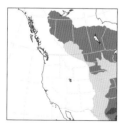

Black-and-white Warbler
Mniotilta varia | BAWW | L 5¼" (13 cm)
- **APPEARANCE** Short tail and long, thin bill impart front-heavy look. All plumages streaked black and white all over. Breeding male has black throat; female and young have mostly white face and throat. Compare with Blackpoll Warbler (p. 436), especially breeding male.
- **VOCALIZATIONS** Song, high and thin, a chanting *wee-see wee-see wee-see wee-see* .... Call note a husky *chint*; flight call a high buzz.
- **BEHAVIORS** More than any other warbler, stays close to branches, even tree trunks. Forages more like a nuthatch (p. 326) than other warblers; body structure (short tail, long bill) accentuates nuthatch-like gestalt. Blackpoll Warbler feeds in treetop foliage, does not creep along trunks.
- **POPULATIONS** Migration, spring and fall, largely to our east; however, is annual on migration in most places in West, making it one of our more common "eastern" warblers.

WOOD-WARBLERS
PARULIDAE

Ovenbird

gleans insects from forest floor; does not overturn leaves like Swainson's Warbler (p. 472)

bold eye ring

Northern Waterthrush

supercilium and underparts washed pale yellow-buff

throat usually streaked

rapidly bobs tail up and down

paler

Black-and-white Warbler

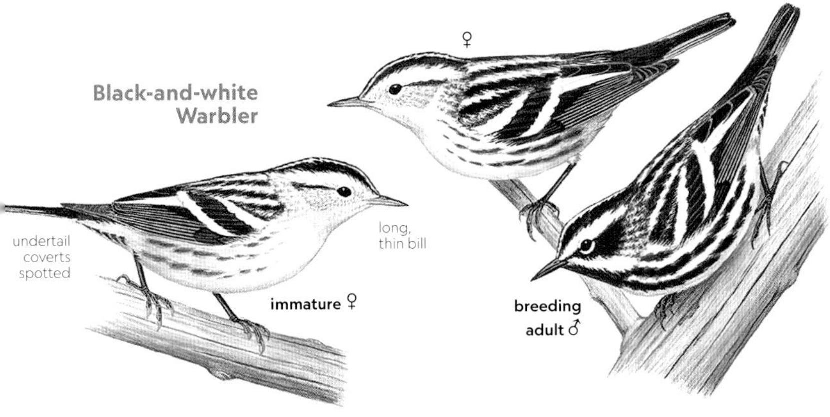

undertail coverts spotted

long, thin bill

immature ♀

♀

breeding adult ♂

Warblers

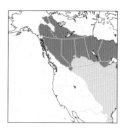

Tennessee Warbler
Leiothlypis peregrina | TEWA | L 4¾" (12 cm)

- **APPEARANCE** Like most other warblers in genus *Leiothlypis* (pp. 426–429), small-bodied and slight-billed. Tennessee is short-tailed and long-winged. Adult male mossy green above with gray cap, white eyebrow, and dark eyeline; underparts, including undertail coverts, whitish. Adult female olive green above, including crown, with white eyebrow and dark eyeline of male; breast yellowish, undertail coverts white. Compare especially with Orange-crowned Warbler; superficially similar to distantly related Philadelphia Vireo (p. 288).
- **VOCALIZATIONS** Tripartite song sounds rushed, builds in amplitude: *tik-tik-tik-twick-twick-twick-twick-TYEW-TYEW-TYEW-TYEW*.... Call note, light and smacking, *tsinck*; flight call a clear, rising *tsee*?
- **BEHAVIORS** Nests close to ground in northern forests. Spring migrants stick to treetops, fall migrants less so. Active; hard to see in canopy, but males declaim loudly on spring migration and breeding grounds.
- **POPULATIONS** Migrates mostly to our east: uncommon but regular in western Great Plains in spring; rare elsewhere in West, mostly fall.

Orange-crowned Warbler
Leiothlypis celata | OCWA | L 5" (13 cm)

- **APPEARANCE** Longer-tailed and shorter-winged than Tennessee, with slightly decurved bill. Trademark orange crown rarely visible. Dark olive-gray above, with thin eyebrow and trace of an eye ring; faintly streaked underparts mostly dusky yellow-gray, with undertail coverts brighter yellow. Tennessee is the classic point of confusion, but beware drab, grayish, late-season Yellow Warbler (p. 436).
- **VOCALIZATIONS** Song a loose trill that usually drops off at end: *titititititi-tlu-tlu-tlu*. Call note a clanking *tswint*, flight call a Tennessee-like *tsee*?
- **BEHAVIORS** Nests in shrubby habitats, with or without forest overstory; on migration and in winter, anywhere there is shrubby vegetation. Active year-round; often forages around eye level; tends feeders.
- **POPULATIONS** Fairly early spring migrant, fairly late fall migrant. Well differentiated subspecies in West include Pacific coast *lutescens*, bright yellow; Channel Is. *sordida*, darker olive and larger-billed; Interior West *orestera*, distinctively gray-headed; and northern nominate *celata*, the dullest subspecies, migrating mostly to our east.

Colima Warbler
Leiothlypis crissalis | COLW | L 5¾" (15 cm)

- **APPEARANCE** Like an oversize Virginia's Warbler (p. 428). Browner overall than Virginia's, with ochre uppertail and undertail coverts; Colima lacks yellow on central breast of Virginia's, and usually shows more rust on crown. Like Virginia's, has white eye ring.
- **VOCALIZATIONS** Trilled song faster than Virginia's; final note of trill of Colima often different from preceding notes. Calls like Virginia's.
- **BEHAVIORS** Nests on ground in hot, sunny oak and oak-pine woods. Gleans leaves like other warblers, but also pecks at galls and branches.
- **POPULATIONS** Away from core breeding grounds in West in Big Bend N.P., also occurs in very small numbers in Davis Mts., where it hybridizes with Virginia's.

Lucy's Warbler
Leiothlypis luciae | LUWA | L 4¼" (11 cm)

■ **APPEARANCE** Our smallest warbler, with short tail and thin, straight, short bill. Gray overall with variable buff-yellow tinge; looks washed-out. Male has rusty swatch on cap and rusty uppertail coverts. Rusty reduced on female; most immatures show yellowish-tawny rumps, a few almost entirely gray. Compare with immature Verdin (p. 302).

■ **VOCALIZATIONS** Song a fast, light, airy trill that changes pitch midway through; songs of Yellow Warbler (p. 436) notoriously variable, with some quite similar to song of Lucy's. Call a high, lisping *tssst;* flight call a high, clear buzz.

■ **BEHAVIORS** Found year-round in hot deserts. Unlike any other warbler breeding in our area, nests in tree holes; sightings in breeding season therefore around watercourses with cottonwoods and willows. Forages in mesquite, greasewood, and other desert shrubs. Active and flitty; twitches tail when foraging.

■ **POPULATIONS** Migrates early in spring; heads back south again mid to late summer. Locally abundant in suitable breeding habitat.

Nashville Warbler
Leiothlypis ruficapilla | NAWA | L 4¾" (12 cm)

■ **APPEARANCE** Short-tailed, straight-billed, and small. Adult in spring has olive green upperparts, yellow underparts (including throat), and blue-gray "helmet" with white eye ring; immature, especially female, in fall has brownish helmet. Closely related Virginia's Warbler never as yellow below. Compare immature Nashville with immature Common Yellowthroat (p. 430), which is dark above and yellow below, with hint of eye ring, but with much longer tail, shorter wings, and long legs.

■ **VOCALIZATIONS** Two-part song starts with disyllabic *s'bee* notes in rapid succession, followed by staccato trill. Call note a sharp *spink;* flight call a clear, rising *tsi?*

■ **BEHAVIORS** Nests in disturbed or early successional forests, often in foothills, with brushy understory. Active; feeds in shrub layer of forest, flitting about flowers and foliage.

■ **POPULATIONS** Two well-differentiated subspecies. Western breeder *ridgwayi,* intermediate in various respects between nominate eastern *ruficapilla* and Virginia's; *ridgwayi* has longer tail and brighter yellow rump than *ruficapilla,* routinely twitches tail, and sings looser, less enunciated song. Migrants east of Rockies mostly *ruficapilla.*

Virginia's Warbler
Leiothlypis virginiae | VIWA | L 4¾" (12 cm)

■ **APPEARANCE** Body structure like that of Nashville. Gray overall with white eye ring. Adult, especially male, sports yellow wash across breast, yellow uppertail and undertail coverts, and small rusty patch on crown; young, especially female, duskier overall with little or no yellow on breast and no rust on crown. Compare also with Lucy's Warbler.

■ **VOCALIZATIONS** Song a loose, amorphous trill that drops in pitch midway through. Call an abrupt *pik,* not unlike call of Lazuli Bunting (p. 450); flight call like Nashville's.

■ **BEHAVIORS** Nests in pine and pine-oak tracts on dry, shrubby, sunny lower slopes. Active; twitches tail as it forages, often in shrub layer.

■ **POPULATIONS** Breeding range may be expanding north and west in places. Has hybridized with Colima Warbler (p. 426) in Davis Mts, Tex.

WOOD-WARBLERS
PARULIDAE

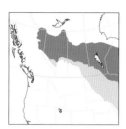

Connecticut Warbler
Oporornis agilis | CONW | L 5¾" (15 cm)

- **APPEARANCE** Large, long-winged, and short-tailed; legs sturdy. Olive above, pale yellow below; dark hood (purple-gray on breeding male, browner in other plumages) has bold and complete eye ring.
- **VOCALIZATIONS** Song syncopated, builds in amplitude: *tee! ch'PEE-tee p'CHIP p'CHEE-CHIP!* Call note a dull *tyip*; flight call high, buzzy, and wavering.
- **BEHAVIORS** Ground-loving species; nests in bogs with broadleaf component. Walks (does not hop) on fallen logs and forest floor.
- **POPULATIONS** Breeds west into our area by way of Boreal Shield across central Canada. Migrates to our east, but casual in late spring to western Great Plains and very rare mid-autumn along Calif. coast.

MacGillivray's Warbler
Geothlypis tolmiei | MGWA | L 5¼" (13 cm)

- **APPEARANCE** Midsize warbler of average proportions. Olive above, yellow below; at all ages, broken eye arcs stand out on dark hood. Hood deep slate blue on adult male, paler and grayer on adult female and young. Compare with Connecticut and Mourning, peripheral in much of range of MacGillivray's, as well as with *orestera* Orange-crowned Warbler (p. 426) and Common Yellowthroat.
- **VOCALIZATIONS** Song breathy and chanting, with ending notes lower: *churry churry churry churr churr*. Call note a smacking *chack*; flight call a short, rising *si?*
- **BEHAVIORS** Nests in broadleaf understory of mountain forests, usually near streams or bogs. Hops on ground and branches (Connecticut walks).
- **POPULATIONS** In parts of B.C., hybridizes, perhaps extensively, with Mourning Warbler, which also may be replacing MacGillivray's in places.

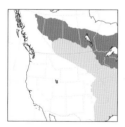

Mourning Warbler
Geothlypis philadelphia | MOWA | L 5¼" (13 cm)

- **APPEARANCE** Same size and shape as MacGillivray's. Hood of male purple-gray, with black around eyes and base of breast. Female, especially young, in fall only weakly hooded; sometimes shows thin eye ring like Connecticut. Compare also with immature Common Yellowthroat.
- **VOCALIZATIONS** Song like MacGillivray's, but less burry. Call note a flat *tchook*, not as smacking as MacGillivray's; flight call like MacGillivray's.
- **BEHAVIORS** Nests at forest edges, especially around roads and logging operations. Hops like MacGillivray's; does not walk.
- **POPULATIONS** Migrates to our east, yet vagrates to Calif. coast every year, mostly fall.

Common Yellowthroat
Geothlypis trichas | COYE | L 5" (13 cm)

- **APPEARANCE** Shape distinctive: short-winged, long-legged, and long-tailed. Adult male has rectangular black face mask, edged bluish above. Other plumages have reduced or no black on face, but all have plain yellow throat and yellow undertail coverts.
- **VOCALIZATIONS** Song chirpy and repetitious: *witchity witchity witchity ....* Call notes include a rough *chap* and rapid chatter; flight call a rough *zzzrt*.
- **BEHAVIORS** Occurs year-round in marshes. Cocks tail like a wren (pp. 330–335). Responds instantly to pishing.
- **POPULATIONS** Hardy; migrates late in fall, with some lingering into winter well north of mapped range.

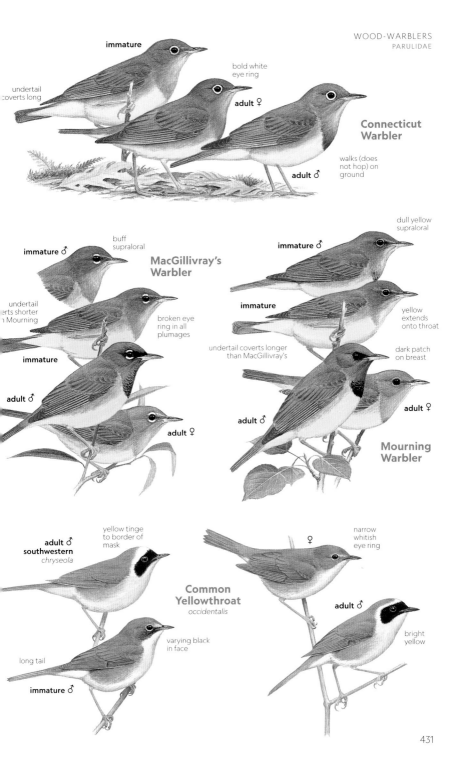

Hooded Warbler
Setophaga citrina | HOWA | L 5¼" (13 cm)

Usually breeds well outside our region, in moist broadleaf forests of Southeast; migration generally well to our east too, but one of the most frequent "eastern" vagrants in the West. Adult, especially male, has black cowl encircling yellow face. All plumages olive green above and yellow below, with extensive white in tail; compare with Wilson's Warbler (p. 442). Feeds in shrub layer or near ground in woods; often fans tail. Song a bright and ringing *s'wee s'wee s'WEE-chee-you;* call note a resonant *tsaap,* flight call buzzy. Probably nests somewhere in the West every year.

American Redstart
Setophaga ruticilla | AMRE | L 5¼" (13 cm)

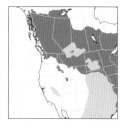

- **APPEARANCE** Delicate and fairly small warbler with long tail. Adult male plumage, not acquired until second fall, mostly black with orange highlights. Young and female mostly gray with yellow on wing and flanks; tail edges are lemon yellow. Second-summer male (around one year of age) begins to show black of adult.
- **VOCALIZATIONS** Variable song comprises a few bright chirps, then a lower buzz: *twee twee twee tzzyeeer.* Call note a smacking *zick;* flight call a clear *ts'weet,* subtly disyllabic.
- **BEHAVIORS** Nests in broadleaf forests, especially around edges and clearings. Foraging, constantly spreads wings and tail, flashing orange or yellow to startle insect prey into flushing.
- **POPULATIONS** Migration mostly to our east, but annual spring and fall, especially in and near foothills of east flank of Rockies. After Northern Waterthrush (p. 424), the second most common "eastern" warbler on migration in the West.

Cape May Warbler
Setophaga tigrina | CMWA | L 5" (13 cm)

- **APPEARANCE** Short-tailed with thin, decurved bill. Breeding male has orangey patch on yellow face, large white wing panel, and black streaking on yellow breast. Other plumages told by yellowish rump, green-edged remiges, and bill shape.
- **VOCALIZATIONS** Song a series of very high notes, often increasing in amplitude: *si see see! SEEE!* Call note, also quite high, a clipped *teep;* flight call a descending *seeew.*
- **BEHAVIORS** Along with Tennessee (p. 426) and Blackpoll (p. 436) Warblers, classified as a spruce budworm specialist. Away from boreal breeding grounds, stops on migration at flowers and, in winter, bird feeders.
- **POPULATIONS** Migration well to our east. Perhaps surprisingly, many records in West away from breeding grounds are of winterers.

Northern Parula
Setophaga americana | NOPA | L 4½" (11 cm)

Something of a distributional doppelganger for the Hooded Warbler: usually breeds well to our east, especially in moist broadleaf woods, but occurs widely in the West, even nesting on rare occasion. Short-tailed and quite small; bill yellow below. All are bluish above with greenish back, white eye arcs, white wing bars, and yellow throat and breast; adult sports variable blue-and-red breastband, fainter on female, nearly absent on immature female. Primary song a rising trill ending in an abrupt *chip;* call note a smacking *tsick,* flight call a clear *zeep.* Habitat generalist on migration, with more in spring in West than fall.

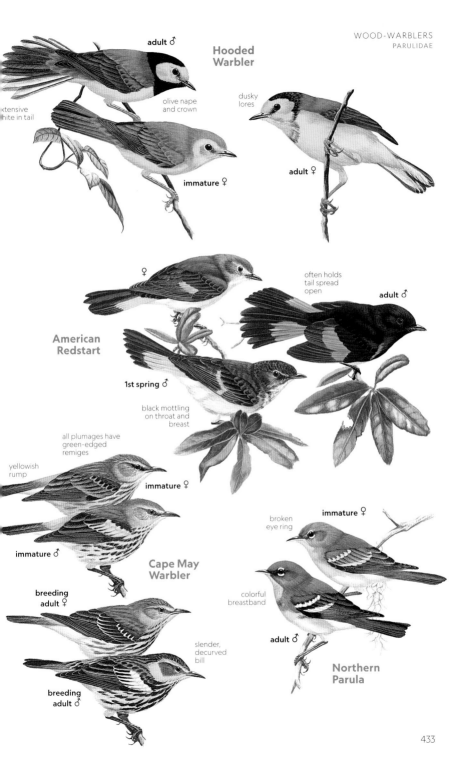

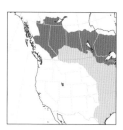

Magnolia Warbler
Setophaga magnolia | MAWA | L 5" (13 cm)

- **APPEARANCE** Built like American Redstart (p. 432): long-tailed and fairly small overall. Tail pattern below unique, with broad white band at base and broad black band at tip. Adult yellow below with thick black streaks; grayish wings have frosty panel. Immature, especially female, has mostly unmarked yellow underparts and blue-gray face with thin white eye ring; young Nashville Warbler (p. 428), in different genus, can closely match plumage of young Magnolia, but body structure and behaviors different.
- **VOCALIZATIONS** Song, not unlike Hooded's (p. 432), clear and ringing, *swee swee swee swee-yee-oh*. Call note, the most distinctive of any *Setopaga* warbler, a muffled *shwrint*; flight call a buzzy *frrzz*.
- **BEHAVIORS** Nests in moist conifer forests, mostly spruce in West; sometimes forages up in canopy, but at least as likely to be in midstory and even ground layers. Active; cocks and spreads tail when feeding.
- **POPULATIONS** Bulk of migration to our east, but a few show up throughout West spring and fall; one of the more common "eastern" vagrants at midlatitudes in western U.S.

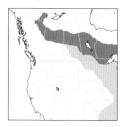

Bay-breasted Warbler
Setopaga castanea | BBWA | L 5½" (14 cm)

- **APPEARANCE** Fairly large for a warbler; long-winged. Adult has extensive chestnut (bay) on flanks and pale patch on neck. Immature has dark legs, thick white wing bars, and pinkish wash on flanks; undertail coverts buffy, breast plain, and overall hue yellow-green. Compare with immature Blackpoll Warbler (p. 436).
- **VOCALIZATIONS** Song fast and high, with notes often paired: *w'si w'si w'si*. Call note a simple *tsip*; flight call a ringing *dzee*.
- **BEHAVIORS** Nests in spruce-fir forests, where it forages mostly from midstory up into canopy. Slow and methodical; probes and gleans leaves, twigs, and lichens.
- **POPULATIONS** Scarce many years, but numbers spike following spruce budworm outbreaks. Migrates well to our east; casual at best away from breeding grounds in West, with strongest accumulation of sightings coastal Calif. and Colo. east of Rockies.

Blackburnian Warbler
Setopaga fusca | BLBW | L 5" (13 cm)

- **APPEARANCE** Subtly long-winged and short-tailed, but structurally average overall for a warbler. Breeding male, the "Firethroat," is unmistakable. Other plumages told by broad yellow eyebrow, yellow throat, and pale "braces" on dark back. Compare with Townsend's Warbler (p. 440); adults, especially males, very different, but drab immatures in fall similar.
- **VOCALIZATIONS** High-pitched primary song stutters up the scale: *ch'ch'ch'ch'-chee-chee-cheeeee*. Call note a bright *chip*; flight call a thin, high *dzi*.
- **BEHAVIORS** Nests and forages in canopy of coniferous forests; feeding strategies like Bay-breasted Warbler, although Blackburnian is more of a gleaner and not quite as deliberate in its actions.
- **POPULATIONS** Western limit of breeding range in Canada in flux; may have withdrawn in recent years. Seems to be more prone to vagrancy than Bay-breasted, with greater accumulation of records in Colo. and Calif.

Yellow Warbler
Setophaga petechia | YEWA | L 5" (13 cm)

■ **APPEARANCE** Short-tailed; the yellowest *Setophaga*, with even the tail completely yellow. All have beady black eye on plain face. Adult male has rusty pinstripes below, more prominent in spring and early summer. Female unmarked yellow all over; some in fall to winter yellow-gray. Orange-crowned (p. 426) has thin eyebrow, short wings, and long tail; female and young Wilson's (p. 442) smaller and longer-tailed with hint of olive cap.

■ **VOCALIZATIONS** Song extremely variable, but the "classic" version comprises chirpy, descending notes in fast series ending with emphatic, rising note: *cheep cheep cheep ch'yeepy CHEET*. Some variants sound like American Redstart (p. 432), Chestnut-sided Warbler, and other warblers. Call note a bright *chip;* flight call a high buzz.

■ **BEHAVIORS** Nests in and near wetlands of all sorts, from riparian groves in hot deserts to willow carrs near timberline. Migrants found in diverse habitats; in limited winter range in West, often in scrubby willows. Feeds by gleaning leaves for arthropods.

■ **POPULATIONS** Boldness of rusty streaking on males variable; southwestern breeders have faintest streaking. In much of range, departs breeding grounds early, by midsummer.

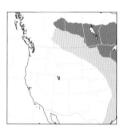

Chestnut-sided Warbler
Setophaga pensylvanica | CSWA | L 5" (13 cm)

■ **APPEARANCE** About the same build as Yellow Warbler, although tail a bit longer. Breeding adult has yellow crown, white eye ring, black "whisker," yellow wing bars, and chestnut flanks. Strangely pale in fall, with blank, wide-eyed look; chestnut reduced or absent, but combination of yellow wing bars, white eye ring, and yellow-green crown imparts distinctive look.

■ **VOCALIZATIONS** Variable song often the inverse of Yellow's: up-slurred notes with terminal down-slurred note. Call note a bit rougher than Yellow's; flight call a fine buzz.

■ **BEHAVIORS** Nests in early successional woods and clearings with dense brush; usually in drier microhabitats than Yellow Warbler. With tail cocked, forages in midstory.

■ **POPULATIONS** Breeders barely reach our area in Canada; bred in very small numbers in Front Range of Colo. in late 20th century, but no confirmed breeding records recently. Migrates mostly to our east, but widespread vagrant across mainland western U.S.

Blackpoll Warbler
Setophaga striata | BLPW | L 5½" (14 cm)

■ **APPEARANCE** Large and long-winged. Feet yellow or yellow-orange. Black cap of breeding male set off from white face; breeding female smudgier. Fall plumage, especially female and young, similar to Bay-breasted Warbler's (p. 434); Blackpoll lacks pink-tinged flanks, is streakier with white undertail coverts and unique yellow feet. Compare also with immature and fall female Yellow-rumped Warbler (p. 438).

■ **VOCALIZATIONS** Primary song a series of very high *zi* notes, rising in amplitude then quickly falling. Call note a loud smack; flight call like Yellow's.

■ **BEHAVIORS** Nests in boreal forest, especially spruce; migrants occur in wooded habitat anywhere. Forages rather methodically, typically from midstory up into canopy.

■ **POPULATIONS** Widespread but generally rare migrant in West. More in spring on western High Plains, but more in fall coastally.

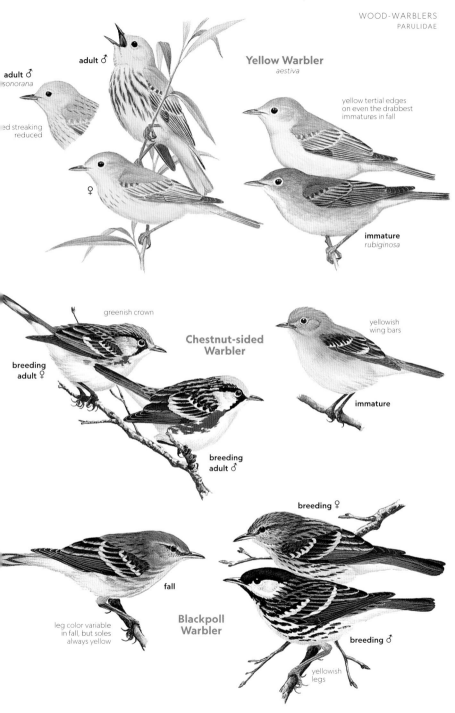

Palm Warbler
Setophaga palmarum | PAWA | L 5½" (14 cm)

- **APPEARANCE** Midsize warbler with long tail. All plumages have obvious eyebrow, streaking below, and yellow undertail coverts. Breeders have obvious reddish cap, reduced or absent fall to winter.
- **VOCALIZATIONS** Song a loose, slow trill. Call note sharp and smacking; flight call a relatively low *zeep*.
- **BEHAVIORS** Terrestrial. Nests on peat in boreal bogs. On migration and in winter, also near ground level, often around brushy woodland edges. Bobs tail constantly.
- **POPULATIONS** Two well-marked subspecies. Western-breeding nominate *palmarum* is widespread but very rare in West on migration; it is one of the more frequently detected "eastern" warblers on Christmas Bird Counts. Eastern breeding *hypochrysea*, much yellower, migrates far to our east, yet supported by scattered records throughout West.

Yellow-rumped Warbler
Setophaga coronata | YRWA | L 5½" (14 cm)

- **APPEARANCE** Large and relatively thick-billed; size of Blackpoll Warbler (p. 436). Highly variable, but all plumages except briefly held juvenile plumage are yellow-rumped ("Butterbutt" is a popular nickname for the species) with throat paler than rest of face. Adult, especially breeding male, brightly marked in black, yellow, and white, but young, especially female, in fall to winter much drabber.
- **VOCALIZATIONS** Song a loose warble, shifting pitch toward end. Geographically variable call note lacks clear tone; flight call a breathy *wheew*.
- **BEHAVIORS** The most generalist warbler in its feeding habits: gleans leaves for caterpillars, habitually hawks for flying insects, and is heavily frugivorous in winter.
- **POPULATIONS** Two well-differentiated subspecies groups in West. "Myrtle" (nominate *coronata*), breeding in northern boreal forest, has thin wing bars and white throat that wraps under cheek; call note a hard, resonant *tchep*. "Audubon's" *(auduboni)*, breeding in Rockies and westward, has large white wing panel and yellow throat that does not wrap under cheek; call note more muffled, a rising *tchit*. Both groups migrate widely across West, moving north early in spring and heading south late in fall. The hardiest warbler; winters well north inland.

Grace's Warbler
Setophaga graciae | GRWA | L 5" (13 cm)

- **APPEARANCE** Small; about the same build as Orange-crowned Warbler (p. 426). Blue-gray above with white wing bars; gray face has thick yellow eyebrow. Bright yellow on throat and breast; belly white with black streaks on flanks. Color scheme roughly similar to that of larger "Audubon's Yellow-rumped Warbler."
- **VOCALIZATIONS** Song a two-part trill, the second part faster than the first; song of Yellow-rumped similar but sounds more leisurely. Call note a down-slurred *chip*, like Yellow Warbler's (p. 436); flight call clipped, high, and lisping.
- **BEHAVIORS** Nests in dry conifer forests, especially ponderosa pine. Forages in canopy, where it is constantly on the move, often calling.
- **POPULATIONS** Range expanding north. Seen only infrequently away from breeding grounds; probably does not wander far from pinewoods on migration.

Black-throated Gray Warbler
Setophaga nigrescens | BTYW | L 5" (13 cm)

- **APPEARANCE** All four species on this page are small, compact warblers with short tails; adult males are black-throated in all four, females and young less so. Black-throated Gray is the most distinctive of the four, gray above and boldly patterned black and white on face, with small yellow patch in front of eye.
- **VOCALIZATIONS** Song, short and buzzy, climbs the scale: *teezle teezle teezle tzeee?* Call note a smacking *tsick*; flight call a clear, ringing *tlee*.
- **BEHAVIORS** Nests and forages in low-stature woods, oak and especially conifer, in dry, sunny landscapes; migrants occur in wooded habitat anywhere.
- **POPULATIONS** Prone to vagrancy; many records in West north of regular breeding range.

Townsend's Warbler
Setophaga townsendi | TOWA | L 5" (13 cm)

- **APPEARANCE** Breeding male like a Black-throated Gray Warbler with the white parts colored yellow. Fall birds, especially young females, similar to Hermit and Black-throated Green Warblers; note dark cheek and yellowish breast of Townsend's.
- **VOCALIZATIONS** Song similar to Black-throated Gray's, notes rising in pitch, buzzier at end. Call note a clinking, lisping *stik*; flight call a sharp *tzee*.
- **BEHAVIORS** Nests and forages in treetops in moist conifer forests; migrants more generalist.
- **POPULATIONS** An elliptical migrant; northbound movement in spring fairly far west, but southbound migrants in fall spread east to western Great Plains.

Hermit Warbler
Setophaga occidentalis | HEWA | L 5½" (14 cm)

- **APPEARANCE** All plumages have yellow face; rest of body lacks yellow and other warm tones, unlike Townsend's and Black-throated Green Warblers.
- **VOCALIZATIONS** Variable song much like Townsend's, but averages buzzier and faster, often with terminal note distinctly different. Calls like Townsend's.
- **BEHAVIORS** Haunts and habits as those of Townsend's; a treetop denizen of conifer tracts.
- **POPULATIONS** Like Townsend's, an elliptical migrant; but doesn't usually get east of Great Basin. Hybridizes with Townsend's around Wash.-Ore. border; these "HETO" hybrids often combine the yellow face of Hermit with the yellow breast of Townsend's.

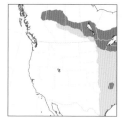

Black-throated Green Warbler
Setophaga virens | BTNW | L 5" (13 cm)

- **APPEARANCE** Rich, unstreaked olive above, with face pattern intermediate between Townsend's and Hermit Warblers: has cheek patch like Townsend's, but weaker and suffused with yellow like Hermit. Underparts mostly white, but with yellowish wash on flanks and vent.
- **VOCALIZATIONS** Two main songs, both comprising short buzzes and sharp whistles: fast *zi zi zi zoo zi* and slower *zooo zeee zoo zoo zeeee*. Calls like those of Townsend's and Hermit.
- **BEHAVIORS** More of a habitat generalist than Townsend's and Hermit; nests in a diversity of conifer forest types, as well as broadleaf tracts.
- **POPULATIONS** Breeding range may be expanding west in our area. Casual migrant across West, with accumulations of records from east flank of Rockies eastward and along Calif. coast.

Rufous-capped Warbler
Basileuterus rufifrons | RCWA | L 5¼" (13 cm)

Mid. Amer. warbler; annual in small numbers to brushy foothills in Southeast Ariz.; scattered records to N. Mex. and West Tex. Tail, long and skinny, flipped about continually. Mostly olive, but with russet face, bold white eyebrow, and yellow below. Size, shape, actions, and habitat recall Bewick's Wren (p. 332). Song a fast, sharp trill.

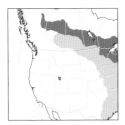

Canada Warbler
Cardellina canadensis | CAWA | L 5¼" (13 cm)

- **APPEARANCE** Small; long-tailed. Upperparts blue-gray, underparts yellow with "necklace" of dark streaks; all have yellow and white "spectacles," fainter on female and young.
- **VOCALIZATIONS** Song, bright and chirpy, starts with introductory *chip*, then goes into disorganized ramble. Call note a smacking *tip*; flight call low and hard, *tyit*.
- **BEHAVIORS** Nests in boreal forest, often aspen, with dense understory. Active but also furtive; forages near ground.
- **POPULATIONS** Declining; classified as threatened in Canada. Casual in West away from breeding grounds.

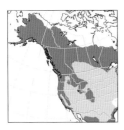

Wilson's Warbler
Cardellina pusilla | WIWA | L 4¾" (12 cm)

- **APPEARANCE** Very small; long-tailed. Olive above, yellow below; beady black eye stares from blank face. Adult males and older adult females have black cap; young wear a weak olive cap.
- **VOCALIZATIONS** Song a series of falling *chi* and *chet* notes. Call note a nasal *wenk*; flight call muffled and low, *schkwrit*.
- **BEHAVIORS** Nests amid shrubby willows and alders in bogs and along streams. With tail cocked, forages close to the ground.
- **POPULATIONS** One of our most common warblers, but recent declines in some areas.

Red-faced Warbler
Cardellina rubrifrons | RFWA | L 5½" (14 cm)

- **APPEARANCE** Midsize warbler with long tail. Adult light gray overall with striking red and black on head; juvenile has duller orange on head.
- **VOCALIZATIONS** Song a bright chirpy warble that often drops in pitch. Call a smacking *tak*; flight call a short, high *tss*.
- **BEHAVIORS** Breeds in montane conifer forests; unusual nest is partially dug into ground. Active; flicks tail while foraging.
- **POPULATIONS** Migratory in our area; departs breeding grounds early, but a few in winter in recent years.

Painted Redstart
Myioborus pictus | PARE | L 5¾" (15 cm)

- **APPEARANCE** Midsize warbler of average proportions. Mostly black with white wing panel and white outer tail feathers; belly of adult brilliant red, gray-black on juvenile.
- **VOCALIZATIONS** Song a bright, amorphous warble. Call note a loud, clear, descending *tyeeoo*, more like Pine Siskin's (p. 378) than any other warbler's.
- **BEHAVIORS** Nests in oak and pine woods; like American Redstart (p. 432), flashes white in wings and tail to startle arthropod prey.
- **POPULATIONS** Most in West leave breeding grounds after nesting, but some winter in Southeast Ariz. and even Southern Calif.

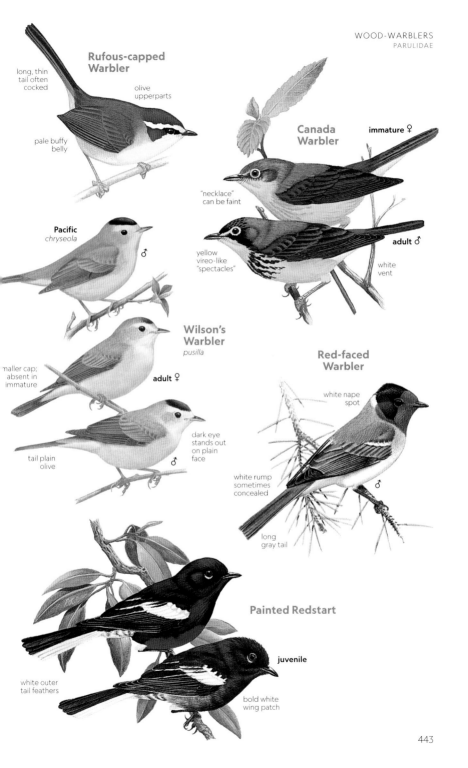

CARDINALS AND KIN | CARDINALIDAE

This New World family comprises about 50 species with bright colors and stout bills. Most are found in forests or forest edges. Note that all our species called "tanager" are in this family, whereas several of our birds called "cardinal" (see p. 452) are in the tanager family.

Hepatic Tanager
Piranga flava | HETA | L 8" (20 cm)

- **APPEARANCE** Large tanager; cutting edge of upper mandible has jagged "tooth." Both sexes monochrome like Summer Tanager, but bill of Hepatic dark and cheek grayish. Male dull brick red; female dusky yellow-orange.
- **VOCALIZATIONS** Song a rich, run-on warble like Black-headed Grosbeak's (p. 448), but delivery slower. Call a low, hollow *tuck*.
- **BEHAVIORS** Nests in dry pine forests. Like other tanagers, feeds slowly and methodically in canopy.
- **POPULATIONS** Returns fairly late in spring. Breeding range slowly advancing north.

Summer Tanager
Piranga rubra | SUTA | L 7¾" (20 cm)

- **APPEARANCE** Large-billed and a bit larger overall than Western Tanager. Bill pale. Adult male rosy-red. Adult female olive-yellow, often washed with dull red. Second-summer male has rosy head and blotchy red and olive-yellow otherwise.
- **VOCALIZATIONS** Song a rich warble, with cadence of Western Tanager but not as hoarse. Call a loud, fast chortle, descending in pitch: *t't'tk'tk'TUCK*.
- **BEHAVIORS** Unlike other tanagers in West, occurs in broadleaf tracts in lowlands. Diet generalist; employs diverse foraging methods.
- **POPULATIONS** Western breeders are subspecies *cooperi*, but many vagrants to coast are eastern *rubra*, which is not as large-billed.

Scarlet Tanager
Piranga olivacea | SCTA | L 7" (18 cm)

Widespread in East; casual vagrant throughout western U.S. Size and shape like Western Tanager. Breeding adult male flaming scarlet with jet-black wings and tail. Other plumages bright yellow-green with dark wings. Song like Western's, but call different: a far-carrying *CHIP! purrr*. Sightings west of Rockies generally late fall, but sightings in western Great Plains often spring.

Western Tanager
Piranga ludoviciana | WETA | L 7¼" (18 cm)

- **APPEARANCE** Smallest tanager regularly occurring in mainland West; bill small. Older males spectacular, yellow and black with red head; some older females are red-headed, and many younger males are only weakly red-headed. Wing bars prominent in all plumages; pale uppertail coverts contrast with darker back (black on adult male, grayer in other plumages).
- **VOCALIZATIONS** Song a short series of harsh rising and falling phrases, often in groups of four. Call an abrupt *p'tick* or *picky-tick*, less explosive than Summer Tanager's.
- **BEHAVIORS** Nests in conifer forests, especially mid-elevation pines. Migrants occur anywhere, sometimes concentrating in flocks during groundings in spring snowstorms.
- **POPULATIONS** Spring migration fairly late; fall migration protracted, with some leaving nesting grounds by midsummer. Increasing in winter in southern coastal Calif.

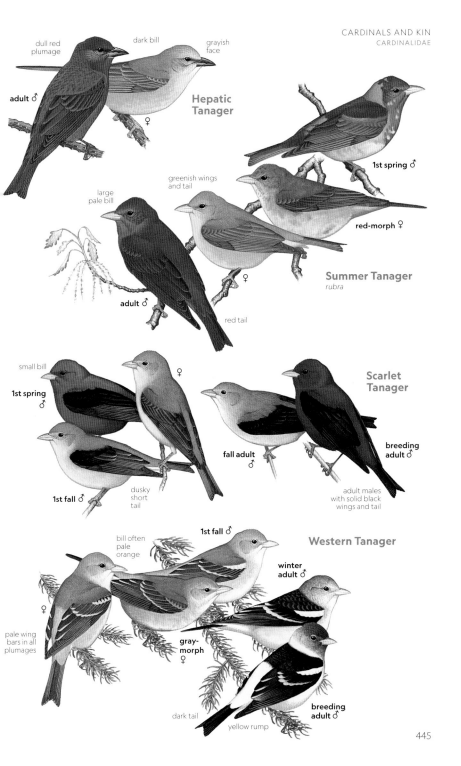

Flame-colored Tanager
Piranga bidentata | FCTA | L 7¼" (18 cm)

Mid. Amer. cardinalid that strays to Southeast Ariz. and Big Bend N.P. in the warmer months. Like Hepatic Tanager (p. 444), has gray bill with "tooth." All plumages have two pale wing bars, white spots on tertials, and streaked back. Adult male, bright orange, is distinctive. Female and young similar to Western Tanager (p. 444), but upper wing bar often white (yellow on Western); note also gray cheek, dark bill, and streaked back and nape of Flame-colored. Song and call similar to Western's. Flame-colored and Western hybridize with some frequency in our area; Southeast Ariz. and West Tex. are the only places where their breeding ranges (barely) overlap.

Introduced to Hawaii

Northern Cardinal
Cardinalis cardinalis | NOCA | L 8¾" (22 cm)

- **APPEARANCE** Slim, long-tailed, and crested. Adult male red, including bill, except for black face mask. Adult female warm brown with ample reddish highlights; also has red bill. Juvenile fairly uniform rufous-brown with dark bill. Compare female and juvenile with grayer, yellow-billed Pyrrhuloxia.
- **VOCALIZATIONS** Song, proclaimed by both sexes, comprises rich, repetitious whistles: *cheer! cheer! cheer!* or *birdy birdy birdy* .... Call note a thin, smacking *tsit*.
- **BEHAVIORS** Occurs year-round in brushy habitats, from cities and suburbs to well-vegetated desert washes and woodland edges. Prominent and conspicuous: sings from exposed perches, forages on lawns, tends feeders, etc.
- **POPULATIONS** Subspecies *superbus*, from southwestern N. Mex. westward, has longer crest, less black on face, slightly more rounded culmen, and brighter tones overall than eastern cardinals, resident in eastern Colo., eastern N. Mex., and West Tex. Also introduced to Hawaii (nominate *cardinalis*, an eastern subspecies), where established on all main islands; Hawaiian cardinals abound in human-dominated landscapes, but also get well into forests to escape competition for food with other granivorous bird species, scarce in Hawaii's forests.

Pyrrhuloxia
Cardinalis sinuatus | PYRR | L 8¾" (22 cm)

- **APPEARANCE** Differs structurally from Northern Cardinal by wispier crest and especially bill structure: upper mandible strongly curved, with hooked tip, imparting a parrotlike gestalt. Adult male gray with splotchy red highlights, including red face mask. Female like female Northern Cardinal, but overall tone grayer; note also yellowish bill, strongly curved above. Juveniles of the two species even more similar, but juvenile Pyrrhuloxia grayer and less colorful, with different bill shape.
- **VOCALIZATIONS** Song like Northern Cardinal's, but not as rich and vigorous. Call note, the opposite: richer and more solid than cardinal's.
- **BEHAVIORS** The "Desert Cardinal," more inclined to arid scrublands than Northern Cardinal; Pyrrhuloxia often occurs in mesquite-dominated desert, sometimes even creosote bush monocultures. Like Northern Cardinal, tends feeders.
- **POPULATIONS** Range slowly expanding north, apparently tracking northward range expansion of mesquite. Occasionally hybridizes with Northern Cardinal; the logo of the Arizona Cardinals football team (bright yellow bill with rounded culmen, brilliant red overall but with black face mask reduced on forehead) is an exemplary hybrid.

Rose-breasted Grosbeak
Pheucticus ludovicianus | RBGR | L 8" (20 cm)

■ **APPEARANCE** Chunky and block-headed with huge bill. Distinguished from Black-headed at all ages by bill color: uniform pale pink on Rose-breasted, bicolored on Black-headed. Adult male Rose-breasted, black and white with inverted red triangle on breast, has rosy wing linings. Brownish female, streaky below with boldly patterned face, has warm buff-yellow wing linings; female Black-headed plainer below, with little streaking. Immature male, with richer buff on breast and reduced streaking, similar to Black-headed, but many show some incoming red feathers on breast. All plumages have white in wing, prominent in flight.

■ **VOCALIZATIONS** Caroled song like American Robin's (p. 348), but faster and sweeter. Call note an abrupt squeak ("wet sneakers on a gym floor"); flight calls varied, most are whistled.

■ **BEHAVIORS** Nests in broadleaf forests, where it forages slowly for arthropods and plant matter. Tends feeders.

■ **POPULATIONS** Migration mostly to our east, but decent pulse through West Tex., N. Mex., and Colo. May to early June.

Black-headed Grosbeak
Pheucticus melanocephalus | BHGR | L 8¼" (21 cm)

■ **APPEARANCE** Same size and shape as Rose-breasted Grosbeak. Breeding male, with extensive orange-umber, distinctive, but other plumages resemble Rose-breasted; all plumages of Black-headed have bicolored bill, dark gray above and paler below. Female has a few fine streaks on buff-suffused breast. Immature male has head pattern like female, but plain, bright orange-umber breast of adult male. All plumages flash white in wing like Rose-breasted, but underwing coverts dusky yellow.

■ **VOCALIZATIONS** Song very similar to that of Rose-breasted, not quite as joyful. Call note of Black-headed less squeaky; varied flight calls like Rose-breasted's.

■ **BEHAVIORS** Nests in broadleaf woods in canyons, on lower slopes, and along rivers. Like Rose-breasted, feeds in canopy.

■ **POPULATIONS** Hybridizes with Rose-breasted where the two species' ranges overlap a bit to our east in the Great Plains. Adult male hybrids, with blended characters, often field identifiable; female and young usually impossible to identify in the field.

Blue Grosbeak
Passerina caerulea | BLGR | L 6¾" (17 cm)

■ **APPEARANCE** Largest *Passerina* bunting (pp. 448–451), with the biggest bill. Breeding male deep indigo, with chestnut wing bars; first-spring (second calendar year) male splotchy indigo and rust. Female and immature cinnamon brown; distinguished from congeners by chestnut wing bars, large bill, and larger size overall.

■ **VOCALIZATIONS** Song a fast, burry warble with little change in pitch, oddly similar to "Western Warbling Vireo" (p. 288). Call note flat and clinking, *plenk;* flight call a low buzz, *szzzzt*.

■ **BEHAVIORS** Nests in semi-open landscapes with scattered tall trees: roadsides, pastures, riparian corridors, etc. Migrants, especially fall, in dense weeds around eye level.

■ **POPULATIONS** Gets back late in spring. Movements in summer complex; late-season nesters, with dependent young into Sept., may have bred elsewhere earlier in the summer.

Lazuli Bunting
Passerina amoena | LAZB | L 5½" (14 cm)

- **APPEARANCE** The four buntings on this page, essentially identical in size and shape, are proportioned like New World sparrows (pp. 386–409). Breeding male Lazuli extensively pale blue with broad white wing bars and orange wash on breast. Female has white wing bars, unstreaked breast with warm brown wash, and off-white belly. Juvenile finely streaked below.
- **VOCALIZATIONS** Song an exuberant eruption of chirps and buzzes, lacking paired elements of Indigo Bunting's. Call note a clipped *spink;* flight call a short buzz.
- **BEHAVIORS** Nests in forest edges and clearings with ample brush in understory. Gleans arthropods from foliage, but also forages for seeds on ground.
- **POPULATIONS** A molt migrant; many in midsummer decamp breeding grounds to molt in Ariz. and N. Mex. during monsoon season.

Indigo Bunting
Passerina cyanea | INBU | L 5½" (14 cm)

- **APPEARANCE** Breeding male deep blue all over. Adult female and juvenile (both sexes) gray-brown with paler throat, blurry streaking on breast, and thin wing bars. In both Lazuli and Indigo Buntings, first-spring male (in second calendar year) and adult male in winter are like breeding adult of corresponding species, but with blotchy brown all over.
- **VOCALIZATIONS** Song comprises slurred elements, often paired, trailing off: *spit spit chew chew spit-it-out-chew.* Calls similar to Lazuli's.
- **BEHAVIORS** Nests in same habitat in West as Lazuli; diet and foraging strategies also the same.
- **POPULATIONS** Outnumbered by Lazuli in zone of overlap in West. The two species hybridize; hybrid males in breeding plumage, mostly blue but with white wing bars and whitish bellies, are moderately distinctive.

Varied Bunting
Passerina versicolor | VABU | L 5½" (14 cm)

- **APPEARANCE** Adult male in good light distinctive, with plum, blueberry, and reddish hues. Brownish female, with thin eye ring, lacks wing bars; unstreaked breast is uniform gray-brown. Upper mandible of Varied more rounded than in other *Passerina* buntings.
- **VOCALIZATIONS** Song bright and steady, lacking buzzy elements of Lazuli and Indigo. Calls similar to those of Lazuli and Indigo.
- **BEHAVIORS** Nests in desert thornscrub: washes and foothills with mesquite, ocotillo, cactus, etc.
- **POPULATIONS** Onset of breeding in subspecies *dickeyae* (Ariz.) influenced by monsoons. Breeders in N. Mex. and West Tex. *(versicolor)* more reddish.

Painted Bunting
Passerina ciris | PABU | L 5½" (14 cm)

- **APPEARANCE** Adult male a riot of color, other plumages mostly plain greenish. Some adult females bright lime-gray, younger birds show just a hint of green; most show a thin but complete eye ring.
- **VOCALIZATIONS** Song a bright warble, sweeter than Lazuli's. Call note more muffled than other *Passerina* buntings; flight call slightly lower-pitched.
- **BEHAVIORS** In limited breeding range in West, often gets into mesquite, but in less arid situations than Varied; also in canopy of shade trees.
- **POPULATIONS** Wanders well north of range annually, often finding its way to feeders; also reaches coast annually.

Dickcissel
Spiza americana | DICK | L 6¼" (16 cm)
- **APPEARANCE** Stout-billed like other cardinalids (pp. 444–453). Breeding male, with yellow and black below, distinctive, but other plumages trickier. All have rufous on wing, gray face with buff-yellow eyebrow and white throat, and at least some yellow on breast.
- **VOCALIZATIONS** Song a few chips followed by short buzzes: *dick dick dick SISS SISS SISS*. Flight call an abrupt, flatulent *ffrrrt*.
- **BEHAVIORS** Nests in tallgrass prairie, including disturbed habitats: pastures, oil fields, roadsides, etc.
- **POPULATIONS** Limits of breeding range intrinsically unstable; common some summers western Great Plains, practically absent others. Small but regular passage in fall through Desert Southwest; annual vagrant to coast.

TRUE TANAGERS | THRAUPIDAE
Despite being restricted to the New World, this is an immense family, with close to 400 species. The thraupids are a challenge to characterize, but some general traits include bright colors, omnivorous diets, and a proclivity for forming small flocks.

Widely established in Hawaii

Red-crested Cardinal
Paroaria coronata | RCCA | L 7½" (19 cm)
- **APPEARANCE** Recalls Northern Cardinal (p. 446), but in different family; Red-crested is smaller overall and smaller-billed. Adult boldly marked red, white, and gray; immature similar, but face and crest a dull gray-red.
- **VOCALIZATIONS** Song, low and rich, an exaggeratedly slow caroling: *tlwee ... tweedle ... sleeur ...*.
- **BEHAVIORS** Inhabits gardens, resorts, and weedy habitats in lowlands; often seen in pairs or family groups bopping around yards near plantings.
- **POPULATIONS** Indigenous to S. Amer.; widely established in Hawaii except Hawaii I., where very rare and greatly outnumbered by Yellow-billed Cardinal.

Yellow-billed Cardinal
Paroaria capitata | YBCA | L 7" (18 cm)

Like Red-crested Cardinal, indigenous to S. Amer. Introduced to Hawaii, where distribution the opposite of Red-crested; Yellow-billed occurs only on Hawaii I., where it outnumbers Red-crested. Color scheme quite similar to Red-crested, but Yellow-billed lacks crest and has orange (not really yellow) bill; Red-crested has silvery bill. Song like Red-crested, but faster. Habits like Red-crested, with special fondness for drier microhabitats.

Established in Hawaii

Saffron Finch
Sicalis flaveola | SAFI | 5–5½" (13–14 cm)
- **APPEARANCE** Compactly built with large bill; wings and tail short. Adult male mostly yellow, but with reddish suffusion on face; female similar, but less vibrantly colored. Immature blotchy gray and yellow, with blurry streaking below.
- **VOCALIZATIONS** Song just a few simple descending chirps. Calls also chirpy, descending like song, but some also ascending.
- **BEHAVIORS** Haunts and habits like the Red-crested and Yellow-billed Cardinals, but more inclined to forming flocks.
- **POPULATIONS** Indigenous to S. Amer.; common on islands of Oahu and Hawaii, less so on Kauai and Maui. Birds established in Hawaii are subspecies *flaveola* from Colombia and Venezuela.

APPENDIX A | RARE BIRDS IN THE WEST

This appendix enumerates 254 bird species so rare in the West that they do not receive full treatment in the main text (pp. 22–453). Species included here have been credibly recorded in the West through summer 2023; the occurrence of almost all is documented by photos (especially in recent years) or specimens (especially in the past). Hundreds of species of escapes from captivity, not established in the wild, are generally not included here.

SPECIES OF CASUAL OCCURRENCE IN THE WEST

Rare birds may be classified as casual or accidental. Casual species occur typically less than annually in a particular region but with sufficient regularity as to be predictable in time and space. We treat 107 species as casual in the West. Note that the breakpoint between casual and accidental is exceedingly fuzzy. The approach here is conservative, with some borderline species treated as accidental rather than casual. The short accounts that follow focus on geographic distribution and field identification. Selected artwork is intended to give a feel for likely plumages shown by vagrants to the West; not all species are pictured.

Tundra Bean-Goose

Taiga Bean-Goose *Anser fabalis* | Eurasian species breeding at generally lower latitudes than closely related Tundra Bean-Goose. Taiga is larger than Tundra, with a longer neck and a longer bill that slopes smoothly and shallowly with forehead. Compare also with juvenile Greater White-fronted Goose (p. 24), much more likely than either bean-goose. Barely annual to Bering Sea region and accidental elsewhere.

Tundra Bean-Goose *Anser serrirostris* | Closely related to Taiga Bean-Goose; breeds at generally higher latitudes than that species. Tundra averages smaller, darker-headed, and shorter-necked than Taiga; shorter bill of Tundra does not slope as smoothly with forehead. Like Taiga Bean-Goose, probably annual to Bering Sea region; reported more frequently away from Alas. than Taiga Bean-Goose.

Whooper Swan

Whooper Swan *Cygnus cygnus* | Widespread Eurasian species similar in size and structure to Trumpeter Swan (p. 28). Bill extensively yellow on adult, dusky pink on juvenile. Very rare but regular, especially in winter, to Aleutian Is.; has even bred there. Scattered records elsewhere in West, mostly near coast; many likely pertain to wild vagrants, but some have been known or presumed escapes from captivity.

Baikal Teal

Baikal Teal *Sibirionetta formosa* | East Asian species rebounding from population crash in 20th century. Both sexes suggest smaller Green-winged Teal (p. 38). Casual to Bering Sea region, mostly fall, with uptick in records recently. Records along U.S. West Coast and Hawaii also increasing, but provenance always a question with sightings away from Alas. of this spectacular duck that is popular in captivity.

Garganey

Garganey *Spatula querquedula* | Old World species annual to N. Amer.; bulk of records in West from Bering Sea region, but widely scattered well inland and to Hawaiian Is. Breeding male has white crescent on dark face; other plumages, more muted, told by dark stripes on face. Many sightings from spring, doubtless reflecting bias of field observers toward easily recognized breeding-plumage males.

Falcated Duck

Falcated Duck *Mareca falcata* | East Asian species reported less than annually in N. Amer.; preponderance of records from Bering Sea region, but also a smattering near coast south to Southern Calif.; accidental Hawaiian Is. Smaller than other *Mareca* ducks, with droopy, disheveled crest. Popular in aviculture, and many records away from Alas. likely or definitely refer to birds of captive origin.

RARE BIRDS IN THE WEST

Eastern Spot-billed Duck *Anas zonorhyncha* I Southeast Asian species that reaches the western Bering Sea region less than annually; accidental to Hawaii. Closely related to Mallard (p. 36), but black bill of Eastern Spot-billed has yellow tip; note also dark stripes on face and white edges to tertials visible on bird at rest. Beware variation in Mallards, including hybrids and aberrant individuals with oddly marked bills and faces.

Eastern Spot-billed Duck

American Black Duck *Anas rubripes* I Part of the Mallard superspecies complex (p. 36), found in eastern Canada and northeastern U.S. Similar to Mexican Duck, but darker; white wing linings of American Black Duck contrast sharply with dark body. Casual across West, but full extent of distribution difficult to ascertain owing to both ID challenges (including hybridization with Mallards) and occasional escapes from captivity.

Common Pochard

Common Pochard *Aythya ferina* I Familiar duck of much of Eurasia, annual in small numbers to Bering Sea region. Bill and forehead sloping like Canvasback, but not as strikingly so; bill pattern (both sexes) more like Redhead (both on p. 38). A few records south along coast to Calif. and to Hawaiian Is. pertain to naturally occurring vagrants; occasionally escapes from captivity.

Smew *Mergellus albellus* I Small Eurasian duck that probably reaches Bering Sea region annually; accidental south to Calif. Closely related to goldeneyes (p. 48). Male, mostly white with a few black markings, unmistakable; female has warm brown crown and bright white cheeks. Scattered records away from Bering Sea region are uncomplicated as far as field ID goes, but the challenge is separating natural vagrants from escapes.

Smew

Least Grebe *Tachybaptus dominicus* I Widespread Neotropical grebe that regularly reaches South Tex. and, more sparingly, Ariz., where present in tiny numbers year-round. Suggests a miniature Pied-billed Grebe (p. 68). Note pointy, all-dark bill and yellow eye, contrasting with gray face. Breeding adults gray with black throat; non-breeders with whitish throat. Also records for Calif., N. Mex., and West Tex.

Least Grebe

Oriental Turtle-Dove *Streptopelia orientalis* I Migratory Asian species found in a variety of habitats. Large and stout; the size of a Rock Pigeon (p. 72). With finely marked patch on hindneck and scaly plumage above, calls to mind Spotted Dove (p. 74). Formerly casual to Bering Sea region, but very few recent records. Accidental farther south, mostly along coast, to Calif.

Oriental Turtle-Dove

Ruddy Ground Dove *Columbina talpacoti* I Widespread in Neotropics; barely enters our area, with preponderance of records from Ariz. Adult Ruddy Ground Dove (pictured on p. 75) differs from Common Ground Dove (p. 74) by dark bill (reddish on Common), plain breast and neck (finely scaled on Common), and face and crown grayer than Common. Most records in West are from winter, but the species has bred in Ariz. and Calif.

Groove-billed Ani *Crotophaga sulcirostris* I Mostly Neotropical species that reaches well into Tex. in summer; very rare but annual in West to southern border states, accidental farther north. All-black with a long tail. Bill stout, decurved, and broad-based; eponymous grooves in bill hard to see. Calls sharp and squeaky, explosive, often in series. Superficially suggests a grackle, but is actually a kind of cuckoo.

Groove-billed Ani

455

APPENDIX A

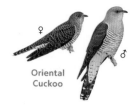

Oriental Cuckoo

Oriental Cuckoo *Cuculus optatus* | Old World species, casual to Bering Sea islands. Very similar to slighter-billed Common Cuckoo (p. 78). Gray-morph female and all males are darker gray above than corresponding plumages of Common Cuckoo, with thicker barring beneath and buffy vent. Hepatic-morph female is more densely barred above than hepatic-morph female Common.

Fork-tailed Swift

Fork-tailed Swift *Apus pacificus* | Asian and Australasian species recorded in Bering Sea region in both late spring and early autumn, but less than annually. Large, dark, and long-tailed, with a conspicuous white rump. Forked tail often held closed. Vagrants usually occur singly, but also sometimes in small flocks. Also known as Pacific Swift.

Purple Gallinule *Porphyrio martinicus* | Mostly Neotropical species that occurs regularly to southeastern U.S. An amazing vagrant, with records to Puget Sound region and even Hawaii. Smaller than a coot; adult very colorful, juvenile mostly buff but with greenish wings and upperparts like adult. Vagrants find their way to ridiculously small ponds, even swimming pools.

Common Crane

Common Crane *Grus grus* | Ecological counterpart in Old World of Sandhill Crane (p. 100), but more closely related to Whooping Crane. Annual in recent years to West, including many well inland. Head and neck of adult strongly marked black, white, and red; bill yellow. Usually seen in flocks of Sandhills.

Whooping Crane 🆎 *Grus americana* | Endangered species with core breeding range, wintering grounds, and migratory corridor just a bit to our east. Even taller than Sandhill (p. 100) and Common Cranes. Plumage mostly white; shows more red on face and head than the smaller species. Migrants barely enter our area in Mont. and Alta., with two in 21st century even to Yukon.

Eurasian Dotterel

Eurasian Dotterel *Charadrius morinellus* | Breeds mostly at high latitudes of Palearctic; formerly bred sparingly to Alas. In breeding plumage, shows white breastband and extensive orange on belly. Nonbreeding adult much more washed-out, but still shows thin white breastband. Formerly casual to West Coast (and accidental to Hawaii), but no records there recently.

Northern Jacana

Wilson's Plover *Charadrius wilsonia* | Widespread but uncommon on Gulf and Atlantic coasts of U.S., but strays to West probably originate from West Mexico populations. Largest of our plovers with a single black breastband; bill long and thick. Flight call a sharply rising utterance, *speee* or *speet!* Even among plovers, closely tied to coastal habitats. Casual to West Coast, with most records from Southern Calif.

Northern Jacana *Jacana spinosa* | Mid. Amer. inhabitant of marshlands that barely reaches U.S. Although traditionally thought of as a South Tex. specialty, preponderance of U.S. records in recent years from Ariz. Adult black and chestnut with neon yellow primaries; immature white below with boldly patterned face. No clear temporal pattern of vagrancy to our area.

Far Eastern Curlew

Far Eastern Curlew 🆎 *Numenius madagascariensis* | Very rare, perhaps annual, stray to Bering Sea region, spring to summer. The largest shorebird on Earth, more immense than even Long-billed Curlew (p. 110); Far Eastern is colder gray-brown overall than Long-billed, with gray-brown legs (blue-gray on Long-billed). Also vagrants to B.C. and Hawaiian Is.

Black-tailed Godwit *Limosa limosa* | Eurasian species, very rare but annual to Bering Sea region; accidental to Hawaii. Bill very long and bicolored as on other godwits, but almost straight; bill more upturned on other godwits. Appears boldly pied in flight, with wing linings white (black in Hudsonian Godwit, p. 112). Vagrants to West are subspecies *melanuroides*, relatively pale and orangey below in spring and summer.

Black-tailed Godwit

Great Knot 🆔 *Calidris tenuirostris* | Found in coastal lowlands of East Asia and Australasia. Casual to Bering Sea region; accidental to Hawaiian Is. Even larger than Red Knot (p. 116), with longer bill. Breeding plumage, with rufous on wings and dense black spotting all over, similar to that of Surfbird (p. 114), but bill structure very different. Known to hybridize with Surfbird.

Great Knot

Broad-billed Sandpiper *Calidris falcinellus* | Migratory Eurasian species. Vagrants have reached Bering Sea region; also one record for Ore. Bill broad at base, hooked down at tip; legs relatively short. In all plumages, facial pattern subtle but distinctive: dark crown bordered by thin white stripe, giving the bird the appearance of an "extra" stripe on face. World population declining, with no records in recent years to N. Amer.

Temminck's Stint

Temminck's Stint *Calidris temminckii* | Old World Arctic tundra and taiga breeder. Regular migrant to Bering Sea region, both spring and fall, with preponderance of records in spring. Bill straight and black; wings relatively short. Legs yellow. Outer tail feathers white. Vocalizations also distinctive: Flight call is a trill or rattle, sometimes prolonged, recalling a longspur flight call.

Long-toed Stint *Calidris subminuta* | Old World counterpart of Least Sandpiper (p. 120). Like Temminck's Stint, Long-toed is regular on migration in Bering Sea region; even more records for Long-toed. Accidental to Hawaii and Calif. Has yellow legs like Temminck's Stint and Least Sandpiper. In all plumages, dark cap of Long-toed set off from pale eyebrow. Flight call similar to Least's: a slurred, rising *t'weee?*

Long-toed Stint

Little Stint *Calidris minuta* | Breeds in Arctic of central Russia and Scandinavia, winters mostly Africa. Not nearly as frequent in Bering Sea region as Red-necked Stint (p. 122); casual along West Coast, and a few have reached Hawaii. Adult in breeding plumage colorful, but not as intensely so as breeding Red-necked; fresh juvenile Little has prominent white V on mantle. Flight call squeakier than Red-necked's.

Jack Snipe *Lymnocryptes minimus* | Eurasian species casual to Bering Sea region, where preponderance of records in fall. Notably smaller than Wilson's (p. 128) and Common Snipes; also shorter-billed than those species. Note also streaked flanks of Jack Snipe; flanks of Wilson's and Common barred. Also records for Wash., Ore., and Calif., but none in past decade.

Jack Snipe

American Woodcock *Scolopax minor* | Distinctive sandpiper that occurs regularly just to our east. Casual west to near foothills of Rockies from Alta. southward, where most records fall through spring; some even in winter. Accidental farther west, to Calif. Rotund, with very long bill and thick, dark transverse barring on crown. A secretive species found in moist, brushy habitats, not mudflats, in East; vagrants to West also find their way to such habitats.

American Woodcock

APPENDIX A

Common Snipe

Terek Sandpiper

Common Sandpiper

Gray-tailed Tattler

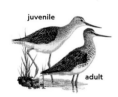

Common Greenshank

Marsh Sandpiper

Common Snipe *Gallinago gallinago* I Widespread Old World counterpart of Wilson's Snipe (p. 128). Regular on migration, both spring and fall, in Bering Sea region; has nested there. Casual to Hawaiian Is.; accidental south to Calif. In flight, Common shows white trailing edge to wing and white underwing panel.

Terek Sandpiper *Xenus cinereus* I Regular to Bering Sea region from spring to late summer. Also several records from West Coast; accidental to Hawaii. Structurally and behaviorally distinct: bill long and upturned; legs short and bright orange; body oddly flattened. Feeds animatedly, running around beaches with crouched posture; both running and standing, bobs tail rapidly.

Common Sandpiper *Actitis hypoleucos* I Old World counterpart of Spotted Sandpiper (p. 128). Regular migrant, more common in spring than fall, to Bering Sea region; casual to Hawaii. Longer-tailed than Spotted Sandpiper. At all seasons, plumage resembles nonbreeding Spotted, but bill and feet of Common not as colorful.

Green Sandpiper *Tringa ochropus* I Widespread Eurasian species; casual, mostly in spring, to Bering Sea region. Similar in ecology and appearance to Solitary Sandpiper (p. 128), but tail pattern different: Green has more white on tail overall with just a bit of weak black barring at tip, especially toward center. Very few records in recent years.

Gray-tailed Tattler *Tringa brevipes* I Old World counterpart of Wandering Tattler (p. 128). Regular on migration, especially in fall, to Bering Sea region; regular in very small numbers, fall through spring, to Hawaiian Is. Accidental to West Coast. Gray-tailed Tattler is paler gray than Wandering, with weaker barring below. Calls of Gray-tailed one or two slurred *tuawee* notes.

Spotted Redshank *Tringa erythropus* I Old World species; casual to Bering Sea region and West Coast; accidental farther inland and to Hawaii. Head, neck, and underparts black in breeding plumage. All plumages show orange legs and thin bill with droopy tip and variable orange at base. Flight call a clipped, two-note *tew-wit!* Few records in recent years to N. Amer.

Common Greenshank *Tringa nebularia* I Old World counterpart of Greater Yellowlegs (p. 130). Very rare but annual to Bering Sea region in spring and late summer; several records from Calif. Similar in build to Greater Yellowlegs, but bill a bit stouter and somewhat more upturned. Legs dull green; in flight, white wedge on back prominent. Flight call, like that of both yellowlegs, a ringing *tyu!*

Marsh Sandpiper *Tringa stagnatilis* I Widespread Eurasian species; casual to Bering Sea region, mostly fall but also several recent records in spring. Slender build accentuated by fine, straight, black bill and long, thin, greenish-yellow legs. Uppertail and lower back white, conspicuously so in flight. Multiple recent records for Calif. and Hawaii.

Long-billed Murrelet *Brachyramphus perdix* I Poorly studied alcid known to breed only in northeastern Asia. Larger overall than Marbled Murrelet (p. 140), with bigger bill. Key difference in black-and-white winter plumage is that Long-billed has entirely dark nape; dark nape of Marbled broken by white collar. Numerous records inland in West (and even in East) in late fall, although almost none in recent years.

Black-tailed Gull *Larus crassirostris* | East Asian species found mostly in coastal lowlands; casual to Pacific coast of N. Amer. from Bering Sea region to Southern Calif. Considered part of the large white-headed gull (LWHG) complex, but a small one. Long bill has dark tip. Broad black tail band distinctive, but note that many LWHGs regular to N. Amer. have broad black tail bands when young.

Common Gull

Common Gull *Larus canus* | Old World gull that is very rare but annual to both East Coast and Bering Sea region. The subspecies that reaches our area is "Kamchatka Gull," *kamtschatschensis*, of eastern Asia, intermediate in size and plumage, especially in first winter, between Short-billed and Ring-billed Gulls (both on p. 154). There are several records of *kamtschatschensis* from Hawaii.

Great Black-backed Gull

Great Black-backed Gull *Larus marinus* | Huge gull of Europe and eastern N. Amer. Annual in very small numbers to eastern flank of Rockies, and a few have made it to West Coast and even Alas. Adult black above with very large bill and pink legs; first-winter with distinctive salt-and-pepper look above. Gradually expanding in East, with concomitant rise in reports of vagrants to West.

Little Tern *Sternula albifrons* | Old World counterpart of Least Tern (p. 168), present in the Americas only at Midway Atoll, where it has bred recently in tiny numbers. Very similar to Least Tern, but Little Tern is a bit larger and proportionately shorter-tailed; calls not as squeaky as those of Least.

White-capped Albatross

White-winged Tern *Chlidonias leucopterus* | Old World "marsh tern" related to Black Tern (p. 168). Numerous records from East, but also casual to West, especially Bering Sea region; records also from Calif. and Hawaii. Upperwings and tail of White-winged paler than in corresponding plumages of Black Tern; breeding White-winged has black underwing linings. No recent records from Bering Sea.

White-capped Albatross *Thalassarche cauta* | Southern Hemisphere species, with primary concentration around New Zealand and Tasmania, where it breeds. Part of the "mollymawk" complex of midsize albatrosses. Roughly similar in plumage to smaller Laysan Albatross (p. 180), but bill of White-capped thicker, more olive-gray. Several records well off West Coast from Wash. to Calif.

Wedge-rumped Storm-Petrel

Wedge-rumped Storm-Petrel *Hydrobates tethys* | Found mostly at low latitudes off the west coast of the Americas, from Mexico to Chile. Small, even for a storm-petrel, with extensive white tail coverts and rump. Sometimes wanders into waters off Southern Calif.; also well inland in Desert Southwest following hurricanes, occasionally numbering in double digits.

Kermadec Petrel *Pterodroma neglecta* | Breeds in South Pacific region, dispersing north during the northern summer; sightings in waters off Hawaiian Is. generally well offshore, spring through fall. Large and variable, ranging from extensively pale below and on head to mostly dark; all morphs flash extensive white in wings. Possibly nested at Kauai in late 20th century. Accidental to Calif.

Kermadec Petrel

Nazca Booby

Nazca Booby *Sula granti* | Close relative of Masked Booby (p. 200), but with smaller range; Nazca breeds mostly Galápagos Is., dispersing north to Mexico. Adult Nazca has slighter, orange-pink bill (yellowish on Masked) and white tail center; juvenile usually lacks white neck collar of Masked. Nazca formerly considered accidental to waters off Calif., but detections really ticking up in past decade, with records to Alas. and Hawaiian Is.

APPENDIX A

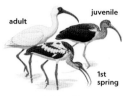
White Ibis

Roseate Spoonbill

White Ibis *Eudocimus albus* | Found mostly in coastal plains of Neotropical wetlands; range reaches north to Gulf Coast and southern Atlantic states; casual vagrant to West, as far northwest as Wash. Bill strongly decurved; adult plumage nearly pure white; juvenile mostly dusky gray-brown. Range expanding and numbers increasing; additional vagrants to West may be expected in coming years.

Roseate Spoonbill *Platalea ajaja* | Widespread Neotropical species that reaches Gulf Coast. Annual vagrant to West, with concentration of records from Southern Calif. and especially Southern Ariz. Remarkable bill and red and pink highlights make the species unmistakable at all ages. Strays to Ariz. probably come from Gulf of California, not Gulf of Mexico.

Swallow-tailed Kite *Elanoides forficatus* | Striking raptor found mostly in Neotropics; breeds north to southeastern coastal plain of U.S. Wanders widely, mostly within East, but also into West; strongest concentration of sightings here in N. Mex. Highly unique in plumage, tail structure, and flight style; the most similar bird, oddly, might be a Magnificent Frigatebird (p. 198) in subadult plumage.

White-tailed Eagle

White-tailed Eagle *Haliaeetus albicilla* | Eurasian counterpart of Bald Eagle (p. 222); breeds regularly to southwestern Greenland. Adult has white tail like Bald Eagle; is also pale-headed, but not as gleamingly white-headed as Bald Eagle. Almost annual to Bering Sea region; famously bred at Attu I. for close to 20 years in late 20th century. Accidental south to Calif. and Hawaii.

Steller's Sea-Eagle

Steller's Sea-Eagle *Haliaeetus pelagicus* | Bald Eagle (p. 222) relative found mostly in coastal lowlands of northeastern Asia; casual to Bering Sea region, accidental to Hawaiian Is. Adult boldly marked with large white wing patches, all-white tail, and white feathering on legs; young, as in Bald Eagle, slowly transition to adult plumage over several years. Very large bill acquires bright yellow-orange hue when bird still quite young.

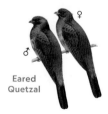
Eared Quetzal

Eared Quetzal *Euptilotis neoxenus* | Like Thick-billed Parrot, essentially endemic to the Sierra Madre Occidental, where it is found in pine-oak forests. Casual to woodlands in "Sky Islands" of Southeast Ariz., where it has attempted to breed; accidental to N. Mex. In same family as Elegant Trogon (p. 242). Eared Quetzal is larger and darker, with dark gray bill (yellow on trogon); also lacks white breastband of trogon. Recorded almost year-round in Ariz., with small spike of sightings in fall.

Great Spotted Woodpecker *Dendrocopos major* | Widespread Palearctic species. Mostly resident, but some disperse a bit in fall to winter; casual to Alas., with records both spring and fall. Suggests Hairy Woodpecker (p. 254), but with large white cheek patch (black on Hairy), long white patch on wing, and extensive red on vent. No records in past decade.

Eurasian Kestrel

Eurasian Kestrel *Falco tinnunculus* | Widespread across much of the Old World; casual to Bering Sea region, accidental to Wash. and Calif. Like a large American Kestrel (p. 258), but with one vertical stripe, not two, on face; male Eurasian has blue tail and orange wings—the opposite of American. Occasionally used for falconry, so sightings far from Bering Sea should be scrutinized for possible captive origin.

Eurasian Hobby *Falco subbuteo* | Something of a distributional doppelganger for Eurasian Kestrel; widespread across much of Old World, casual to Bering Sea, accidental farther south on coast. Adult distinctive, but juvenile suggests a small Peregrine Falcon (p. 260). As with Eurasian Kestrel, possibility of escapes from falconry should be considered for birds away from Alas.

Eurasian Hobby

Monk Parakeet *Myiopsitta monachus* | Indigenous to S. Amer. south of Amazon Basin; well established in parts of eastern U.S. Underparts mostly pale gray; remiges blue. Not considered established in West, but escapes often seen in southern border states; occasionally they nest in the wild. Used to be seen with some regularity in Pacific Northwest states, but fewer sightings in recent years.

Thick-billed Parrot 🆔 *Rhynchopsitta pachyrhyncha* | A marvelous Mexican parrot, the treasure of the Sierra Madre Occidental. Formerly wandered to Chiricahua Mts., Ariz., sometimes in flocks, probably annually; perhaps bred. A large, long-tailed parrot; mostly green, with red forehead and "shoulders." Repatriation efforts in Ariz. in late 20th century were unsuccessful.

Nutting's Flycatcher

Nutting's Flycatcher *Myiarchus nuttingi* | Mid. Amer. species resident north to northern Mexico. A few records to Calif., Ariz., and West Tex.; has bred Mohave Co., Ariz. Very similar to Ash-throated Flycatcher (p. 268); the two were formerly treated as one species. Nutting's more richly colored, and the *inside* of its mouth is bright orange—get photos! Call of Nutting's a short, rising whistle.

Fork-tailed Flycatcher *Tyrannus savana* | Widespread in Neotropics; some populations are austral migrants that reach the East every year. Spectacularly long-tailed adult sometimes confused with Pin-tailed Whydah (p. 352); shorter-tailed immature, dark above and pale below, could be passed off as Eastern Kingbird (p. 272). Casual to West, with scattered records to coast and north to Alta.

Fork-tailed Flycatcher

Tufted Flycatcher

Tufted Flycatcher *Mitrephanes phaeocercus* | Mostly Mid. Amer. species that has been annual in recent years to Southeast Ariz., where it has bred; also scattered records to West Tex. Rather like a small *Contopus* flycatcher (and thought to be a close relative), but with deep cinnamon on face and below; distinctly crested. Like *Contopus* flycatchers, tends to perch out in the open on dead or bare snags.

Eastern Wood-Pewee *Contopus virens* | Eastern counterpart of Western Wood-Pewee (p. 274); breeding range regularly reaches west almost to northeastern Mont. Differs weakly in plumage from Western, but songs different: Song of Eastern a pure-tone whistle, *PEEEE-uh-weeee*. Both species give weak whistles on migration. Eastern very rare but annual to foothills of Rockies; casual farther west.

Eastern Wood-Pewee

White-eyed Vireo

White-eyed Vireo *Vireo griseus* | Common in thickets in southeastern U.S. Bright and small; head gray with yellow around bill and eye. White eye of adult distinctive, but juvenile's eye is dark. Distinctive song *(chick! uh-duh-WEEE-oh chick!)* sometimes sung by vagrants in spring. Rare vagrant to western U.S. and southwestern Canada, with most records from southern Rockies and southwestern U.S.

APPENDIX A

Yellow-throated Vireo

Brown Shrike

Middendorff's
Grasshopper Warbler

Siberian House-Martin

Willow Warbler

Wood Warbler

Yellow-throated Vireo *Vireo flavifrons* | Breeds in leafy hardwood forests, often in river bottoms, of eastern U.S. Large and stocky, with eye-catching yellow throat and breast; note conspicuous yellow "spectacles," white wing bars, and white belly. Song like Plumbeous Vireo (p. 286), but even more exaggeratedly slow. Rare vagrant to western U.S. and southwestern Canada, especially southern Rockies and Calif.

Brown Shrike *Lanius cristatus* | Migratory Asian species, casual to Bering Sea region, accidental to West Coast (B.C., Calif., Alas. Panhandle). Warm brown plumage and lack of white in wings separate it from N. Amer. shrikes, but several rarer Eurasian shrikes (such as a remarkable Red-backed x Turkestan Shrike, *Lanius collurio* x *L. phoenicuroroides*, Mendocino Co., Calif., 2015) are possible vagrants to N. Amer.

Middendorff's Grasshopper Warbler *Helopsaltes ochotensis* | Migratory East Asian species, found mostly at low elevations around coasts. Casual to Alas. spring to fall. A skulker; mostly unmarked brown-buff all over, with pale supercilium. Ample tail is broad with whitish tips. Precise taxonomic status unclear, but placement within grassbird family Locustellidae enjoys widespread support.

Lanceolated Warbler *Locustella lanceolata* | Migratory species that breeds across central Eurasia from southern Finland to Russian Far East. Casual to Bering Sea region. Remarkably, at least four singing males held territories on Buldir I. in the Aleutians in the summer of 2007, and at least one of them bred. Accidental to Calif.; no records from N. Amer. in recent years.

Western House-Martin *Delichon urbicum* and **Siberian House-Martin** *Delichon lagopodum* | Until recently treated as conspecific. Both are small and deep blue above, with deeply notched tail; rump white. Both species have probably occurred in N. Amer. (*lagopodum* to Bering Sea region, *urbicum* to Atlantic Canada), but records under review at the present time.

Willow Warbler *Phylloscopus trochilus* | Highly migratory Old World species; breeds across northern Eurasia, winters in southern Africa. Casual in fall to Bering Sea region. Related and similar to Arctic Warbler (p. 316), but smaller and a bit more colorful; supercilium faint and wing bar almost absent. All records from 21st century; almost annual now, with bulk of records from Gambell, St. Lawrence I., Alas. Accidental south to Calif.

Common Chiffchaff *Phylloscopus collybita* | Migratory Old World species similar to Willow Warbler, but grayer and darker; Chiffchaff has black legs and bill (more orange-pink on Willow Warbler) and shorter wings than Willow Warbler. Accumulation of records in N. Amer. mirrors that of Willow Warbler; unrecorded here until 21st century, but now somewhat regular, with records both spring and fall.

Wood Warbler *Phylloscopus sibilatrix* | Mostly European breeder that winters in equatorial Africa. Casual in fall to Bering Sea region. A rather warmly colored "phyllosc," with yellow supercilium, yellowish throat, greenish upperparts, and white belly; legs yellow-orange. Almost all records in N. Amer. are from 21st century, including one extremely well-watched vagrant in Los Angeles, Oct. 2022.

Dusky Warbler *Phylloscopus fuscatus* | Asian species that wanders to N. Amer. more than any other "phyllosc" except Arctic Warbler (p. 316), a regular breeder in Alas. Like a browner, oversize Arctic Warbler without even a hint of a wing bar. Very rare migrant in Bering Sea region, with more records in fall than spring. Casual farther south, with quite an accumulation of records, all fall, from San Francisco area southward.

Dusky Warbler

Yellow-browed Warbler *Phylloscopus inornatus* | Migratory Eurasian species that breeds east almost to Bering Sea. Casual in fall (mostly Sept.) to Bering Sea region; accidental farther south to B.C. and Calif. Rather boldly marked for a "phyllosc," with prominent supercilium and strong wing bars. All records for N. Amer. from 21st century.

Yellow-browed Warbler

Kamchatka Leaf Warbler *Phylloscopus examinandus* | Very similar to closely related Arctic Warbler (p. 316). Best distinction in the field is the call note, disyllabic in Kamchatka Leaf Warbler but monosyllabic in Arctic Warbler. Was treated as conspecific with Arctic Warbler until early 2010s; status in N. Amer. still being worked out. There are a few records for Bering Sea region, both spring and fall.

Blue Mockingbird *Melanotis caerulescens* | Endemic to Mexico, where generally nonmigratory. Several records, all relatively recent, to southern border states from Calif. to West Tex. (and farther east). Dark blue all over with broad black ear patch. Identification rarely a challenge, but separating naturally occurring vagrants from escapes is more vexing.

Blue Mockingbird

Wood Thrush *Hylocichla mustelina* | Breeds in hardwood, mostly lowland forests in the East. Larger and plumper than *Catharus* thrushes. Rich rufous-brown above; white underparts extensively spotted with bold black spots. Very rare but annual west to foothills of Rockies spring and fall. Casual farther west, mostly fall; one made it all the way to St. Paul I. in Bering Sea, Oct. 2014.

Aztec Thrush *Ridgwayia pinicola* | Mexican species found in conifer-oak forests. Adult male distinctively patterned in black and white; female similarly patterned but browner. Juvenile streaked and scalloped. Nonmigratory in most of range, but some wander north; a few reach Ariz. year-round, with concentration of records in late summer. A few 20th-century records for West Tex.

Aztec Thrush

Gray-streaked Flycatcher *Muscicapa griseisticta* | Migratory East Asian species; casual to Bering Sea region in spring and fall, with stronger concentration of records from fall. Very long-winged. Plain brown above; white below with prominent streaking. Along with the next three species, an Old World Flycatcher, family Muscicapidae.

Gray-streaked Flycatcher

Dark-sided Flycatcher *Muscicapa sibirica* | Migratory East Asian species; casual to Bering Sea region, with records both in late spring and early autumn. Similar to Gray-streaked Flycatcher, but shorter-winged and a bit smaller overall. Sooty brown above; paler below with dusky wash on breast and flanks. Head and face marked with thin eye ring, pale throat, and thin white "necklace."

Red-flanked Bluetail *Tarsiger cyanurus* | Mostly Asian species. Casual to Bering Sea region both spring and fall, with preponderance of records in fall; also a scattering of records coastally from B.C. to Southern Calif., and even a couple well inland (Idaho, Wyo.). In all plumages, tail is blue and flanks orangey.

Red-flanked Bluetail

APPENDIX A

Asian Stonechat

Siberian Accentor

Olive-backed Pipit

Hawfinch

Eurasian Bullfinch

Oriental Greenfinch

Asian Stonechat *Saxicola maurus* | Taxonomically vexed Old World flycatcher. Stonechats are small, short-tailed birds with a pale rump and pale wing patch; breeding males striking, other plumages more subtle. Vagrants to Alas., casual to Bering Sea region, treated by some authorities as Amur Stonechat, *Saxicola stejnegeri*, but the name Siberian Stonechat is also applied—it's complicated!

Siberian Accentor *Prunella montanella* | Asian species. The only member of its family, Prunellidae, ever recorded in N. Amer. Like a brightly colored sparrow, but thin-billed. Black-and-peach face and throat distinctive. Almost annual to Bering Sea region, fall to early winter. Also a cluster of scattered records from southwestern Canada and northwestern lower 48 states: B.C., Alta., Wash., Idaho, and Mont.

Gray Wagtail *Motacilla cinerea* | Migratory Eurasian-African species; annual in very small numbers to Bering Sea region, both spring and fall, but preponderance of records in fall. Accidental south to Calif. Longer-tailed and a bit larger than similar Eastern Yellow Wagtail (p. 356). On Gray Wagtail, yellow extends to rump; white in tail and wings prominent in flight. Breeding male black-throated, other plumages pale-throated. Flight call a sharp smacking sound, usually uttered in twos and threes in rapid succession.

Olive-backed Pipit *Anthus hodgsoni* | Asian species; closely related and similar to even rarer Tree Pipit (p. 472). Its olive back has faint dark streaks; the mostly white underparts have prominent black streaks. Legs pink. Annual in tiny numbers in spring and especially fall in Bering Sea region; also a small cluster of records farther south (Calif., Nev., Baja California).

Hawfinch *Coccothraustes coccothraustes* | Old World counterpart of Evening Grosbeak (p. 360). Almost annual to Bering Sea region, both in late spring and late summer. Irruptive; for example, was widely noted in Bering Sea region in June 2023. Casual in Alas. away from Bering Sea. Like Evening Grosbeak, huge-billed, chunky overall; clad in copper and silver, with prominent white wing panels.

Common Rosefinch *Carpodacus erythrinus* | Widespread migratory Palearctic species. Annual in tiny numbers, spring and fall, to Bering Sea region; accidental south to Calif. Recalls House Finch (p. 372), but a bit smaller, with culmen particularly rounded; male more extensively red, female has greenish hues. Despite similarity to House Finch, Common Rosefinch is apparently more closely related to the Hawaiian honeycreepers.

Eurasian Bullfinch *Pyrrhula pyrrhula* | Widespread in diverse forest types in Palearctic. Nearly annual to Bering Sea region in spring and fall; also multiple records away from Bering Sea region in Alas. Male distinctively patterned in pink, black, white, and gray; female similarly but more subtly marked. Close relative of larger Pine Grosbeak (p. 360).

Oriental Greenfinch *Chloris sinica* | East Asian species of woodland edges. Casual to Bering Sea region in spring; also a few fall records there. Accidental to B.C., Ore., and Calif. Rather like a large goldfinch, but bill is a bit larger; fairly dark overall, with bright yellow in wings. Adult (both sexes) unmarked below; juvenile streaked below.

Little Bunting *Emberiza pusilla* | Migratory species that breeds widely in Old World taiga, where quite common. Annual in tiny numbers to Alas., mostly Bering Sea region and mostly in fall. Also scattered records on U.S. West Coast and even inland (Ariz., eastern Ore.). Stub-billed, short-legged, and short overall. Has bright orangey highlights on head and wings; finely streaked below.

Little Bunting

Eastern Towhee *Pipilo erythrophthalmus* | Eastern counterpart of Spotted Towhee (p. 408), breeding west regularly to central Great Plains. Instead of heavily spotted upperparts, has just a single white spot ("pocket handkerchief") on wing. Hybrids with Spotted Towhee present year-round at extreme eastern edge of our area in U.S. and south-central Canada. Apparent pure Easterns are casual west to foothills of Rockies and accidental farther west; field ID probably not possible in some instances.

Worm-eating Warbler *Helmitheros vermivorum* | Breeds in broadleaf forests to our east, especially on drier, south-facing slopes. Long-billed, short-tailed. Head and breast washed warm buff; face has prominent black stripes. Annual vagrant to West, reaching coast most years; most records in West in southern half of lower 48 states.

Worm-eating Warbler

Louisiana Waterthrush *Parkesia motacilla* | Breeds along streams in forests in East. Similar to Northern Waterthrush (p. 424), but Louisiana is larger-billed, longer-winged, and shorter-tailed. Eyebrow broad, long, and white; white throat is unmarked. Not even annual in most places west of Continental Divide, but annual in tiny numbers in Southeast Ariz., midsummer through winter; most there find their way to very small ponds, often in woods, where they may be present for weeks at a time.

Louisiana Waterthrush

Golden-winged Warbler *Vermivora chrysoptera* | Breeds to our east in early successional habitats. Male has striking black-and-white head; body mostly gray with yellow wing panel; scheme of adult female same as adult male, but colors and contrast muted. Casual vagrant west of Rockies; somewhat more records for this species in West than for Blue-winged.

Blue-winged Warbler *Vermivora cyanoptera* | Like Golden-winged Warbler, breeds in early successional habitats to our east. Mostly yellow, with a thin black line through eye; the blue-gray wings have white wing bars. Closely related to Golden-winged Warbler; hybrids and backcrosses frequently noted. Casual west of Rockies in U.S. and southern Canada, with records to West Coast.

Golden-winged Warbler

Prothonotary Warbler *Protonotaria citrea* | Breeds to our east in bottom-land swamps. Underparts and entire head radiant yellow; long-billed and short-tailed. Plain blue-gray wings lack wing bars; tail flashes much white in flight. A few show up every year in West, mostly in southern half of lower 48 states; records about equally split between spring and fall migration seasons.

Prothonotary Warbler

Crescent-chested Warbler *Oreothlypis superciliosa* | Mostly nonmigratory species of forests in Mexico and northern Mid. Amer.; casual to Southeast Ariz. Bluish-gray and olive above, with prominent white supercilium. Bright yellow below; adult male has prominent rusty spot on breast, weaker in other plumages. Formerly considered a parula, but now treated as closer to genus *Leiothlypis*.

Crescent-chested Warbler

APPENDIX A

Black-throated Blue Warbler

Yellow-throated Warbler

immature ♀
Prairie Warbler

Fan-tailed Warbler

Slate-throated Redstart:

Yellow Grosbeak

Kentucky Warbler *Geothlypis formosa* | Breeds in moist bottomland forests of Southeast. Dark olive above; bright yellow below, including throat. Face mostly dark, but with yellow above and behind eye, creating "spectacled" appearance. Annual vagrant west of Rockies, with more records during spring migration than fall migration.

Black-throated Blue Warbler *Setophaga caerulescens* | Breeds to our east in mixed conifer–northern hardwoods forests. Buffy-green female and black-and-blue male very different, but both show pale at base of primaries, forming a "pocket handkerchief." A few make it west of Rockies every year, with more sightings in fall than spring, with concentration of records in Oct.

Pine Warbler *Setophaga pinus* | Breeds and winters in pinewoods to our east. Large-billed, short-winged, and long-tailed. All have prominent wing bars; yellowish underparts have blurry side streaks. A few make it every year to West, with concentration of records in southern Rockies; some are found on migration, but just as many occur in the West in winter.

Yellow-throated Warbler *Setophaga dominica* | Breeds to our east in lowland forests. Suggests Grace's Warbler (p. 438), but larger and longer-billed; note gray upperparts, bright yellow underparts, and high-contrast black-and-white on face. A few make it each year to west of Rockies, fall to spring, mostly in southwestern U.S.; one in Alas. Panhandle Sept. 2015 was seriously lost.

Prairie Warbler *Setophaga discolor* | Breeds to our east in early successional habitats. Small and slight; breeding adult has black and olive on yellow face, bright yellow underparts with black streaks on sides of breast, and reddish streaks on back. Annual vagrant to west of Rockies, with decent accumulation of records on West Coast; one on Middleton I., Gulf of Alaska, Sept. 1990, was remarkable.

Fan-tailed Warbler *Basileuterus lachrymosus* | Mid. Amer. species of forested foothills and mountains; resident in much of range, but migratory in northern Sierra Madre Occidental. Fairly large warbler with ample tail, often fanned. Dark above with yellow crown; yellowish below. Face combines broken eye ring with white spot in front of eye. Casual to Ariz., accidental to N. Mex. and West Tex.

Slate-throated Redstart *Myioborus miniatus* | Widespread Neotropical species, occurring from Bolivia to northern Mexico. Mostly resident, but northern populations partially migratory, and the species is a casual stray to Southeast Ariz., southern N. Mex., and West Tex. Similar to Painted Redstart (p. 442), but lacks white on face and wings and has reduced white in tail.

Yellow Grosbeak *Pheucticus chrysopeplus* | Mostly resident in forests of Mexico and Guatemala; partially migratory in northern part of range. Annual in recent years in tiny numbers to southern Ariz. spring to summer; has gotten as far north as Colo. Huge-billed; mostly yellow, but with black on wings and tail. Closely related to Black-headed and Rose-breasted Grosbeaks (both on p. 448).

SPECIES OF ACCIDENTAL OCCURRENCE IN THE WEST

Accidental species, the rarest of the rare, occur far less than annually. Many of the 147 accidentals in the West have been documented fewer than 10 times. However, documentation of accidentals has surged in the 21st century, spurred by the proliferation of digital cameras and the explosion of interest in birding. A decent number of the species here, with steadily accumulating records of late, could be treated as casuals.

Lesser White-fronted Goose *Anser erythropus* | Migratory Palearctic species, with one record for Attu I., Alas. Kept in captivity, and sometimes escapes to wild.

Pink-footed Goose *Anser brachyrhynchus* | Western European species, increasing in winter in East. Recent records for Colo. and B.C.

Barnacle Goose *Branta leucopsis* | European species, increasing in winter in Northeast. Scattered records in West a mix of vagrants and escapes.

Mottled Duck *Anas fulvigula* | Eastern counterpart of Mexican Duck; hybridizes with Mallard (both on p. 36). Accidental in our area west to foothills of the Rockies.

Common Scoter *Melanitta nigra* | Eurasian species, split from Black Scoter (p. 44) in early 2010s. Records, all recent, from Ore. and Calif. (also Que.).

Masked Duck *Nomonyx dominicus* | Neotropical duck, related to Ruddy Duck (p. 50); formerly regular to South Tex. One record for West Tex., July 1976.

Chuck-will's-widow *Antrostomus carolinensis* | Large goatsucker that breeds to our east; has strayed to eastern N. Mex. and, remarkably, twice to coastal Calif.

Eastern Whip-poor-will *Antrostomus vociferus* | Eastern counterpart of Mexican Whip-poor-will (p. 82), with records as far west as Ore. and Calif.

Gray Nightjar *Caprimulgus jotaka* | Asian goatsucker, with one old record (May 1977), a carcass at Buldir I. in the Aleutians.

White-collared Swift *Streptoprocne zonaris* | Widespread Neotropical species, with several records for East; also one to Calif., May 1982.

White-throated Needletail *Hirundapus caudacutus* | Widespread Australasian species, with several records for Bering Sea region, spring and summer.

Common Swift *Apus apus* | Highly migratory Old World species, with scattered records in N. Amer.; several records from the Bering Sea region in early summer.

Mexican Violetear *Colibri thalassinus* | Mid. Amer. hummingbird, annual in tiny numbers to East; scattered records for West, north to Alta. and west to coast.

Amethyst-throated Mountain-gem *Lampornis amethystinus* | Mid. Amer. hummingbird that has strayed to West Tex. (and, remarkably, Que.).

Bumblebee Hummingbird *Selasphorus heloisa* | Tiny Mexican endemic with two 19th-century records (both specimens) from Ariz.

Xantus's Hummingbird *Basilinna xantusii* | Endemic to Baja California, but has strayed north to Southern Calif. and even B.C.

Cinnamon Hummingbird *Amazilia rutila* | Mid. Amer. species, with single records for both Ariz. and N. Mex.

Rufous-necked Wood-Rail *Aramides axillaris* | Widely but disjunctly distributed in Neotropics, mostly lowlands; one made it to central N. Mex., summer 2013.

King Rail *Rallus elegans* | Large freshwater species found to our east; several records west to eastern Colo. and N. Mex.

Clapper Rail *Rallus crepitans* | Eastern saltmarsh counterpart of Ridgway's Rail (p. 96); one record for south-central N. Mex., 2009 and 2010, presumably of the same individual.

Common Moorhen *Gallinula chloropus* | Wide-spread Old World species, with a record for Shemya I., Alas., Oct. 2010.

Eurasian Coot *Fulica atra* | Widespread Old World species casual to eastern Canada; a few records for Bering Sea region.

APPENDIX A

Sungrebe *Heliornis fulica* | Only New World representative of the finfoot family. Occurs widely in Neotropical wetlands; one reached central N. Mex., Nov. 2008.

Limpkin *Aramus guarauna* | Mostly Neotropical species that breeds north to Fla. Range rapidly expanding, with two recent records (2023) for Colo.

Hooded Crane *Grus monacha* | East Asian species, with several records for lower 48 states, including Idaho, maybe referring to same individual or flock; also Alas.

Black-winged Stilt *Himantopus himantopus* | Old World counterpart of Black-necked Stilt (p. 102); several records for Bering Sea and Hawaiian Is.

Eurasian Oystercatcher *Haematopus ostralegus* | Old World species, casual to Nfld.; one strayed to Buldir I., Alas., spring 2012.

Northern Lapwing *Vanellus vanellus* | Large Eurasian plover, almost annual to Northeast. Records of lone birds in West for Aleutians and Midway Atoll.

European Golden-Plover *Pluvialis apricaria* | Mostly northwestern Eurasian species almost annual to Nfld. Several records in West for Alas. and one for N. Mex.

Little Ringed Plover *Charadrius dubius* | Widespread Eurasian species, with a couple of spring records for western Aleutians; also one from Midway Atoll.

Greater Sand-Plover *Charadrius leschenaultii* | Old World species, with two records from N. Amer., including one in West (Calif., 2001).

Little Curlew *Numenius minutus* | East Asian species closely related to Eskimo Curlew (p. 473); several records for Alas. and West Coast.

Spoon-billed Sandpiper CR *Calidris pygmea* | Critically endangered species; breeds Russian Far East, but no recent records for N. Amer.; 20th-century sightings in Alas. and B.C.

Purple Sandpiper *Calidris maritima* | North Atlantic counterpart of Rock Sandpiper (p. 118); vagrants to West have included several far inland.

Solitary Snipe *Gallinago solitaria* | Asian species, with several recent records from Bering Sea region.

Pin-tailed Snipe *Gallinago stenura* | Breeds mostly Russia; several records for Bering Sea region, spring and late summer.

Oriental Pratincole *Glareola maldivarum* | Asian species; related to shorebirds but feeds aerially. Two records, both mid-1980s, for Bering Sea region.

Swallow-tailed Gull *Creagrus furcatus* | Found mostly off west coast of S. Amer.; a few records for Calif.; also one to Wash.

Pallas's Gull *Ichthyaetus ichthyaetus* | Distinctive Central Asian breeder (a large black-headed gull), with a single record for N. Amer. (Shemya I., Alas., May 2019).

Belcher's Gull *Larus belcheri* | Found on west coast of S. Amer.; vagrants to N. Amer. include one in Southern Calif., 1997-98.

Kelp Gull *Larus dominicanus* | Very widespread in Southern Hemisphere. Scattered records in N. Amer. include vagrants to Calif. and Colo.

Bridled Tern *Onychoprion anaethetus* | Fairly widespread in warm waters worldwide, but accidental to West; records for Hawaii and Calif.

Inca Tern *Larosterna inca* | Found along Pacific coast of S. Amer.; one wandered around the Hawaiian Is., 2021-22.

Great Crested Tern *Thalasseus bergii* | Widespread in tropical waters of Old World; two records for Hawaii.

Sandwich Tern *Thalasseus sandvicensis* | Taxonomically fraught species with wide distribution, including East; records for West in Colo., Calif., and Hawaii.

Chatham Albatross *Thalassarche eremita* | Breeds only on the Chatham Is., off New Zealand, but wanders to west coast of S. Amer.; one or two records off Calif.

Salvin's Albatross *Thalassarche salvini* | Close relative of Chatham Albatross, but with wider breeding distribution; has wandered to Calif. and even Alas.

Light-mantled Albatross *Phoebetria palpebrata* | Beautiful species of south polar region; one, far out of range, made it to Calif., July 1994.

Wandering Albatross *Diomedea exulans* | Enormous species of Southern Ocean; has wandered north to waters off Ore. and Calif. Recently split into four species by several authorities; experts still evaluating the two records for the West.

Ringed Storm-Petrel *Hydrobates hornbyi* | Occurs off west coast of S. Amer., but one made it to waters off Southern Calif., Aug. 2005.

Swinhoe's Storm-Petrel *Hydrobates monorhis* | Breeds western Pacific north to Russia; one likely made it to Alas., Aug. 2003.

Ainley's Storm-Petrel *Hydrobates cheimomnestes* | Poorly known species that nests in winter only on Guadalupe I., off Mexico. One record off Ventura Co., Calif., Dec. 2021.

Northern Giant-Petrel *Macronectes halli* | Huge, albatross-like relative of Northern Fulmar (p. 186). Southern Hemisphere species; one made it to waters off Wash., Dec. 2019.

Gray-faced Petrel *Pterodroma gouldi* | Breeds New Zealand, but several records off Calif., midsummer through early winter.

Providence Petrel *Pterodroma solandri* | Disperses well north after breeding off Australia; has reached waters off Alas. (including, once, a flock), B.C., and Wash.

Herald Petrel *Pterodroma heraldica* | Ranges widely in tropical Pacific Ocean, occasionally reaching Hawaiian waters, generally well offshore.

Stejneger's Petrel *Pterodroma longirostris* | Breeds off Chile, winters northwest all the way to Japan; several fall records for Hawaii and Calif.

Tahiti Petrel *Pseudobulweria rostrata* | Widespread in equatorial Pacific; several fall records for Hawaii.

Jouanin's Petrel *Bulweria fallax* | Indian Ocean species, with records for Midway Atoll and Southern Calif.

White-chinned Petrel *Procellaria aequinoctialis* | Widespread in Southern Hemisphere oceans; several fall records off Calif.

Parkinson's Petrel *Procellaria parkinsoni* | New Zealand breeder that ranges widely in South Pacific; one fall record off Calif.

Streaked Shearwater *Calonectris leucomelas* | Western Pacific species. Multiple fall records off Calif. and Ore. coasts; one in Wyo., June 2006, was amazing.

Cory's Shearwater *Calonectris diomedea* | Breeds eastern Atlantic, regularly dispersing to East Coast. Several fall records for Calif., all fairly recent.

Great Shearwater *Ardenna gravis* | Widespread in Atlantic Ocean. A number of records, almost all 21st century, from Alas. to Calif.

Bryan's Shearwater CR *Puffinus bryani* | World's smallest shearwater. Very rare, breeding off Japan. Accidental to Midway Atoll.

Lesser Frigatebird *Fregata ariel* | Occurs mostly tropical Indian and western Pacific Oceans. Records in West for Hawaiian Is., Calif., and, remarkably, Wyo.

Northern Gannet *Morus bassanus* | North Atlantic relative of boobies. One long-lived individual has been in the San Francisco area since 2012.

Anhinga *Anhinga anhinga* | Related to cormorants; widespread in Neotropics and reaches Southeast. Accidental to West Tex., southern Rockies, and Calif.

Yellow Bittern *Ixobrychus sinensis* | Widespread in southern Asia; one strayed all the way to Attu I., Alas., May 1989.

Gray Heron *Ardea cinerea* | Familiar Old World counterpart of Great Blue Heron (p. 206). Multiple records, almost all recent, from Bering Sea region.

Intermediate Egret *Ardea intermedia* | Occurs widely from India east to southern Japan, as well as much of Australasia; two records for Aleutians.

Chinese Egret *Egretta eulophotes* | Southeast Asian species; one old record (June 1974) from Aleutians.

Little Egret *Egretta garzetta* | Old World counterpart of Snowy Egret (p. 208); annual to East in very small numbers. One record for Aleutians (May 2000).

Chinese Pond-Heron *Ardeola bacchus* | Southeast Asian species, with several records, spring and summer, for Bering Sea region.

APPENDIX A

Chinese Sparrowhawk *Accipiter soloensis* | Southeast Asian species. One in Aleutians in 1985 was either this species or Japanese Sparrowhawk, *A. gularis*.

Eurasian Goshawk *Accipiter gentilis* | Until very recently, treated as conspecific with American Goshawk (p. 220). One record for West (Shemya I., Alas., 2001).

Black Kite *Milvus migrans* | Widespread Old World species, with several records for Hawaiian Is.; one at St. Paul I. in Bering Sea was found in Jan. (2017).

White-tailed Hawk *Geranoaetus albicaudatus* | Patchily distributed Neotropical species that gets well into Tex. One strayed to north-central N. Mex. (June 2017).

Long-legged Buzzard *Buteo rufinus* | Afro-Asian species; one overwintered on St. Paul I. in Bering Sea, 2018–19.

Oriental Scops-Owl *Otus sunia* | Found in South and Southeast Asia; two records, both 1970s, for Aleutians.

Northern Boobook *Ninox japonica* | Southeast Asian species, with two records for Aleutians within a year of one another, 2007 and 2008.

Eurasian Hoopoe *Upupa epops* | Extraordinary species—colorful, crested, long-billed, and terrestrial—of Old World. Four records, all from fall, for Alas.

Ringed Kingfisher *Megaceryle torquata* | Widespread Neotropical species that gets well into Tex. One visited Graham Co., Ariz., 2018–19.

Eurasian Wryneck *Jynx torquilla* | Unusual member of woodpecker family, widespread in Old World. Two fall records for Alas.

Aplomado Falcon *Falco femoralis* | Neotropical raptor. Formerly occurred west to Ariz.; recent sightings in West Tex. and N. Mex. may derive in part from released birds.

Gray-collared Becard *Pachyramphus major* | Mid. Amer. species, found in diverse forest types; two records for Chiricahua Mts., Ariz.

Small-billed Elaenia *Elaenia parvirostris* | S. Amer. flycatcher with widespread records in N. Amer., including one off San Francisco, Calif., Sept. 2022.

Great Kiskadee *Pitangus sulphuratus* | Striking Neotropical flycatcher; occurs north to Tex., where range expanding. Records for West in Ariz., N. Mex., and Colo.

Piratic Flycatcher *Legatus leucophaius* | Widespread intra-Neotropical migrant that has wandered north to U.S. Records for West in N. Mex. and West Tex.

Variegated Flycatcher *Empidonomus varius* | S. Amer. species that overshoots to East occasionally. One record in our area for Wash. (Sept. 2008).

Couch's Kingbird *Tyrannus couchii* | Mid. Amer. and South Tex. species; range expanding northward. Scattered records West Tex. to Calif.; also one for Colo.

Acadian Flycatcher *Empidonax virescens* | Breeds in broadleaf forests to our east; accidental to Colo., N. Mex., and West Tex.

Pine Flycatcher *Empidonax affinis* | Found in mountain forests of Mexico and Guatemala; two records for Ariz.

Black-capped Vireo *Vireo atricapilla* | Breeding range restricted mostly to Tex. Vagrants in West to N. Mex., Ariz., and, remarkably, B.C.

Red-backed Shrike *Lanius collurio* | Western Palearctic species that winters to southern Africa. Several records for Alas.; one each for B.C., Hawaii.

Thick-billed Warbler *Arundinax aedon* | This species and the next three are in Old World family Acrocephalidae (reed warblers). One record for Alas. (Sept. 2017).

Icterine Warbler *Hippolais icterina* | Mostly European breeder that winters well south in Africa. Two records for Bering Sea region.

Sedge Warbler *Acrocephalus schoenobaenus* | Breeds mostly Europe and winters mostly Africa. Two fall records for St. Lawrence I., Alas.

Blyth's Reed Warbler *Acrocephalus dumetorum* | Breeds mostly Central Asia, winters mostly India. Three fall records for St. Lawrence I., Alas.

Pallas's Grasshopper Warbler *Helopsaltes certhiola* | Eastern Palearctic species, with a single fall record for St. Lawrence I., Alas.

River Warbler *Locustella fluviatilis* | Along with the previous two species, in Old World family Locustellidae (grassbirds). One fall record for St. Lawrence I., Alas.

Brown-chested Martin *Progne tapera* | Widespread S. Amer. swallow with several records for East. Only record for West is of one in Ariz., Feb. 2006.

Pallas's Leaf Warbler *Phylloscopus proregulus* | Tiny eastern Palearctic species. One record for St. Lawrence I., Alas., Sept. 2006.

Lesser Whitethroat *Sylvia curruca* | Currently placed in same family as Wrentit (p. 318). One record for St. Lawrence I., Alas., Sept. 2002.

Gray Silky-flycatcher *Ptiliogonys cinereus* | Mid. Amer. relative of Phainopepla (p. 324). One visited El Paso, Tex., winter 1994–95.

Sinaloa Wren *Thryophilus sinaloa* | West Mexico endemic, except for recent influx into southern Ariz.; almost annual there in tiny numbers in past 15 years.

Long-billed Thrasher *Toxostoma longirostre* | Permanent resident in northeastern Mexico and South Tex., expanding northward. Has strayed to N. Mex. and even Colo.

Brown-backed Solitaire *Myadestes occidentalis* | Incredible songster of mountain forests of Mid. Amer. Two records for Ariz.

Naumann's Thrush *Turdus naumanni* | Eastern Palearctic species. Scattered records for Alas., spring and fall; vagrants also to Yukon, B.C., and Wash.

Fieldfare *Turdus pilaris* | Familiar species across Palearctic; in addition to records for Alas. (spring), also B.C. and Mont.

Redwing *Turdus iliacus* | Like Fieldfare, familiar across Palearctic. Spring and fall records for Alas.; also B.C. and Wash.

Song Thrush *Turdus philomelos* | Mostly western Palearctic species, with two recent records for Alas. and one for Wash.

Clay-colored Thrush *Turdus grayi* | Neotropical species that reaches South Tex. Has strayed to West Tex., N. Mex., and Ariz.

White-throated Thrush *Turdus assimilis* | Neotropical species that ranges well north in Mexican highlands. Two recent records for Ariz.

Asian Brown Flycatcher *Muscicapa dauurica* | Asian species, with several records for Bering Sea region, both spring and fall. This species and the next nine are in the family Muscicapidae (Old World flycatchers).

Spotted Flycatcher *Muscicapa striata* | Widespread Eurasian-African species, with single record for N. Amer. (St. Lawrence I., Alas., Sept. 2014).

Siberian Blue Robin *Larvivora cyane* | East Asian species, with two records from Bering Sea region.

Rufous-tailed Robin *Larvivora sibilans* | Asian species, with several records, all 21st century, from Bering Sea region.

Narcissus Flycatcher *Ficedula narcissina* | Breeds mostly Japan, with two records, both spring, for Attu I., Alas.; both were of spectacular males.

Mugimaki Flycatcher *Ficedula mugimaki* | Eastern Palearctic species, with a single record for Aleutians (Shemya I., Alas., May 1985).

Taiga Flycatcher *Ficedula albicilla* | Asian species, with scattered records, mostly spring, from Bering Sea region; also one fall record for Calif.

Common Redstart *Phoenicurus phoenicurus* | Breeds mostly western Palearctic, winters well south in Africa. One fall record (Oct. 2013) for St. Paul I., Alas.

Rufous-tailed Rock-thrush *Monticola saxatilis* | Breeds southern Europe to Central Asia, winters Africa; single record for Utqiagvik, Alas., June 2021.

Pied Wheatear *Oenanthe pleschanka* | Breeds mostly in Asian interior; single record for Seward Peninsula, July 2017.

Lavender Waxbill *Glaucestrilda caerulescens* | West African species. Introduced to Hawaii, but not considered established; recently increasing on Hawaii I., however.

Eurasian Tree Sparrow *Passer montanus* | Old World species introduced in Midwest; birds in West (e.g., B.C., Wash.) could derive from Midwestern population.

Citrine Wagtail *Motacilla citreola* | Mostly Central Asian breeder, with records in West from Alas., B.C., and Calif.

Tree Pipit *Anthus trivialis* | Breeds from Europe to Central Asia, winters to southern Africa; scattered records in Bering Sea region, both late spring and early fall.

Pechora Pipit *Anthus gustavi* | Breeds mostly northern Russia, with scattered records for Bering Sea region, both spring and fall.

Pallas's Rosefinch *Carpodacus roseus* | East Asian species, with single record for St. Paul I., Alas., Sept. 2015.

Asian Rosy-Finch *Leucosticte arctoa* | Asian species, with a single record for Adak I., Alas., Dec. 2011.

Eurasian Siskin *Spinus spinus* | Found in both Europe and East Asia. Several records for Aleutians, spring and also winter. Sometimes kept in captivity, and escapes possible anywhere.

Pine Bunting *Emberiza leucocephalos* | Eastern Palearctic species with scattered records, all in fall, for Bering Sea region.

Yellow-browed Bunting *Emberiza chrysophrys* | East Asian species, with a single record for St. Lawrence I., Alas., Sept. 2007.

Yellow-throated Bunting *Emberiza elegans* | East Asian species, with a single record for Attu I., Alas., Sept. 2007.

Yellow-breasted Bunting [CR] *Emberiza aureola* | Asian species, with scattered records, spring and fall, for Bering Sea region.

Gray Bunting *Emberiza variabilis* | Range-restricted East Asian species, with several spring records for western Aleutians.

Pallas's Bunting *Emberiza pallasi* | Asian species, with several records for Bering Sea region, spring and fall.

Reed Bunting *Emberiza schoeniclus* | Old World species with generally more southwestern distribution than Little Bunting (p. 465).

Worthen's Sparrow [EN] *Spizella wortheni* | Endangered species of grasslands in northern Mexico, with single, puzzling record for N. Mex., June 1884.

Henslow's Sparrow *Centronyx henslowii* | Declining species of grasslands to our east; vagrants have strayed west to Colo. and N. Mex.

Black-vented Oriole *Icterus wagleri* | Mid. Amer. species that ranges well north in Mexico, with records for West for Ariz. and West Tex.

Swainson's Warbler *Limnothlypis swainsonii* | Breeds southeastern U.S. Scattered records to N. Mex. and Colo., and one to Ariz.

Cerulean Warbler *Setophaga cerulea* | Breeds eastern U.S. and Ont., with small clusters of records for southern Rockies region and especially Southern Calif.

Tropical Parula *Setophaga pitiayumi* | Widespread Neotropical species that reaches South Tex. Has strayed to Calif., Ariz., N. Mex., West Tex., and Colo.

Golden-cheeked Warbler [EN] *Setophaga chrysoparia* | Breeds only central Tex., but one was found off San Francisco (Sept. 1971)! Also a few records for West Tex.

Golden-crowned Warbler *Basileuterus culicivorus* | Widespread Neotropical species that irregularly reaches South Tex. Records in West for N. Mex. and Colo.

Blue-black Grassquit *Volatinia jacarina* | Neotropical tanager with one recent record for Ariz.

APPENDIX B | EXTINCT AND LIKELY EXTINCT BIRDS IN THE WEST

Human-caused extinctions of birds in the West can be grouped into three periods: (1) those caused by First Peoples arriving via the Bering Sea region more than 10,000 years ago and then swiftly populating the Americas; (2) those caused by Polynesian mariners who reached the Hawaiian Islands around 1,000 years ago; and (3) those caused by settlers of mostly European ancestry beginning close to 500 years ago.

Many birds, including the huge teratorns, went extinct after the arrival of First Peoples, but direct evidence for human agency in such extinctions is scant at best. Nevertheless, it is the general consensus of the scientific community that First Peoples substantially altered the landscapes and even climate of the Americas. The impacts of seafaring Polynesians are likewise poorly documented, but are widely conjectured to have been catastrophic for global biodiversity, resulting in the extinction of hundreds and possibly thousands of bird species, especially flightless rails. Much better documented are extinctions in the West beginning with the Cook expedition's arrival in Hawaii in 1778. All the species in the enumeration below went extinct after that date.

No bird species endemic to the mainland West have gone extinct since European settlement began. That is in contrast to the East, which has lost several endemic species to extinction, like the Great Auk. However, some species with mostly eastern ranges did occur in the West (for example, Passenger Pigeon and Eskimo Curlew) and are included here. All the rest listed below were endemic to the Hawaiian Islands.

Hawaiian endemics that have gone extinct since 1778 include two rails in the genus *Zapornia*, five Oos in the extinct endemic family Mohoidae (related to waxwings), three solitaires in the genus *Myadestes*, and 22 honeycreepers (all in the finch family)—for a total of 32 extinct Hawaiian endemics. The Labrador Duck, Passenger Pigeon, Eskimo Curlew, and Carolina Parakeet—primarily eastern species of peripheral or accidental occurrence in the West—bring the total to 36. In a problematic gray area is the functionally extinct Hawaiian Crow, which occurs now only in captivity.

Labrador Duck, *Camptorhynchus labradorius*
(may have occurred west to Yukon)

Passenger Pigeon, *Ectopistes migratorius*
(widespread records from mainland West)

Laysan Rail, *Zapornia palmeri*

Hawaiian Rail, *Zapornia sandwichensis*

Eskimo Curlew, *Numenius borealis*
(bred in Nvt., possibly west to Alas.)

Carolina Parakeet, *Conuropsis carolinensis*
(a few records, poorly attested, from West)

Hawaiian Crow, *Corvus hawaiiensis*
(does not occur in the wild)

Kauai Oo, *Moho braccatus*

Oahu Oo, *Moho apicalis*

Bishop's Oo, *Moho bishopi*

Hawaii Oo, *Moho nobilis*

Kioea, *Chaetoptila angustipluma*

Kamao, *Myadestes myadestinus*

Amaui, *Myadestes woahensis*

Olomao, *Myadestes lanaiensis*

Poo-uli, *Melamprosops phaeosoma*

Oahu Alauahio, *Paroreomyza maculata*

Kakawahie, *Paroreomyza flammea*

Kona Grosbeak, *Chloridops kona*

Lesser Koa-Finch, *Rhodacanthis flaviceps*

Greater Koa-Finch, *Rhodacanthis palmeri*

Ula-ai-hawane, *Ciridops anna*

Laysan Honeycreeper, *Himatione fraithii*

Hawaii Mamo, *Drepanis pacifica*

Black Mamo, *Drepanis funerea*

Ou, *Psittirostra psittacea*

Lanai Hookbill, *Dysmorodrepanis munroi*

Kauai Nukupuu, *Hemignathus hanapepe*

Oahu Nukupuu, *Hemignathus lucidus*

Maui Nukupuu, *Hemignathus affinis*

Lesser Akialoa, *Akialoa obscura*

Kauai Akialoa, *Akialoa stejnegeri*

Oahu Akialoa, *Akialoa ellisiana*

Maui-nui Akialoa, *Akialoa lanaiensis*

Greater Amakihi, *Viridonia sagittirostris*

Oahu Akepa, *Loxops wolstenholmei*

Maui Akepa, *Loxops ochraceus*

GLOSSARY

ABA American Birding Association, publisher of *Birding* magazine and the *ABA Checklist*, updated multiple times annually at aba.org/aba-checklist.

ABA Area North America north of Mexico and Hawaii, plus offshore waters.

accidental Generally denotes a species recorded five or fewer times overall in a region or fewer than three times in the past 30 years.

advertising call A loud, far-carrying vocalization, typically given in courtship.

AOS American Ornithological Society, publisher of the *Checklist of North and Middle American Birds*, updated at least annually at checklist.american ornithology.org/taxa.

auriculars Feather tract covering the ear (aural) openings, often bordered with contrasting stripes or lines; also called "ear coverts."

axillaries Feathers that make up the bird's "armpit," or "wingpit," visible in the field only in flight.

backcross A second-generation (or later) hybrid, typically between one hybrid parent and one "pure" parent.

bare parts Those areas of a bird's body completely without feathers, typically the eyes, bill, and feet.

body plan A bird's overall build or structure.

bosque A grove of broadleaf trees, especially cottonwoods, along a river or pond.

breeding plumage The coat of feathers worn by many, but not all, birds when they are courting and raising young.

brood parasitism Mating system wherein a female lays its eggs in the nest of another species.

call note Short vocalization generally used to communicate alarm, contact, begging, or other messages between birds.

carpal bar A band on the inner wing formed by contrasting secondary coverts.

casual Generally denotes a species recorded 10 or fewer times overall in a region, with at least three records in the past 30 years.

cavity nester A population or species that nests in holes, natural or artificial.

character A morphological trait.

clade A group of species or other taxa that share a common ancestor.

class In the Linnaean hierarchy, a clade of orders; the class Aves are the birds.

cline A gradual change in certain characteristics of individuals of the same population or species, evident in a geographic progression from one population to the next.

colony A densely packed, sometimes cooperative community of nesting birds.

commensalism A two-species relationship in which one benefits and the other is unaffected.

common Denotes a species easily found at a certain place and time of year.

common name A standardized English name set by the AOS.

congener A similar or closely related species.

convergent evolution The process whereby similar environments promote development of similar traits among distantly related species or groups of species.

cooperative breeding Mating system wherein individuals help to raise young that are not their direct offspring.

coverts Feathers that cover other feather tracts, especially on a bird's wing or tail.

critically endangered A species considered by the IUCN to be in likely danger of imminent extinction.

crown The top of a bird's head, extending from the forehead to the nape.

cryptic Describing a bird that is difficult to detect in its environment due to morphological features and/or behavior.

cryptic species Two or more populations that look nearly identical but have considerable genetic differences and do not interbreed.

GLOSSARY

culmen The ridge of the upper mandible from the base to the tip.

decurved Of the bill, curving downward.

digiscoping Taking a photograph by holding a small camera or cellphone against the eyepiece of a spotting scope.

display A behavior presented in courtship or threat, often, but not always, delivered with song or other vocalizations.

diurnal Active during the day.

eclipse plumage The dull or plain plumage of most ducks, worn only in late summer.

endangered A species that is considered by the IUCN to be in likely danger of eventual extinction if corrective measures are not taken.

endemic Occurring only within a given geographic region.

extinction The permanent and complete disappearance of all the members of a population of an organism, especially a species.

extirpation The extinction of a population, especially a species, from a region.

eye crescent A partial eye ring visible either above or below—or both above and below—the eye.

eye ring A circle of feathers around a bird's eye.

eye stripe Synonym of *eyeline*, but typically thicker.

eyeline A thin stripe of feathers extending horizontally in front of and behind the eye.

family In the Linnaean hierarchy, a clade of genera; ends in -*idae* for birds.

feather tract A group of feathers that typically share a function, for example, powering flight (primaries) or protecting another feather tract (coverts).

feral A formerly domesticated animal now living in the wild; also refers to a formerly domesticated population now established in the wild.

field mark A visible characteristic used to help identify a bird.

flanks The area of a bird's underparts on either side of its belly, usually visible just below the folded wing.

fledge To leave the nest for the first time.

fledgling A young bird that has left the nest but often still depends on its parents for food and other care.

flight A coordinated passage of birds on the wing, often on migration.

flight call Typically simple, often loud vocalization given by a bird on the wing, especially on nocturnal migration.

flight feathers The remiges (primaries, secondaries, tertials) and rectrices.

flush To flee from disturbance, usually by flying.

forecrown Feathering just above the eye.

frontal shield Unfeathered plate directly above the upper mandible.

frugivorous Feeding on fruit.

gape The juncture of the upper and lower mandible.

generalist Having broad engagement with the environment, for example, by using multiple habitats or eating diverse food types.

genotype An organism's genetic material.

genus (pl.: genera) In the Linnaean hierarchy, a clade of species.

gestalt An overall impression of the shape and size of a bird.

gonys A ridge formed where two segments of the lower mandible join.

gorget The brilliant patch of throat feathers on a hummingbird.

granivorous Feeding on seeds.

greater coverts The outermost and usually largest tract of feathers overlaying the secondaries.

guild An assemblage of species with similar ecologies and life histories.

GLOSSARY

hybrid The offspring of breeding between individuals of different species.

immature An individual that has not attained full adult plumage; can indicate juvenile or, especially, subsequent plumages.

indigenous Pertaining to a bird's geographic range prior to human introductions.

inner primaries The primaries farthest from the tip of the wing.

insectivorous Feeding on insects.

intergrade An individual showing a complex array of traits of two (or rarely more) species.

irruption An irregular seasonal movement often related to variation in food resources.

IUCN International Union for Conservation of Nature, an organization that provides guidance on understanding biological diversity and the threats it faces.

juvenile A bird in its first plumage, grown in around the time of fledging; in many species, the juvenile plumage is the first of two or more immature plumages.

kettle A group of soaring birds, especially hawks or vultures, often on migration.

kingdom In the Linnaean hierarchy, a clade of phyla.

kleptoparasitism Theft of food.

lek A communal courtship display.

lesser coverts The innermost tract of feathers overlaying the secondaries.

leucistic Having whiter than normal plumage due to a partial lack of melanin pigments.

lores The area between the upper base of a bird's bill and its eyes.

lower mandible The lower half of the bill, also called just the "mandible."

lumping Reclassifying two or more bird species into a single species.

malar Feather tract extending from the base of a bird's bill downward along either side of its throat.

mantle Describes the back, scapulars, and upperwing coverts as a whole, often applied to gulls.

median coverts The feather tract between the greater and lesser coverts.

median crown stripe A contrasting line of feathers down the center of the top of the head.

microhabitat A pocket of differentiated habitat within a larger habitat type; for example, a wetter, leafier area of a forest.

migrant A bird in the process of moving from one place to another, often between breeding and nonbreeding territories.

mimicry The incorporation of other species' sounds into an individual bird's songs or calls.

mobbing Attacking or harassing a predator, sometimes as a group.

molt The process of growing new feathers to replace old ones; almost all birds in the West have a complete molt in late summer or fall.

monophyletic Referring to a group of organisms, especially a group of species, that share a common ancestor not shared with any other such group.

morph A population of birds with a consistent color difference from other populations in the same species.

morphology The physical, often structural, attributes of a bird; may exclude color and pattern.

nail The hard, hooked, often darkened tip of the upper mandible.

nape The back of a bird's head, below the crown and above the hindneck.

naturalized Referring to a population or species introduced, typically by human means, to a region but currently established and self-sustaining in the wild.

nectivorous Feeding on nectar.

nominate subspecies A subspecies whose name is the same as the name of its species; for example, *Zonotrochia leucophrys leucophrys*.

nonbreeding plumage A bird's coat of feathers when not courting and raising young.

GLOSSARY

nonindigenous Pertaining to the parts of a bird's geographic range where it was introduced, deliberately or unintentionally, by humans.

obligate Describing an individual, population, or species restricted to a particular niche or life strategy.

occasional Denotes a species that is found less than annually but nevertheless with a repeatable and predictable pattern of occurrence within a region.

orbital ring Circle of bare skin immediately surrounding the eye; compare *eye ring*.

order In the Linnaean hierarchy, a clade of families; ends in *-iformes* for birds.

outer primaries The primaries closest to the tip of the wing.

outer rectrices The rectrices farthest from the center of the tail.

panel A contrasting swatch of color, often paler than its surroundings, visible on the wing.

parasitism The practice of exploiting another individual or organism for food or other resources.

passage transient A bird in the process of migrating from one place to another.

passerine A species in the order Passeriformes, often called songbirds or perching birds.

patagium The leading edge of the inner wing; if this area is conspicuously dark, it forms a patagial bar.

pelagic Of the open ocean.

phenotype The observable characters of a bird's genotype; examples include sex, color, and many behaviors.

phylum In the Linnaean hierarchy, a clade of classes.

pishing The human practice of mimicking the sound of birds' alarm calls (*psh psh psh*) to attract their attention.

plumage A bird's coat of feathers.

population A group of organisms; does not correspond neatly to species or subspecies.

postocular stripe A stripe extending back from the eye.

precocial Pertaining to a species whose young develop very quickly following hatching.

primaries The outermost, longest, and most powerful flight feathers of a bird.

primary coverts Feathers arranged in rows that overlay the bases of the primaries.

rare Refers to a species that occurs annually, but in very low numbers, sometimes only individually; scarcer than uncommon but more numerous than casual.

record Verified report of a bird's occurrence at a given time and place.

rectrices (sing.: rectrix) The strong flight feathers of the tail.

regular Denotes consistent, predictable occurrences over a particular time frame, like a year or season.

remiges (sing.: remex) The flight feathers—primaries, secondaries, and tertials—of the wing.

resident A population or species present in a region for a season or year-round.

roost The place a bird goes to rest during periods of inactivity.

rufous Reddish.

rump The part of a bird's body directly above the uppertail coverts.

scapulars The tract of feathers that overlays the area where the wing attaches to the body; sometimes called "shoulders."

scavenger An individual or species that forages on dead organic matter or human refuse.

scientific name The name for an organism, often derived from Latin or Greek and consisting of two words: the genus name, then the species name.

secondaries The inner, fairly long, fairly powerful flight feathers of a bird.

secondary coverts Feather tracts that overlay the secondaries.

GLOSSARY

sexual dimorphism Referring to male and female morphologies that are appreciably different to the human eye.

solicitation whistle A shrill sound intended to arouse the interest of another bird, typically of the same species and often a biological close relative.

song Patterned vocalizations given by males and many females, often to defend territory or attract mates.

specialties Species or subspecies known to occur in a certain area and sought after by birders.

species A population of organisms capable of producing fertile offspring with one another, reproductively isolated from other such populations.

spectacles Pale eye rings connected above the bill, giving the appearance of glasses.

spectrogram A visual representation of a bird sound, with frequency plotted on the vertical axis and time on the horizontal axis.

splitting Reclassifying birds within a single species as multiple species.

subadult A bird that has not attained full adult plumage; synonym of *immature*.

subspecies Within a species, groups of closely related yet morphologically distinct individuals from different geographic regions.

supercilium A tract of feathers above a bird's eye, extending from the base of the bill to behind the eye; synonym of "eyebrow."

superspecies A group of closely related species, often so close that introgression is rampant within the group.

syrinx A bird's vocal organ.

taxon (pl.: taxa) A taxonomic grouping.

taxonomy The scientific classification of organisms based on evolutionary relationships.

tertials The innermost flight feathers of a bird's wing, technically the innermost secondaries; sometimes distinct in color and shape from the rest of the secondaries.

torpor A condition of lowered metabolism, often brought on very rapidly, typically in response to environmental stresses like heat or cold.

uncommon Denotes a species or population found in small numbers, but reliably so, at a certain place and time of year.

undertail coverts Feather tract covering the bases of the rectrices from below.

underwing coverts Tracts of feathers on the underside of the wing that cover the bases of the primaries and secondaries; "wing linings" is a synonym.

undifferentiated (of plumage) Referring to a bland, unpatterned, or otherwise unremarkable appearance.

upper mandible The upper half of the bill, also called the "maxilla."

uppertail coverts Feather tract covering the bases of the rectrices from above.

upperwing coverts Tracts of feathers on the upperside of the wing that cover the bases of the primaries and secondaries.

vagrant A bird in a location typically not inhabited or visited by its species.

vent Part of a bird where the belly meets the undertail coverts.

very rare Denotes a species found in extremely low numbers, but usually reliably so, at a certain place and time of year.

visitant A species that appears at a particular time of the year, especially winter.

wing bar A stripe of contrasting color across the middle of a bird's wing.

ILLUSTRATIONS CREDITS

All maps created in partnership with the Cornell Lab of Ornithology, using eBird data. Visit *science.ebird.org* to explore interactive maps and data.

The following artists contributed the illustrations in this second edition: Jonathan Alderfer, David Beadle, Peter Burke, Andrew Guttenberg, Marc R. Hanson, Cynthia J. House, H. Jon Janosik, Donald L. Malick, Killian Mullarney, Marquette Mutchler, Michael O'Brien, John P. O'Neill, Kent Pendleton, Diane Pierce, John C. Pitcher, H. Douglas Pratt, David Quinn, Chuck Ripper, N. John Schmitt, Thomas R. Schultz, and Daniel S. Smith.

Front cover—Pendleton; back cover—(BOTH) Guttenberg; 2—Pratt; 4—Philip Georgakakos; 6—Ken Etzel; 7—TJ Lenahan; 12—Bill Vriesema; 15—Ben Sonnenberg; 17 (UP)—Annette Shaff/Shutterstock; 17 (LO LE)—Jaymi Heimbuch/Minden Pictures; 17 (LO RT)—Marky Mutchler; 18—Lieutenant Elizabeth Crapo, NOAA Corps; 19—Eryn Lynum; 20 (UP)—photo of Virginia Rose used with permission of Birdability; 20 (LO)—Luke Franke; 21 (LE)—Anne Craig; 21 (RT)—ergioboccardo/iStock/Getty Images; 23—House; 25—House; except Greater White-fronted Goose by Schultz; Graylag Goose by Alderfer; 27—House; except Brant by Alderfer; Cackling Goose (*taverneri, hutchinsii,* and flight) and Canada Goose (standing *moffitti* and "Lesser") by Schmitt; 29—House; except Hawaiian Goose and Black Swan by Guttenberg; Trumpeter Swans and Tundra Swans (heads) by Mullarney; 31—House; except Egyptian Goose by Schmitt; Muscovy Duck by Guttenberg; 33—House; except Cinnamon Teal (female) by Schmitt; teal hybrid by Guttenberg; 35—House; except wigeon hybrid by Schultz; 37—Guttenberg; except Mallard and Mexican Duck by House; 39—House; except Green-winged Teal (swimming female) by Schmitt; 41—House; 43—House; except Common Eider (flight) by Alderfer; 45—House; except Stejneger's Scoter by Alderfer; 47—House; except Harlequin Duck (swimming and standing adult males) by Alderfer; 49—House; except goldeneye hybrid by Alderfer; 51—House; 53—Pendleton; except Northern Bobwhite *(taylori)* by Schmitt; 55-7—Pendleton; 59—Pendleton; except Rock Ptarmigan *(evermanni)* by Schmitt; 61—Pendleton; except Greater Sage-Grouse (displaying male) and Gunnison Sage-Grouse by Alderfer; Sooty Grouse (hooting male and displaying male *sitkensis*) by Schmitt; 63—Pendleton; except Sharp-tailed Grouse (displaying male) by Alderfer; 65—Pendleton; except Kalij Pheasant and Indian Peafowl by Guttenberg; 67—Guttenberg; except Himalayan Snowcock and Chukar (standing figures by Pendleton; Chukar (flight) by Schmitt; 69—Alderfer; 71—Janosik; except Western Grebe (winter adult) and Clark's Grebe (winter adult) by Alderfer; grebe hybrid by Guttenberg; 73—Pratt; except Chestnut-bellied Sandgrouse by Guttenberg; Band-tailed Pigeon (juvenile) by Alderfer; Eurasian Collared-Dove by Schmitt and Alderfer; 75—Schmitt and Alderfer; except Zebra Dove by Guttenberg; Inca Dove by Malick; Common Ground Dove and Ruddy Ground Dove by Alderfer; 77—Schmitt and Alderfer; except Greater Roadrunner by Schmitt; 79—Pratt; except Common Cuckoo by Quinn; 81—Schmitt; except Common Nighthawk by Schultz; Common Poorwill (tail) by Ripper; 83—Schmitt; except Buff-collared Nightjar (tail) by Ripper; Mexican Whip-poor-will (tail) by Webb; 85—Schmitt; except Mariana Swiftlet by Guttenberg; 87—Schmitt and Alderfer; 89—Schmitt and Alderfer; except Lucifer Hummingbird (flight and perched) by Webb; 91—Schmitt and Alderfer; except Calliope Hummingbird by Alderfer; 93—Alderfer; 95—Schmitt and Alderfer; 97—Schultz; except Sora (standing and head) by Hanson; 99—Hanson; except Common Gallinule (breeding and juvenile) by Schultz; Hawaiian Coot by Guttenberg; 101—Schultz; except Yellow Rail (standing figures) by Hanson; Sora (chick) by Guttenberg; Sandhill Crane (standing figures) by Pierce; 103—Janosik; except Black-necked Stilt *(knudseni)* by Guttenberg; American Oystercatcher (standing) by Schultz; 105—Alderfer; 107—Smith; except Killdeer (standing figures) by Mullarney; Common Ringed Plover (standing figures) and Semipalmated Plover (standing figures) by Pitcher; 109—Smith; except Piping Plover (standing figures and head) and Snowy Plover (standing figures) by Pitcher; Lesser Sand-Plover (standing figures) and Mountain Plover (standing figures and ventral flight) by Mullarney; 111—Smith; except Upland Sandpiper (standing figures) by Mullarney; Bristle-thighed Curlew (standing figure) by Schmitt; 113—Alderfer; except Bar-tailed Godwit (flight) and Marbled Godwit (flight) by Smith; 115—Pitcher; except Black Turnstone (flight) by Alderfer; 117—Schultz; except Red Knot (flight) and Stilt Sandpiper (standing figures) by Alderfer; Stilt Sandpiper (flight) and Curlew Sandpiper (flight) by Smith; 119—Schultz; except Rock Sandpiper (standing figures) by Pitcher; all flight figures by Smith; 121—Pitcher; except all flight figures by Smith; 123—Pitcher; except Semipalmated Sandpiper (flight) and Western Sandpiper (flight) by Smith; 125—Mullarney; except all flight figures by Smith; 127—Alderfer; 129—Pitcher; except Wilson's Snipe by Alderfer; 131—Pitcher; except Lesser Yellowlegs (flight) and Greater Yellowlegs (flight) by Smith; Willet by O'Brien; 133—Mullarney; 135-7—Schultz; 139—Alderfer; except Dovekie (breeding adults and swimming winter) and Common Murre (standing) by Ripper; Common Murre (swimming) by Schmitt; 141—Ripper; except Black Guillemot (running, swimming, and flying) by Alderfer; Pigeon Guillemot (flight) by Schmitt; 143—Alderfer; except Ancient Murrelet by Schmitt and Alderfer; 145—Alderfer; 147—Alderfer; except Cassin's Auklet by Schmitt; 149-63—Schultz; 165—Guttenberg; except Brown Noddy by Schultz; 167—Schultz; except Gray-backed Tern by Guttenberg; 169-73—Schultz; 175—Janosik; except White-tailed Tropicbird (adult) by Guttenberg; 177-9—Quinn; 181-7—Alderfer; 189—Alderfer; except Juan Fernandez Petrel and White-necked Petrel by Mutchler; 191—Alderfer; except Bonin Petrel and Black-winged Petrel by Mutchler; 193—Alderfer; except Short-tailed Shearwater (with dark underwing) and Sooty Shearwater (flight) by Hanson; 195—Alderfer; 197—Alderfer; except Christmas Shearwater by Mutchler; 199—Janosik; except Wood Stork by Pierce; Great Frigatebird (flying adult female) by Schultz; Great Frigatebird (all others) by Guttenberg; 201-3—Alderfer; 205—Janosik; except Neotropic Cormorant and Brown Pelican (breeding adult) by Schultz; 207—Burke; except Great Blue Heron by Pierce; 209—Pierce; except Snowy Egret (flight) and Little Blue Heron (flight) by Schultz; 211—Pierce; except Tricolored Heron (flight), Reddish Egret, and Cattle Egret (flight) by Schultz; 213—Burke; except Green Heron (standing) by Pierce; Green Heron (flight) by Schultz; 215—Burke; except White-faced Ibis (flight) and Glossy x White-faced hybrid by Guttenberg; California Condor by Malick; 217—Malick; except ventral flying figures by Pendleton; 219—Malick; except Golden Eagle (flying adult) and Bald Eagle by Pendleton; Northern Harrier by Schmitt; 221—Malick; except three flying juveniles by Schmitt; 223—Malick; except

ILLUSTRATION CREDITS

Bald Eagle (3rd year), Mississippi Kite (flying adults), and Common Black Hawk by Schmitt; Mississippi Kite (immature) by Pendleton; 225—Schmitt; 227—Schmitt; except Hawaiian Hawk by Guttenberg; Short-tailed Hawk (perched) and Swainson's Hawk (perched) by Malick; 229—Schmitt; except Zone-tailed Hawk (perched) by Malick; Red-tailed Hawk (light-morph *calurus* in flight) by Guttenberg; 231—Malick; except Rough-legged Hawk (perched and flying juvenile) and Ferruginous Hawk (perched adult) by Malick; 233—5—Malick; 237—Malick; 239—Malick; 241—Malick; except Long-eared Owl (flight) and Northern Saw-whet Owl (*brooksi*) by Schmitt; 243—Malick; except Elegant Trogon (male, female, and tails) by O'Neill; Elegant Trogon (juvenile) and Green Kingfisher (flight) by Schmitt; 245—9—Malick; 251—Malick; except American Three-toed Woodpecker (*dorsalis*) by Schultz; 253—Malick; except Nuttall's Woodpecker (juvenile) by Schmitt; 255—Malick; except Hairy Woodpecker (*orius*) by Schultz; 257—Malick; except Northern Flicker (intergrade) by Schultz; 259—Malick; except Crested Caracara (flight) by Pendleton; American Kestrel (dorsal flight) and Merlin (*suckleyi* flight) by Schmitt; 261—Malick; except all flight figures by Schmitt; 263—5—Schmitt; 267—Alderfer; except Northern Beardless-Tyrannulet by Beadle; Sulphur-bellied Flycatcher by Schultz; 269—Burke; 271—Alderfer; 273—Schultz; except Thick-billed Kingbird (1st fall) by Alderfer; Scissor-tailed Flycatcher by Pratt; 275—81—Beadle; 283—Pratt; 285—Schmitt; except Bell's Vireo by Pratt; Gray Vireo by Schultz; 287—Schultz; 289—Beadle; except Red-eyed Vireo by Pratt; Warbling Vireo (on nest) by Guttenberg; 291—Guttenberg; 293—Pratt; except Loggerhead Shrike (flight) and Northern Mockingbird (flight) by Schmitt; 295—Pratt; except Blue Jay (flight) by Schmitt; 297—Schmitt; except Mexican Jay (juvenile) by Pratt; 299—Pratt; except Yellow-billed Magpie (juvenile) by Schmitt; 301—Schmitt; except Chihuahuan Raven (flight) and Common Raven (flight and calling) by Alderfer; 303—O'Brien; except Verdin by O'Neill; 305—307— O'Brien; 309—Beadle; 311—Pratt; except Millerbird by Guttenberg; Violet-green Swallow (dorsal flight and perched adult male) by Schmitt; 313—Schmitt; except Purple Martin (*arboricola*) by Schultz; 315—Pratt; except Cliff Swallow (head detail and *pyrrhonota* flight) by Guttenberg; Cave Swallow (flight) by Beadle; 317—O'Neill; except Japanese Bush-Warbler by Guttenberg; Arctic Warbler by Quinn; 319—Guttenberg; except Red-whiskered Bulbul by Pratt; Wrentit by O'Brien; 321—Guttenberg; 323—Schmitt; except Red-billed Leiothrix by Guttenberg; 325—Pratt; except Bohemian Waxwing (perched adult and flight) and Cedar Waxwing (flight) by Quinn; 327—Pratt; except White-breasted Nuthatch and Brown Creeper by Schultz; 329—Pratt; except Blue-gray Gnatcatcher (*obscura*) by Schultz; 331—Pratt; except Pacific Wren (*pacificus*) by Beadle; Winter Wren by Schmitt; 333—Schmitt; except Carolina Wren and Bewick's Wren (*eremophilus*) by Pratt; 335—Pratt; except Rock Wren and Cactus Wren (*sandiegensis* and nest) by Schmitt; 337—Pratt; except Curve-billed Thrasher by Schmitt; Brown Thrasher by Schultz; 339—Schmitt; 341—Pratt; except Sage Thrasher and Northern Mockingbird (flight) by Schmitt; European Starling (flight) by Guttenberg; Common Myna by Alderfer; 343—Pratt; except Mountain Bluebird immature by Guttenberg; 345—Guttenberg; except Townsend's Solitaire by Pratt; 347—Schultz; 349—Pratt; except Eyebrowed Thrush and Dusky Thrush by Quinn; 351—Quinn; except White-rumped Shama by Guttenberg; 353—Schmitt; except Olive Warbler by Pratt; 355—Guttenberg; except Scaly-breasted Munia by Schmitt; 357—Pratt; except House Sparrow by Schmitt; 359—Quinn; except flight figures by Guttenberg; 361—Pierce; 363–9– Guttenberg; 371—Schmitt; except Gray-crowned Rosy-Finch (*littoralis* and male *tephrocotis*) by Beadle; Gray-crowned Rosy-Finch (*umbrina*) by Pierce; 373—Schultz; 375—Guttenberg; except Common Redpoll and Hoary Redpoll by Pierce; 377—Pierce; except Cassia Crossbill by Schultz; 379—Pierce; except Pine Siskin (green morph) by Quinn; 381—Pierce; except wing tip figures by Schultz; 383—Pierce; except wing tip figures by Schultz; 385—Pierce; except Snow Bunting (1st winter female) by Alderfer; Rustic Bunting by Quinn; 387—Schultz; 389—Schultz; except Northern Red Bishop by Schmitt; Five-striped Sparrow by Pierce; 391—Pierce; 393—Schultz; 395—Schultz; except American Tree Sparrow by Pierce; 397—Pierce; except Dark-eyed Junco (*cismontanus*) by Guttenberg; Dark-eyed Junco (*shufeldti* and *mearnsi*) by Beadle; Dark-eyed Junco (flight) by Schmitt; 399—Pierce; except White-crowned Sparrow by Schultz; 401—Pierce; except Bell's Sparrow (*canescens*) by Beadle; three tails by Schultz; 403—Schmitt; except Vesper Sparrow by Schultz; Baird's Sparrow by Pierce; 405—Pierce; except Song Sparrow (juvenile) and Lincoln's Sparrow by Schultz; 407—Burke; except Rufous-crowned Sparrow by Schultz; 409—Burke; 411—Schultz; except Yellow-breasted Chat (male) by Burke; Yellow-headed Blackbird (perched figures) by Pratt; all flight figures by Guttenberg; 413—Schultz; 415—17—Burke; 419—Schultz; except Red-winged Blackbird (males) by Pratt; Tricolored Blackbird (males) by Schmitt; all flight figures by Guttenberg; 421—Pratt; except Brown-headed Cowbird by Schmitt; all flight figures by Guttenberg; 423—Pratt; except Bronzed Cowbird (*loyei*) by Beadle; all flight figures by Guttenberg; 425—Schultz; except Ovenbird by Pratt; 427—Pratt; 429—Schultz; except Lucy's Warbler and Nashville Warbler (*ruficapilla*) by Pratt; 431—Schultz; 433—Pratt; 435—Pratt; except Magnolia Warbler by Schultz; Bay-breasted Warbler (fall male) by Beadle; 437—Pratt; except Yellow Warbler by Schultz; 439—Pratt; except Palm Warbler and Yellow-rumped Warbler (fall *coronata*) by Schultz; 441—Pratt; except Black-throated Gray Warbler (immature) by Schultz; 443—Pratt; except Rufous-capped Warbler by Burke; Wilson's Warbler by Schultz; Red-faced Warbler by Beadle; 445—Burke; 447—Pierce; except Flame-colored Tanager by Burke; Northern Cardinal (*superbus*) by Beadle; 449—Pierce; 451—Schultz; except Lazuli x Indigo hybrid by Guttenberg; 453—Guttenberg; except Dickcissel by Pierce; 454—Mullarney; except Whooper Swan and Falcated Duck by House; 455—Alderfer; except Common Pochard and Smew by House; Oriental Turtle-Dove by Schmitt and Alderfer; Groove-billed Ani by Pratt; 456—Schmitt; except Oriental Cuckoo by Quinn; Common Crane by Pierce; Eurasian Dotterel by Mullarney; Northern Jacana by Janosik; 457—Alderfer; except Great Knot by Schultz; Temminck's and Long-toed Stints by Pitcher; 458—Pitcher; except Common Snipe by Alderfer; Marsh Sandpiper by Quinn; 459—Alderfer; except Common and Great Black-backed Gulls by Schultz; Kermadec Petrel by Mutchler; 460—Pierce; except White-tailed Eagle and Steller's Sea-Eagle by Schmitt; Eared Quetzal by O'Neill; Eurasian Kestrel by Malick; 461—Pratt; except Eurasian Hobby by Schmitt; Nutting's Flycatcher by Burke; Tufted Flycatcher and Eastern Wood-Pewee by Beadle; 462—Quinn; except Yellow-throated Vireo by Pratt; 463—Quinn; except Blue Mockingbird by Alderfer; Aztec Thrush by Schmitt; 464—Quinn; except Eurasian Bullfinch and Oriental Greenfinch by Pierce; 465—Pratt; except Little Bunting by Quinn; Louisiana Waterthrush by Schultz; Crescent-chested Warbler by Beadle; 466—Pratt; except Fan-tailed Warbler by Burke; Yellow Grosbeak by Pierce; 481—Jeff Gordon.

ABOUT THE AUTHOR

TED FLOYD is the longtime editor of *Birding* magazine, the award-winning flagship publication of the American Birding Association (ABA). He has written five bird books, including the *Smithsonian Field Guide to the Birds of North America* (HarperCollins, 2008) and *How to Know the Birds* (National Geographic, 2019). Floyd is also the author of more than 200 popular articles, technical papers, and book chapters on birds and nature, and he is a frequent speaker at bird festivals and ornithological society conferences worldwide. He has served on several nonprofit boards and is a recipient of the ABA Claudia Wilds Award for Distinguished Service

ACKNOWLEDGMENTS

More than 40 years ago, the National Geographic Society's *Field Guide to the Birds of North America* ushered in a new era of knowledge and enjoyment for bird lovers in the United States and Canada. That first edition (1983) was instantly hailed for being accurate, authoritative, and useful. Six subsequent editions (1987, 1999, 2002, 2007, 2011, 2017) built on that initial success, adding new species and expanded text, while keeping abreast of rapid changes in ornithological nomenclature and taxonomic relationships. Regional texts, pocket guides, and an encyclopedic *Complete Birds* followed.

Literally hundreds of birders, ornithologists, and editors have contributed to the excellence of the galaxy of National Geographic bird books, but two are deserving of special recognition. Jon L. Dunn and Jonathan Alderfer, who guided so many National Geographic birding books from conception to completion, were uncompromising in their precision, thoroughness, and attention to detail. Through their eyes, birders everywhere have learned to see the world of birds more clearly, more accurately, and, ultimately, more lovingly. All of us have been made better birders, directly or indirectly, through their legacy.

Many individuals have contributed their time and knowledge to the field guides preceding this one and its companions. All of us involved in the production of the present volume have benefited greatly from their many and varied contributions, which we acknowledge with great appreciation. Individuals who assisted in the preparation of this book's predecessors are listed in the acknowledgments of each edition.

The author thanks the following individuals, who provided facts, data, and other information with direct bearing on new content in this second edition of the *National Geographic Field Guide to the Birds of the United States and Canada—West*: Jody Allair, Elisabeth Ammon, Katie Andersen, Eliana Ardila, Yousif Attia, Amar Ayyash, Jen Ballard, Lauryn Benedict, Ned Bohman, Joe Buchanan, Joanna Burger, Peter Burke, the other Peter Burke, Jay Carlisle, Jared Clarke, Dominic Couzens, Lisa Crampton, Diana Doyle, Jennie Duberstein, Pete Dunne, Bill Evans, Shawneen Finnegan, Leslie Flint, Kimball Garrett, Tony Gaston, Dan Gibson, Bob Gill, Nathan Goldberg, Asher Gorbet, Catherine Hamilton, Steve Hampton, Lauren Harter, David Hartley, Justine Hausheer, Marshall Iliff, Jean Iron, Dave Irons, Alvaro Jaramillo, Kenn Kaufman, Aaron Lang, Cin-Ty Lee, Paul Lehman, Tony Leukering, Mark Lockwood, Patrick Maurice, Guy McCaskie, Chris McCreedy, Mia McPherson, Liz Medes, Kathy Mihm-Dunning, Larry Neel, Greg Neise, Adrianna Nelson, Duane Nelson, Kristy Nelson, Ann Nightingale, Christian Nunes, Michael O'Brien, Peter Olsoy, Ed Pandolfino, Nathan Pieplow, Donna Pomeroy, Doug Pratt, John Puschock, Liam Ragan, Van Remsen, Terry Rich, Margaret Rubega, Jennifer Rycenga, Rebecca Safran, Cathie Sandell, Bill Schmoker, Ioana Seritan, Ali Sheehey, Steve Shunk, David Sibley, Dessi Sieburth, Kelly Smith, Dave Sonneborn, Kate Stone, Justin Streit, Claire Stuyck, Nate Swick, Mandy Talpas, Jeanne Tinsman, David Tønnessen, Lara Tseng, Philip Unitt, Mark VanderVen, Eric VanderWerf, Debbie Van Dooremolen, Carolyn R. Van Hemert, Alex Wang, Claire Wayner, David Wilcove, Sheri Williamson, Jay Withgott, Chris Wood, and Rick Wright.

The species accounts here are drawn primarily from the author's 30+ years of extensive field experience in the region, but several online references were indispensable and consulted practically daily during the writing of this guide. Foremost among these is the Macaulay Library *(ebird.org/catalog)*, with 56 million photos, 2.1 million audio recordings, and 271,000 videos of birds. For general guidance about behavior, ecology, and systematics, the author made use of the miraculously thorough Birds of the World *(birdsoftheworld.org)*.

State and provincial resources were regularly consulted, but two stand out for their supreme value in preparing the text for this book: Daniel D. Gibson and Jack W. Withrow's "Inventory of the Species and Subspecies of Alaska Birds, Second Edition" (*Western Birds*, vol. 46, pp. 94–185), published in 2015, and Robert L. Pyle and Peter Pyle's *Birds of the Hawaiian Islands: Occurrence, History, Distribution, and Status*, version 2, published online in 2017 by the B. P. Bishop Museum. Morphometric data for bird

ACKNOWLEDGMENTS

species occurring only in the Hawaiian Islands region are based on Helen and André F. Raine's *ABA Field Guide to Birds of Hawaii* (Scott & Nix, 2020) and are used with permission.

The art here, created and refined over the years for National Geographic's birding field guides, represents the work of numerous artists, identified in the illustrations credits (pp. 479–480) and informed by numerous museum collections, cited in previous field guides. We continue to thank these artists and their research sources for this important collection of illustrations. The Denver Museum of Nature & Science and its Science Liaison, Jeff Stephenson, deserve special thanks for guiding the author's own research during the past two decades—including for this field guide.

The birding gods smiled on us when Andrew Guttenberg, one of the greatest ornithological illustrators of his generation, agreed to join the project. New illustrations for this edition have been created by Andrew, whose knowledge and aesthetics have added so much to this book, not only in the form of new art but also in layout arrangement, annotation composition, and deep grasp of ornithology. Thanks also to Marky Mutchler, who helped compose the art annotations, paying particular attention to the technical aspects of avian anatomy, and who also contributed several excellent illustrations. Both Andrew and Marky also read the text and offered many corrections and clarifications.

Behind the scenes, a dedicated team of editors, designers, interns, assistants, and project managers tended to matters great and small, guiding this project from its earliest conception to the final finished product. Editorial assistant Margo Rosenbaum managed a mountain of Google Docs, Excel spreadsheets, and more; text editor Rebecca Heisman applied word-for-word and even character-by-character precision to a thorough review of the entire manuscript; graphic designer Carol Norton created page layouts, clean and clear, that present a considerable density of text and art in a manner that is aesthetically appealing and, more important, easily understood; cartographer Debbie Gibbons created state-of-the-art maps that combine the latest science with a novel and powerful way of visualizing birds' dynamic ranges in both space and time; and senior editor Susan Tyler Hitchcock hatched the wonderful plan for this and companion field guides, ambitious and sometimes crazy and ultimately deeply satisfying for all involved.

Last but certainly not least, project manager Adrienne Izaguirre brought order to chaos; it is not hyperbole to say that this field guide would have imploded for sure without Adrienne's preternatural discipline, almost freakish attention to detail, and unsurpassed excellence overall. Although it wasn't in her job description, Adrienne provided invaluable expertise in everything from copyediting to cartography; along the way, she made countless corrections and more than a handful of original contributions to the species accounts and other technical content in this book.

Special thanks to those essential to our partnership with the Cornell Lab of Ornithology, especially Miyoko Chu and Tom Auer. We also thank Steve Mlodinow, one of the foremost experts on bird ranges in the West and beyond, for his exacting review of the maps and text, bringing to bear his exhaustive knowledge of when and where birds occur in terms of both historical patterns and very recent range shifts.

At National Geographic Books, thanks go to Lisa Thomas, who identified the need to reinvigorate National Geographic's line of bird field guides and invited the author and his colleagues to join her in doing so. Throughout the negotiations and beyond, as we have created this book and its companion volumes, the author has been ably represented by Russell Galen of Scovil Galen Ghosh Literary Agency, Inc.—a wise mentor and tireless advocate whose experience has guided many decisions along the way.

Peter Pyle and Michael Retter read the entire manuscript, catching thousands of typos, omissions, misstatements, and other infelicities. Even more important, they read for clarity and coherence, the twin touchstones for the success of any field guide or other natural history reference. Jenna Anthony, Jodhan Fine, Rajan Rao, Alex Wang, and Katie Warner also read large portions of the manuscript and provided many corrections and clarifications regarding the bird faunas of their areas of expertise: the Canadian Rockies, the Desert Southwest, California, the Hawaiian Islands, and the Salish Sea region, respectively. As we prepared the book, we appreciated the help of Michael O'Connor, copy editor Jen Hess, and proofreader Mary Stephanos.

A field guide is a technical work, to be sure, but it is also intended for a contemporary readership. To put things in perspective, well over half the human population of the United States and Canada wasn't even born when the first edition of the *National Geographic Field Guide to the Birds of North America* was published, in 1983. In this regard, the author frequently solicited the counsel of Zoomers Hannah Floyd and Andrew Floyd in matters of metaphor and idiom for a modern audience. Maya Izaguirre, representing Gen Alpha, born about the time the work on this book began, was too young to weigh in on style and sensibility, but she enlivened many a Zoom meeting with her constant cheer and companionableness.

Learning about birds is, more than ever, a crowd-sourced and communitarian undertaking. It is not an exaggeration to say that literally thousands of bird lovers have contributed in one way or another to the content in this new edition of the *National Geographic Field Guide to the Birds of the United States and Canada—West*. To one and all: Thank you!

—*Ted Floyd*

INDEX

The page number for the main entry for each species is listed in **boldface** type and refers to text page opposite the illustration.

A

ABA (American Birding Association) 14, 21
Acanthis
 flammea **374**
 hornemanni **374**
Accentor, Siberian **464**
Accidental species 467–472, 473
Accipiter
 atricapillus **220**
 cooperii **220**
 gentilis **470**
 soloensis **470**
 striatus **220**
Accipitridae (family) 218–231
Acridotheres tristis **340**
Acrocephalidae (family) 310–311
Acrocephalus
 dumetorum **470**
 familiaris **310**
 schoenobaenus **470**
Actitis
 hypoleucos **458**
 macularius **128**
Aechmophorus
 clarkii **70**
 occidentalis **70**
Aegithalidae (family) 316–317
Aegolius
 acadicus **240**
 funereus **240**
Aerodramus bartschi **84**
Aeronautes saxatalis **84**
Aethia
 cristatella **144**
 psittacula **144**
 pusilla **144**
 pygmaea **144**
Agapornis roseicollis **264**
Agelaius
 phoeniceus **418**
 tricolor **418**
Aimophila ruficeps **406**
Aix
 galericulata **30**
 sponsa **30**
Akekee **368**
Akepa
 Hawaii **368**
 Maui 473
 Oahu 473
Akialoa
 Kauai 473
 Lesser 473
 Maui-nui 473
 Oahu 473
Akialoa
 ellisiana 473
 lanaiensis 473
 obscura 473
 stejnegeri 473
Akiapolaau **364**

Akikiki **362**
Akohekohe **368**
Alaska 13
Alauahio
 Maui **362**
 Oahu 473
Alauda arvensis **308**
Alaudidae (family) 308–309
Albatross
 Black-footed 12, **180**
 Chatham **468**
 Laysan **180**
 Light-mantled **468**
 Salvin's **468**
 Short-tailed **180**
 Wandering **469**
 White-capped **459**
Alcedinidae (family) 242–243
Alcidae (family) 138–147
Alectoris chukar **66**
Alle alle **138**
Alopochen aegyptiacus **30**
Amakihi
 Greater 473
 Hawaii **366**
 Kauai **366**
 Oahu **366**
Amandava amandava **354**
Amaui 473
Amazilia rutila **467**
Amazona
 finschi **264**
 oratrix **264**
 viridigenalis **264**
American Birding Association (ABA) 14, 21
American Ornithological Society (AOS) checklist 7
Ammodramus savannarum **388**
Ammospiza
 leconteii **402**
 nelsoni **402**
Amphispiza bilineata **390**
Amphispizopsis quinquestriata **388**
Anas
 acuta **38**
 crecca **38**
 diazi **36**
 fulvigula **467**
 laysanensis **36**
 platyrhynchos **36**
 rubripes **455**
 wyvilliana **36**
 zonorhyncha **455**
Anatidae (family) 22–51
Anatomy 10–11, 17
Anhinga **469**
Anhinga anhinga **469**
Ani, Groove-billed **455**
Anianiau **366**
Anous
 ceruleus **164**
 minutus **164**
 stolidus **164**
Anser
 albifrons **24**

 brachyrhyncus **467**
 caerulescens **24**
 canagicus **22**
 erythropus **467**
 fabalis **454**
 rossii **24**
 serrirostris **454**
Anthus
 cervinus **358**
 gustavi **472**
 hodgsoni **464**
 rubescens **358**
 spragueii **358**
 trivialis **472**
Antigone canadensis **100**
Antrostomus
 arizonae **82**
 carolinensis **467**
 ridgwayi **82**
 vociferus **467**
AOS (American Ornithological Society) checklist 7
Apapane **364**
Aphelocoma
 californica **296**
 insularis **296**
 wollweberi **296**
 woodhouseii **296**
Apodidae (family) 82–85
Apus
 apus **467**
 pacificus **456**
Aquila chrysaetos **218**
Aramides axillaris **467**
Aramus guarauna **468**
Aratinga nenday **262**
Archilochus
 alexandri **88**
 colubris **88**
Ardea
 alba **208**
 cinerea **469**
 herodias **206**
 intermedia **469**
Ardeidae (family) 206–213
Ardenna
 bulleri **194**
 carneipes **194**
 creatopus **194**
 gravis **469**
 grisea **192**
 pacifica **192**
 tenuirostris **192**
Ardeola bacchus **469**
Arenaria
 interpres **114**
 melanocephala **114**
Artemisiospiza
 belli **400**
 nevadensis **400**
Arundinax aedon **470**
Asio
 flammeus **240**
 otus **240**
Athene cunicularia **238**

INDEX

Auklet
  Cassin's **146**
  Crested **144**
  Least **144**
  Parakeet **144**
  Rhinoceros **146**
  Whiskered **144**
Auriparus flaviceps **302**
Avadavat, Red **354**
Avocet, American **102**
Aythya
  affinis **40**
  americana **38**
  collaris **40**
  ferina **455**
  fuligula **40**
  marila **40**
  valisineria **38**

B

Baeolophus
  atricristatus **306**
  inoratus **306**
  ridgwayi **306**
  wollweberi **306**
Bartramia longicauda **110**
Basileuterus
  culicivorus **472**
  lachrymosus **466**
  rufifrons **442**
Basilinna
  leucotis **94**
  xantusii **467**
Becard
  Gray-collared **470**
  Rose-throated **266**
Behavioral ecology 17–18
Binoculars 19, 20
Birdsong 8, 17, 18
Bishop, Northern Red **352**
Bittern
  American **206**
  Least **206**
  Yellow **469**
Blackbird
  Brewer's **420**
  Red-winged **418**
  Rusty **420**
  Tricolored **418**
  Yellow-headed **410**
Bluebird
  Eastern **342**
  Mountain 15, **342**
  Western 18, **342**
Bluetail, Red-flanked **463**
Bluethroat 13, **350**
Bobolink **410**
Bobwhite, Northern **52**
Bombycilla
  cedrorum **324**
  garrulus **324**
Bombycillidae (family) 324–325
Bonasa umbellus **56**
Boobook, Northern **470**
Booby
  Blue-footed **200**
  Brown **200**
  Masked **200**
  Nazca **459**
  Red-footed **200**
Botaurus lentiginosus **206**
Brachyramphus
  brevirostris **140**
  marmoratus **140**
  perdix **458**
Brambling **360**
Brant **26**
Branta
  bernicla **26**
  canadensis **26**
  hutchinsii **26**
  leucopsis **467**
  sandvicensis **28**
Brotogeris
  chiriri **262**
  versicolurus **262**
Bubo
  scandiacus **234**
  virginianus **234**
Bubulcus ibis **210**
Bucephala
  albeola **46**
  clangula **48**
  islandica **48**
Bufflehead **46**
Bulbul
  Red-vented **318**
  Red-whiskered **318**
Bullfinch, Eurasian **464**
Bulweria
  bulwerii **190**
  fallax **469**
Bunting
  Gray **472**
  Indigo **450**
  Lark **390**
  Lazuli 18, **450**
  Little **465**
  McKay's **384**
  Painted **450**
  Pallas's **472**
  Pine **472**
  Reed **472**
  Rustic **384**
  Snow **384**
  Varied **450**
  Yellow-breasted **472**
  Yellow-browed **472**
  Yellow-throated **472**
Bush-Warbler, Japanese **316**
Bushtit **316**
Buteo
  albonotatus **228**
  brachyurus **228**
  jamaicensis **228**
  lagopus **230**
  lineatus **224**
  plagiatus **224**
  platypterus **224**
  regalis **230**
  rufinus **470**
  solitarius **226**
  swainsoni **226**
Buteogallus anthracinus **222**
Butorides virescens **212**
Buzzard, Long-legged **470**

C

Cairina moschata **30**
Calamospiza melanocorys **390**
Calcariidae (family) 380–385
Calcarius
  lapponicus **380**
  ornatus **382**
  pictus **380**
Calidris
  acuminata **124**
  alba **118**
  alpina **118**
  bairdii **120**
  canutus **116**
  falcinellus **457**
  ferruginea **116**
  fuscicollis **120**
  himantopus **116**
  maritima **468**
  mauri **122**
  melanotos **124**
  minuta **457**
  minutilla **120**
  ptilocnemis **118**
  pugnax **124**
  pusilla **122**
  pygmea **468**
  ruficollis **122**
  subminuta **457**
  subruficollis **124**
  temminckii **457**
  tenuirostris **457**
  virgata **114**
Calliope calliope **350**
Callipepla
  californica **54**
  gambelii **54**
  squamata **54**
Calonectris
  diomedea **469**
  leucomelas **469**
Calothorax lucifer **88**
Calypte
  anna **90**
  costae **90**
Camptorhynchus labradorius 473
Camptostoma imberbe **266**
Campylorhyncus brunneicapillus **334**
Canachites canadensis **56**
Canada, western 13
Canary
  Island **374**
  Yellow-fronted **374**
Canvasback **38**
Caprimulgidae (family) 80–83
Caprimulgus jotaka **467**
Caracara, Crested **258**
Caracara plancus **258**
Cardellina
  canadensis **442**
  pusilla **442**
  rubrifrons **442**
Cardinal
  Northern **446**
  Red-crested **452**
  Yellow-billed **452**
Cardinalidae (family) 444–453

INDEX

Cardinalis
   cardinalis **446**
   sinuatus **446**
Carpodacus
   erythrinus **464**
   roseus **472**
Casual species 454–466
Catbird, Gray **336**
Cathartes aura **216**
Cathartidae (family) 214–217
Catharus
   fuscescens **346**
   guttatus **346**
   minimus **346**
   ustulatus **346**
Catherpes mexicanus **334**
Centrocercus
   minimus **60**
   urophasianus **60**
Centronyx
   bairdii **402**
   henslowii **472**
Cepphus
   columba **140**
   grylle **140**
Cerorhinca monocerata **146**
Certhia americana **326**
Certhiidae (family) 326–327
Cettiidae (family) 316–317
Chaetoptila angustipluma 473
Chaetura
   pelagica **84**
   vauxi **84**
Chamaea fasciata **318**
Charadriidae (family) 104–109
Charadrius
   dubius **468**
   hiaticula **106**
   leschenaultii **468**
   melodus **108**
   mongolus **108**
   montanus **108**
   morinellus **456**
   nivosus **108**
   semipalmatus **106**
   vociferus **106**
   wilsonia **456**
Chasiempis
   ibidis **290**
   sandwichensis **290**
   sclateri **290**
Chat, Yellow-breasted **410**
Chickadee
   Black-capped **302**
   Boreal **304**
   Chestnut-backed **304**
   Gray-headed **304**
   Mexican **304**
   Mountain **302**
Chiffchaff, Common **462**
Chlidonias
   leucopterus **459**
   niger **168**
Chloridops kona 473
Chloris sinica **464**
Chloroceryle americana **242**

Chlorodrepanis
   flava **366**
   stejnegeri **366**
   virens **366**
Chondestes grammacus **390**
Chordeiles
   acutipennis **80**
   minor **80**
Christmas Bird Count 21
Chroicocephalus
   philadelphia **150**
   ridibundus **150**
Chuck-will's-widow **467**
Chukar **66**
Ciconiidae (family) 198–199
Cinclidae (family) 342–343
Cinclus mexicanus **342**
Circus hudsonius **218**
Ciridops anna 473
Cistothorus
   palustris **332**
   stellaris **332**
Clangula hyemalis **46**
Coccothraustes
   coccothraustes **464**
   vespertinus **360**
Coccyzus
   americanus **78**
   erythropthalmus **78**
Colaptes
   auratus **256**
   chrysoides **256**
Colibri thalassinus **467**
Colinus virginianus **52**
Columba livia **72**
Columbidae (family) 72–77
Columbina
   inca **74**
   passerina **74**
   talpacoti **455**
Condor, California **214**
Contopus
   cooperi **274**
   pertinax **274**
   sordidulus **274**
   virens **461**
Conuropsis carolinensis 473
Coot
   American **98**
   Eurasian **467**
   Hawaiian **98**
Copsychus malabaricus **350**
Coragyps atratus **216**
Cormorant
   Brandt's **202**
   Double-crested **202**
   Neotropic **204**
   Pelagic **202**
   Red-faced **202**
Cornell Lab of Ornithology 6, 15, 19
Corthylio calendula **322**
Corvidae (family) 292–301
Corvus
   brachyrhynchos **300**
   corax **300**
   cryptoleucus **300**
   hawaiiensis 473
Coturnicops noveboracensis **100**

Cowbird
   Bronzed **422**
   Brown-headed **420**
Crane
   Common **456**
   Hooded **468**
   Sandhill **100**
   Whooping **456**
Creagrus furcatus **468**
Creeper
   Brown **326**
   Hawaii **368**
Crithagra mozambica **374**
Crossbill
   Cassia **376**
   Red **376**
   White-winged **376**
Crotophaga sulcirostris **455**
Crow
   American **300**
   Hawaiian 473
Cuckoo
   Black-billed **78**
   Common **78**
   Oriental **456**
   Yellow-billed **78**
Cuculidae (family) 76–79
Cuculus
   canorus **78**
   optatus **456**
Curlew
   Bristle-thighed **110**
   Eskimo 473
   Far Eastern **456**
   Little **468**
   Long-billed **110**
Cyanecula svecica **350**
Cyanocitta
   cristata **294**
   stelleri **294**
Cygnus
   buccinator **28**
   columbianus **28**
   cygnus **454**
   olor **28**
Cynanthus latirostris **94**
Cypseloides niger **82**
Cyrtonyx montezumae **52**

D

Delichon
   lagopodum **462**
   urbicum **462**
Dendragapus
   fuliginosus **60**
   obscurus **60**
Dendrocopos major **460**
Dendrocygna
   autumnalis **22**
   bicolor **22**
Dickcissel **452**
Diomedea exulans **469**
Diomedeidae (family) 180–181
Dipper, American 17, **342**
Dolichonyx oryzivorus **410**
Dotterel, Eurasian **456**

INDEX

Dove
- African Collared- 72, 73
- Common Ground **74**
- Eurasian Collared- **72**
- Inca **74**
- Mourning **76**
- Oriental Turtle- **455**
- Ruddy Ground- **455**
- Spotted **74**
- White-winged **76**
- Zebra **74**

Dovekie **138**

Dowitcher
- Long-billed **126**
- Short-billed **126**

Drepanis
- *coccinea* **364**
- *funerea* 473
- *pacifica* 473

Dryobates
- *albolarvatus* **254**
- *arizonae* **254**
- *nuttallii* **252**
- *pubescens* **252**
- *scalaris* **252**
- *villosus* **254**

Dryocopus pileatus **256**

Duck
- American Black **455**
- Eastern Spot-billed **455**
- Falcated **454**
- Harlequin **46**
- Hawaiian **36**
- Labrador 473
- Laysan **36**
- Long-tailed **46**
- Mandarin **30**
- Masked **467**
- Mexican **36**
- Mottled **467**
- Muscovy **30**
- Ring-necked **40**
- Ruddy **50**
- Tufted **40**
- Whistling- (see Whistling-Duck)
- Wood **30**

Dumetella carolinensis **336**

Dunlin **118**

Dysmorodrepanis munroi 473

E

Eagle
- Bald **222**
- Golden **218**
- Steller's Sea- **460**
- White-tailed **460**

ebird 15, 16, 20

Ectopistes migratorius 473

Egret
- Cattle **210**
- Chinese **469**
- Great **208**
- Intermediate **469**
- Little **469**
- Reddish **210**
- Snowy **208**

Egretta
- *caerulea* **208**
- *eulophotes* **469**
- *garzetta* **469**
- *rufescens* **210**
- *thula* **208**
- *tricolor* **210**

Eider
- Common **42**
- King **42**
- Spectacled **42**
- Steller's **42**

Elaenia, Small-billed **470**

Elaenia parvirostris **470**

Elanoides forficatus **460**

Elanus leucurus **218**

Elepaio
- Hawaii **290**
- Kauai **290**
- Oahu **290**

Emberiza
- *aureola* **472**
- *chrysophrys* **472**
- *elegans* **472**
- *leucocephalos* **472**
- *pallasi* **472**
- *pusilla* **465**
- *rustica* **384**
- *schoeniclus* **472**
- *variabilis* **472**

Emberizidae (family) **384**–**385**

Empidonax
- *affinis* **470**
- *alnorum* **276**
- *difficilis* **280**
- *flaviventris* **276**
- *fulvifrons* **280**
- *hammondii* **278**
- *minimus* **278**
- *oberholseri* **278**
- *traillii* **276**
- *virescens* **470**
- *wrightii* **280**

Empidonomus varius **470**

Eremophila alpestris **308**

Estrilda astrild **354**

Estrildidae (family) **354**–**355**

Ethics of birding 21

Eudocimus albus **460**

Eugenes fulgens **86**

Euodice cantans **354**

Euphagus
- *carolinus* **420**
- *cyanocephalus* **420**

Euplectes franciscanus **352**

Euptilotis neoxenus **460**

Extinct and likely extinct birds 473

F

Falco
- *columbarius* **258**
- *femoralis* **470**
- *mexicanus* **260**
- *peregrinus* **260**
- *rusticolus* **260**
- *sparverius* **258**
- *subbuteo* **461**
- *tinnunculus* **460**

Falcon
- Aplomado **470**
- Peregrine **260**
- Prairie **260**

Falconidae (family) 258–261

Ficedula
- *albicilla* **471**
- *mugimaki* **471**
- *narcissina* **471**

Fieldfare **471**

Finch
- Asian Rosy- **472**
- Black Rosy- **370**
- Brown-capped Rosy- **370**
- Cassin's **372**
- Gray-crowned Rosy- **370**
- Greater Koa- 473
- House **372**
- Laysan **362**
- Lesser Koa- 473
- Nihoa **362**
- Purple **372**
- Saffron **452**

Flicker
- Gilded **256**
- Northern **256**

Flight anatomy 10

Flycatcher
- Acadian **470**
- Alder **276**
- Ash-throated **268**
- Asian Brown **471**
- Brown-crested **268**
- Buff-breasted **280**
- Dark-sided **463**
- Dusky **278**
- Dusky-capped **268**
- Fork-tailed **461**
- Gray **280**
- Gray-streaked **463**
- Great Crested **268**
- Hammond's **278**
- Least **278**
- Mugimaki **471**
- Narcissus **471**
- Nutting's **461**
- Olive-sided **274**
- Pine **470**
- Piratic **470**
- Scissor-tailed **272**
- Spotted **471**
- Sulphur-bellied **266**
- Taiga **471**
- Tufted **461**
- Variegated **470**
- Vermilion **282**
- Western **280**
- Willow **276**
- Yellow-bellied **276**

Francolin
- Black **66**
- Erckel's **66**
- Gray **66**

Francolinus
- *francolinus* **66**
- *pondicerianus* **66**

Fratercula
- *cirrhata* **146**
- *corniculata* **146**

Fregata
- *ariel* **469**
- *magnificens* **198**
- *minor* **198**

INDEX

Fregatidae (family) 198–199
Frigatebird
  Great **198**
  Lesser **469**
  Magnificent **198**
Fringilla montifringilla **360**
Fringillidae (family) 360–379
Fulica
  alai **98**
  americana **98**
  atra **467**
Fulmar, Northern **186**
Fulmarus glacialis **186**

G

Gadwall **34**
Gallinago
  delicata **128**
  gallinago **458**
  solitaria **468**
  stenura **468**
Gallinula
  chloropus **467**
  galeata **98**
Gallinule
  Common **98**
  Purple **456**
Gallus gallus **66**
Gannet, Northern **469**
Garganey **454**
Garrulax
  canorus **320**
  pectoralis **320**
Gavia
  adamsii **178**
  arctica **176**
  immer **178**
  pacifica **176**
  stellata **176**
Gaviidae (family) 176–179
Gelochelidon nilotica **168**
Geococcyx californianus **76**
Geopelia striata **74**
Geothlypis
  formosa **466**
  philadelphia **430**
  tolmiei **430**
  trichas **430**
Geranoaetus albicaudatus **470**
Glareola maldivarum **468**
Glaucestrilda caerulescens **471**
Glaucidium
  brasilianum **236**
  gnoma **236**
Gnatcatcher
  Black-capped **328**
  Black-tailed **328**
  Blue-gray **328**
  California **328**
Godwit
  Bar-tailed **112**
  Black-tailed **457**
  Hudsonian **112**
  Marbled **112**
Goldeneye
  Barrow's **48**
  Common **48**
Goldfinch
  American **378**
  Lawrence's **378**
  Lesser **378**
Goose
  Barnacle **467**
  Cackling **26**
  Canada **26**
  Egyptian **30**
  Emperor **22**
  Graylag 24, **25**
  Greater White-fronted **24**
  Hawaiian **28**
  Lesser White-fronted **467**
  Pink-footed **467**
  Ross's **24**
  Snow **24**
  Taiga Bean- **454**
  Tundra Bean- **454**
Goshawk
  American **220**
  Eurasian **470**
Grackle
  Common **422**
  Great-tailed **422**
Grassquit, Blue-black **472**
Grebe
  Clark's **70**
  Eared **68**
  Horned **68**
  Least **455**
  Pied-billed **68**
  Red-necked **70**
  Western **70**
Greenfinch, Oriental **464**
Greenshank, Common **458**
Grosbeak
  Black-headed **448**
  Blue **448**
  Evening **360**
  Kona **473**
  Pine **360**
  Rose-breasted **448**
  Yellow **466**
Grouse
  Dusky **60**
  Greater Sage- **60**
  Gunnison Sage- **60**
  Ruffed **56**
  Sharp-tailed **62**
  Sooty 12, **60**
  Spruce **56**
Gruidae (family) 100–101
Grus
  americana **456**
  grus **456**
  monacha **468**
Guillemot
  Black **140**
  Pigeon **140**
Gull
  anatomy 11
  Belcher's **468**
  Black-headed **150**
  Black-tailed **459**
  Bonaparte's **150**
  California **158**
  Common **459**
  Franklin's **152**
  Glaucous **162**
  Glaucous-winged **162**
  Great Black-backed **459**
  Heermann's **152**
  Herring 11, **158**
  Iceland **160**
  Ivory **148**
  Kelp **468**
  Laughing **152**
  Lesser Black-backed **160**
  Little **150**
  Pallas's **468**
  Ring-billed **154**
  Ross's **150**
  Sabine's **148**
  Short-billed **154**
  Slaty-backed **162**
  Swallow-tailed **468**
  Western **156**
  Yellow-footed **156**
Gygis alba **164**
Gymnogyps californianus **214**
Gymnorhinus cyanocephalus **294**
Gyrfalcon **260**

H

Haematopodidae (family) 102–103
Haematopus
  bachmani **102**
  ostralegus **468**
  palliatus **102**
Haemorhous
  cassinii **372**
  mexicanus **372**
  purpureus **372**
Haliaeetus
  albicilla **460**
  leucocephalus **222**
  pelagicus **460**
Harrier, Northern **218**
Hawaii 14
Hawfinch **464**
Hawk
  Broad-winged **224**
  Common Black **222**
  Cooper's **220**
  Ferruginous **230**
  Gray **224**
  Harris's **224**
  Hawaiian **226**
  Red-shouldered **224**
  Red-tailed **228**
  Rough-legged **230**
  Sharp-shinned **220**
  Short-tailed **226**
  Swainson's **226**
  White-tailed **470**
  Zone-tailed **228**
Heliomaster constantii **86**
Heliornis fulica **468**
Helmitheros vermivorum **465**
Helopsaltes
  certhiola **470**
  ochotensis **462**
Hemignathus
  affinis 473
  hanapepe 473
  lucidus 473
  wilsoni **364**

487

INDEX

Heron
  Chinese Pond- **469**
  Gray **469**
  Great Blue **206**
  Green **212**
  Little Blue **208**
  Tricolored **210**
Himantopus
  himantopus **468**
  mexicanus **102**
Himatione
  fraithii 473
  sanguinea **364**
Hippolais icterina **470**
Hirundapus caudacutus **467**
Hirundinidae (family) 310–315
Hirundo rustica **314**
Histrionicus histrionicus **46**
Hobby, Eurasian **461**
Honeycreeper
  Hawaiian 14, 362–369, 473
  Laysan 473
Hookbill, Lanai 473
Hoopoe, Eurasian **470**
Horornis diphone **316**
Hummingbird
  Allen's **92**
  anatomy 11
  Anna's **90**
  Berylline **94**
  Black-chinned **88**
  Broad-billed **94**
  Broad-tailed **92**
  Bumblebee **467**
  Calliope 4, 5, **90**
  Cinnamon **467**
  Costa's 12, **90**
  Lucifer **88**
  Rivoli's **86**
  Ruby-throated **88**
  Rufous 11, **92**
  Violet-crowned **94**
  White-eared **94**
  Xantus's **467**
Hwamei **320**
Hydrobates
  castro **186**
  cheimomnestes **469**
  furcatus **184**
  homochroa **184**
  hornbyi **469**
  leucorhous **182**
  melania **184**
  microsoma **184**
  monorhis **469**
  socorroensis **182**
  tethys **459**
  tristrami **186**
Hydrobatidae (family) 182–187
Hydrocoloeus minutus **150**
Hydroprogne caspia **168**
Hylocichla mustelina **463**

I
Ibis
  Glossy **214**
  White **460**
  White-faced **214**
Ichthyaetus ichthyaetus **468**

Icteria virens **410**
Icteridae (family) 410–423
Icteriidae (family) 410–411
Icterus
  bullockii **416**
  cucullatus **414**
  galbula **416**
  parisorum **416**
  pustulatus **414**
  spurius **414**
  wagleri **472**
Ictinia mississippiensis **222**
Identification
  anatomy 10–11, 17
  apps 19–20
  behavioral ecology 17–18
  feeding behavior 18
  introduction 17–18
Iiwi **364**
iNaturalist 19, 20
Interior West 12–13
Ixobrychus
  exilis **206**
  sinensis **469**
Ixoreus naevius **348**

J
Jacana, Northern **456**
Jacana spinosa **456**
Jaeger
  Long-tailed **136**
  Parasitic **136**
  Pomarine **134**
Jay
  Blue **294**
  California Scrub- **296**
  Canada **292**
  Island Scrub- **296**
  Mexican **296**
  Pinyon **294**
  Steller's 2, 5, **294**
  Woodhouse's Scrub- **296**
Junco
  Dark-eyed **396**
  Yellow-eyed **396**
Junco
  hyemalis **396**
  phaeonotus **396**
Junglefowl, Red **66**
Jynx torquilla **470**

K
Kakawahie 473
Kamao 473
Kestrel
  American **258**
  Eurasian **460**
Killdeer **106**
Kingbird
  Cassin's **270**
  Couch's **470**
  Eastern **272**
  Thick-billed **272**
  Tropical **270**
  Western **270**
Kingfisher
  Belted 18, **242**
  Green **242**
  Ringed **470**

Kinglet
  Golden-crowned **322**
  Ruby-crowned **322**
Kioea 473
Kiskadee, Great **470**
Kite
  Black **470**
  Mississippi **222**
  Swallow-tailed **460**
  White-tailed **218**
Kittiwake
  Black-legged **148**
  Red-legged **148**
Knot
  Great **457**
  Red **116**

L
Lagopus
  lagopus **58**
  leucura **58**
  muta **58**
Lampornis
  amethystinus **467**
  clemenciae **86**
Laniidae (family) 292–293
Lanius
  borealis **292**
  collurio **470**
  cristatus **462**
  ludovicianus **292**
Lapwing, Northern **468**
Laridae (family) 148–173
Lark, Horned **308**
Larosterna inca **468**
Larus
  argentatus **158**
  belcheri **468**
  brachyrhynchus **154**
  californicus **158**
  canus **459**
  crassirostris **459**
  delawarensis **154**
  dominicanus **468**
  fuscus **160**
  glaucescens **162**
  glaucoides **160**
  heermanni **152**
  hyperboreus **162**
  livens **156**
  marinus **459**
  occidentalis **156**
  schistisagus **162**
Larvivora
  cyane **471**
  sibilans **471**
Laterallus jamaicensis **100**
Laughingthrush, Greater Necklaced **320**
Legatus leucophaius **470**
Leiothlypis
  celata **426**
  crissalis **426**
  luciae **428**
  peregrina **426**
  ruficapilla **428**
  virginiae **428**
Leiothrichidae (family) 320–323
Leiothrix, Red-billed **322**
Leiothrix lutea **322**

INDEX

Leucophaeus
  atricilla **152**
  pipixcan **152**
Leucosticte
  arctoa **472**
  atrata **370**
  australis **370**
  tephrocotis **370**
Limnodromus
  griseus **126**
  scolopaceus **126**
Limnothlypis swainsonii **472**
Limosa
  fedoa **112**
  haemastica **112**
  lapponica **112**
  limosa **457**
Limpkin **468**
Locustella
  fluviatilis **471**
  lanceolata **462**
Lonchura
  atricapilla **354**
  punctulata **354**
Longspur
  Chestnut-collared **382**
  Lapland **380**
  Smith's **380**
  Thick-billed **382**
Loon
  Arctic **176**
  Common **178**
  Pacific **176**
  Red-throated **176**
  Yellow-billed **178**
Lophodytes cucullatus **48**
Lophura leucomelanos **64**
Lovebird, Rosy-faced **264**
Loxia
  curvirostra **376**
  leucoptera **376**
  sinesciuris **376**
Loxioides bailleui **362**
Loxops
  caeruleirostris **368**
  coccineus **368**
  mana **368**
  ochraceus 473
  wolstenholmei 473
Lymnocryptes minimus **457**

M

Macronectes halli **469**
Magpie
  Black-billed **298**
  Yellow-billed **298**
Magumma parva **366**
Mallard **36**
Mamo
  Black 473
  Hawaii 473
Maps
  approach to range maps 15–16
  regions 12–14
Mareca
  americana **34**
  falcata **454**
  penelope **34**
  strepera **34**

Martin
  Brown-chested **471**
  Purple **312**
  Siberian House- **462**
  Western House- **462**
Meadowlark
  Chihuahuan **412**
  Eastern **412**
  Western **412**
Megaceryle
  alcyon **242**
  torquata **470**
Megascops
  asio **234**
  kennicottii **234**
  trichopsis **232**
Melamprosops phaeosoma 473
Melanerpes
  aurifrons **246**
  carolinus **246**
  erythrocephalus **244**
  formicivorus **244**
  lewis **244**
  uropygialis **246**
Melanitta
  americana **44**
  deglandi **44**
  nigra **467**
  perspicillata **44**
  stejnegeri **44**
Melanotis caerulescens **463**
Meleagris gallopavo **56**
Melospiza
  georgiana **404**
  lincolnii **404**
  melodia **404**
Melozone
  aberti **406**
  crissalis **406**
  fusca **406**
Merganser
  Common **50**
  Hooded **48**
  Red-breasted **50**
Mergellus albellus **455**
Mergus
  merganser **50**
  serrator **50**
Merlin **258**
Merlin Bird ID app 19–20
Micrathene whitneyi **236**
Millerbird **310**
Milvus migrans **470**
Mimidae (family) 336–341
Mimus polyglottos **340**
Mitrephanes phaeocercus **461**
Mniotilta varia **424**
Mockingbird
  Blue **463**
  Northern **340**
Moho
  apicalis 473
  bishopi 473
  braccatus 473
  nobilis 473
Mohoidae (family) 473
Molothrus
  aeneus **422**
  ater **420**

Molt 10
Monarchidae (family) 290–291
Monticola saxatilis **471**
Moorhen, Common **467**
Morus bassanus **469**
Motacilla
  alba **356**
  cinerea **464**
  citreola **472**
  tschutschensis **356**
Motacillidae (family) 356–359
Mountain-gem
  Amethyst-throated **467**
  Blue-throated **86**
Munia
  Chestnut **354**
  Scaly-breasted **354**
Murre
  Common **138**
  Thick-billed **138**
Murrelet
  Ancient **142**
  Craveri's **142**
  Guadalupe **142**
  Kittlitz's **140**
  Long-billed **458**
  Marbled 13, **140**
  Scripps's **142**
Muscicapa
  dauurica **471**
  griseisticta **463**
  sibirica **463**
  striata **471**
Muscicapidae (family) 350–351
Myadestes
  lanaiensis 473
  myadestinus 473
  obscurus **344**
  occidentalis **471**
  palmeri **344**
  townsendi **344**
  woahensis 473
Mycteria americana **198**
Myiarchus
  cinerascens **268**
  crinitus **268**
  nuttingi **461**
  tuberculifer **268**
  tyrannulus **268**
Myioborus
  miniatus **466**
  pictus **442**
Myiodynastes luteiventris **266**
Myiopsitta monachus **461**
Myna, Common **340**

N

Nannopterum
  auritum **202**
  brasilianum **204**
National Audubon Society 21
Needletail, White-throated **467**
Night-Heron
  Black-crowned **212**
  Yellow-crowned **212**
Nighthawk
  Common **80**
  Lesser **80**

INDEX

Nightjar
　Buff-collared **82**
　Gray **467**
Ninox japonica **470**
Noddy
　Black **164**
　Blue-gray **164**
　Brown **164**
Nomonyx dominicus **467**
Nucifraga columbiana **298**
Nukupuu
　Kauai 473
　Maui 473
　Oahu 473
Numenius
　americanus **110**
　borealis 473
　madagascariensis **456**
　minutus **468**
　phaeopus **110**
　tahitiensis **110**
Nutcracker, Clark's **298**
Nuthatch
　Pygmy **326**
　Red-breasted **326**
　White-breasted **326**
Nyctanassa violacea **212**
Nycticorax nycticorax **212**

O

Oceanites oceanicus **182**
Oceanitidae (family) 182–183
Odontophoridae (family) 52–55
Oenanthe
　oenanthe **350**
　pleschanka **471**
Olomao 473
Omao **344**
Onychoprion
　aleuticus **166**
　anaethetus **468**
　fuscatus **166**
　lunatus **166**
Oo
　Bishop's 473
　Hawaii 473
　Kauai 473
　Oahu 473
Oporornis agilis **430**
Oreomystis bairdi **362**
Oreortyx pictus **52**
Oreoscoptes montanus **340**
Oreothlypis superciliosa **465**
Oriole
　Baltimore **416**
　Black-vented **472**
　Bullock's **416**
　Hooded **414**
　Orchard **414**
　Scott's **416**
　Streak-backed **414**
Osprey **216**
Otus sunia **470**
Ou 473
Ovenbird **424**
Owl
　Barn **232**
　Barred 17, **238**
　Boreal **240**
　Burrowing **238**
　Eastern Screech- **234**
　Elf **236**
　Ferruginous Pygmy- **236**
　Flammulated **232**
　Great Gray **238**
　Great Horned **234**
　Long-eared **240**
　Northern Hawk **236**
　Northern Pygmy- **236**
　Northern Saw-whet **240**
　Oriental Scops- **470**
　Short-eared **240**
　Snowy **234**
　Spotted 17, **238**
　Western Screech- **234**
　Whiskered Screech- **232**
Oxyura jamaicensis **50**
Oystercatcher
　American **102**
　Black **102**
　Eurasian **468**

P

Pachyramphus
　algaiae **266**
　major **470**
Pacific region 12
Padda oryzivora **354**
Pagophila eburnea **148**
Palila **362**
Palmeria dolei **368**
Pandion haliaetus **216**
Pandionidae (family) 216–217
Parabuteo unicinctus **224**
Parakeet
　Carolina 473
　Mitred **262**
　Monk **461**
　Nanday **262**
　Red-masked **262**
　Rose-ringed **264**
　White-winged **262**
　Yellow-chevroned **262**
Paridae (family) 302–307
Parkesia
　motacilla **465**
　noveboracensis **424**
Paroaria
　capitata **452**
　coronata **452**
Paroreomyza
　flammea 473
　maculata 473
　montana **362**
Parrot
　Lilac-crowned **264**
　Red-crowned **264**
　Red-lored 264, 265
　Thick-billed **461**
　Yellow-headed **264**
Parrotbill, Maui **364**
Partridge, Gray **64**
Parula
　Northern **432**
　Tropical **472**
Parulidae (family) 424–443

Passer
　domesticus **356**
　montanus **471**
Passerculus sandwichensis **388**
Passerella iliaca **394**
Passerellidae (family) 386–409
Passeridae (family) 356–357
Passerina
　amoena **450**
　caerulea **448**
　ciris **450**
　cyanea **450**
　versicolor **450**
Patagioenas fasciata **72**
Pavo cristatus **64**
Peafowl, Indian **64**
Pelecanidae (family) 204–205
Pelecanus
　erythrorhynchos **204**
　occidentalis **204**
Pelican
　American White **204**
　Brown **204**
Perdix perdix **64**
Perisoreus canadensis **292**
Peterson, Roger Tory 12
Petrel
　Black-winged **190**
　Bonin **190**
　Bulwer's **190**
　Cook's **190**
　Gray-faced **469**
　Hawaiian **188**
　Herald **469**
　Jouanin's **469**
　Juan Fernandez **188**
　Kermadec **459**
　Mottled **188**
　Murphy's **186**
　Northern Giant- **469**
　Parkinson's **469**
　Providence **469**
　Stejneger's **469**
　Tahiti **469**
　White-chinned **469**
　White-necked **188**
Petrochelidon
　fulva **314**
　pyrrhonota **314**
Peucaea
　botterii **386**
　carpalis **386**
　cassinii **386**
Peucedramidae (family) 352–353
Peucedramus taeniatus **352**
Pewee
　Eastern Wood- **461**
　Greater **274**
　Western Wood- **274**
Phaethon
　aethereus **174**
　lepturus **174**
　rubricauda **174**
Phaethontidae (family) 174–175
Phainopepla **324**
Phainopepla nitens **324**
Phalacrocoracidae (family) 202–205
Phalaenoptilus nuttallii **80**

INDEX

Phalarope
  Red **132**
  Red-necked **132**
  Wilson's **132**
Phalaropus
  fulicarius **132**
  lobatus **132**
  tricolor **132**
Phasianidae (family) 56–67
Phasianus colchicus **64**
Pheasant
  Kalij **64**
  Ring-necked **64**
Pheucticus
  chrysopeplus **466**
  ludovicianus **448**
  melanocephalus **448**
Phoebastria
  albatrus **180**
  immutabilis **180**
  nigripes **180**
Phoebe
  Black **282**
  Eastern **282**
  Say's **282**
Phoebetria palpebrata **468**
Phoenicurus phoenicurus **471**
Phylloscopidae (family) 316–317
Phylloscopus
  borealis **316**
  collybita **462**
  examinandus **463**
  fuscatus **463**
  inornatus **463**
  proregulus **471**
  sibilatrix **462**
  trochilus **462**
Pica
  hudsonia **298**
  nuttalli **298**
Picidae (family) 244–257
Picoides
  arcticus **250**
  dorsalis **250**
Pigeon
  Band-tailed **72**
  Passenger 473
  Rock **72**
Pinicola enucleator **360**
Pintail, Northern **38**
Pipilo
  chlorurus **408**
  erythrophthalmus **465**
  maculatus **408**
Pipit
  American **358**
  Olive-backed **464**
  Pechora **472**
  Red-throated **358**
  Sprague's **358**
  Tree **472**
Piranga
  bidentata **446**
  flava **444**
  ludoviciana **444**
  olivacea **444**
  rubra **444**
Pitangus sulphuratus **470**
Platalea ajaja **460**

Plectrophenax
  hyperboreus **384**
  nivalis **384**
Plegadis
  chihi **214**
  falcinellus **214**
Ploceidae (family) 352–353
Plover
  American Golden- **104**
  Black-bellied **104**
  Common Ringed **106**
  European Golden- **468**
  Greater Sand- **468**
  Lesser Sand- **108**
  Little Ringed **468**
  Mountain **108**
  Pacific Golden- **104**
  Piping **108**
  Semipalmated **106**
  Snowy **108**
  Wilson's **456**
Pluvialis
  apricaria **468**
  dominica **104**
  fulva **104**
  squatarola **104**
Pochard, Common **455**
Podiceps
  auritus **68**
  grisegena **70**
  nigricollis **68**
Podicipedidae (family) 68–71
Podilymbus podiceps **68**
Poecile
  atricapillus **302**
  cinctus **304**
  gambeli **302**
  hudsonicus **304**
  rufescens **304**
  sclateri **304**
Polioptila
  caerulea **328**
  californica **328**
  melanura **328**
  nigriceps **328**
Polioptilidae (family) 328–329
Polysticta stelleri **42**
Poo-uli 473
Pooecetes gramineus **402**
Poorwill, Common **80**
Porphyrio martinicus **456**
Porzana carolina **96**
Prairie-Chicken
  Greater **62**
  Lesser **62**
Pratincole, Oriental **468**
Procellaria
  aequinoctialis **469**
  parkinsoni **469**
Procellariidae (family) 186–197
Progne
  subis **312**
  tapera **471**
Protonotaria citrea **465**
Prunella montanella **464**
Psaltiparus minimus **316**
Pseudobulweria rostrata **469**
Pseudonestor xanthophrys **364**
Psiloscops flammeolus **232**

Psittacara
  erythrogenys **262**
  mitratus **262**
Psittacidae (family) 262–265
Psittacula krameri **264**
Psittaculidae (family) 264–265
Psittirostra psittacea 473
Ptarmigan
  Rock **58**
  White-tailed **58**
  Willow **58**
Pternistis erckelii **66**
Pterocles exustus **72**
Pteroclidae (family) 72–73
Pterodroma
  cervicalis **188**
  cookii **190**
  externa **188**
  gouldi **469**
  heraldica **469**
  hypoleuca **190**
  inexpectata **188**
  longirostris **469**
  neglecta **459**
  nigripennis **190**
  sandwichensis **188**
  solandri **469**
  ultima **186**
Ptilogonatidae (family) 324–325
Ptilogonys cinereus **471**
Ptychoramphus aleuticus **146**
Puaiohi **344**
Puffin
  Horned **146**
  Tufted **146**
Puffinus
  bryani **469**
  nativitatis **196**
  newelli **196**
  opisthomelas **196**
  puffinus **196**
Pycnonotidae (family) 318–319
Pycnonotus
  cafer **318**
  jocosus **318**
Pyrocephalus rubinus **282**
Pyrrhula pyrrhula **464**
Pyrrhuloxia **446**

Q

Quail
  California **54**
  Gambel's **54**
  Montezuma **52**
  Mountain **52**
  Scaled **54**
Quetzal, Eared **460**
Quiscalus
  mexicanus **422**
  quiscala **422**

R

Rail
  Black **100**
  Clapper **467**
  Hawaiian 473
  King **467**
  Laysan 473
  Ridgway's **96**

INDEX

Rufous-necked Wood- **467**
Virginia **96**
Yellow **100**
Rallidae (family) 96–101
Rallus
  crepitans **467**
  elegans **467**
  limicola **96**
  obsoletus **96**
Ramosomyia violiceps **94**
Range maps, approach to 15–16
Rare birds in the West 454–472
Raven
  Chihuahuan **300**
  Common **300**
Recurvirostra americana **102**
Recurvirostridae (family) 102–103
Redhead **38**
Redpoll
  Common **374**
  Hoary **374**
Redshank, Spotted **458**
Redstart
  American **432**
  Common **471**
  Painted **442**
  Slate-throated **466**
Redwing **471**
Regulidae (family) 322–323
Regulus satrapa **322**
Remizidae (family) 302–303
Rhodacanthis
  flaviceps 473
  palmeri 473
Rhodostethia rosea **150**
Rhynchophanes mccownii **382**
Rhynchopsitta pachyrhyncha **461**
Ridgwayia pinicola **463**
Riparia riparia **312**
Rissa
  brevirostris **148**
  tridactyla **148**
Roadrunner, Greater **76**
Robin
  American **348**
  Rufous-backed **348**
  Rufous-tailed **471**
  Siberian Blue **471**
Rock-thrush, Rufous-tailed **471**
Rosefinch
  Common **464**
  Pallas's **472**
Rubythroat, Siberian **350**
Ruff **124**
Rynchops niger **172**

S

Salpinctes obsoletus **334**
Sanderling **118**
Sandgrouse, Chestnut-bellied **72**
Sandpiper
  anatomy 11
  Baird's **120**
  Broad-billed **457**
  Buff-breasted **124**
  Common **458**
  Curlew **116**
  Green **458**
  Least **120**
  Marsh **458**
  Pectoral **124**
  Purple **468**
  Rock **118**
  Semipalmated 11, **122**
  Sharp-tailed **124**
  Solitary **128**
  Spoon-billed **468**
  Spotted **128**
  Stilt **116**
  Terek **458**
  Upland **110**
  Western **122**
  White-rumped **120**
  Wood **130**
Sapsucker
  Red-breasted 13, **248**
  Red-naped 13, **248**
  Williamson's **250**
  Yellow-bellied 13, **248**
Saucerottia beryllina **94**
Saxicola maurus **464**
Sayornis
  nigricans **282**
  phoebe **282**
  saya **282**
Scaup
  Greater **40**
  Lesser **40**
Scolopacidae (family) 110–133
Scolopax minor **457**
Scoter
  Black **44**
  Common **467**
  Stejneger's **44**
  Surf **44**
  White-winged **44**
Seek app 19–20
Seiurus aurocapilla **424**
Selasphorus
  calliope **90**
  heloisa **467**
  platycercus **92**
  rufus **92**
  sasin **92**
Serinus canaria **374**
Setophaga
  americana **432**
  caerulescens **466**
  castanea **434**
  cerulea **472**
  chrysoparia **472**
  citrina **432**
  coronata **438**
  discolor **466**
  dominica **466**
  fusca **434**
  graciae **438**
  magnolia **434**
  nigrescens **440**
  occidentalis **440**
  palmarum **438**
  pensylvanica **436**
  petechia **436**
  pinus **466**
  pitiayumi **472**
  ruticilla **432**
  striata **436**
  tigrina **432**
  townsendi **440**
  virens **440**
Shama, White-rumped **350**
Shearwater
  Black-vented **196**
  Bryan's **469**
  Buller's **194**
  Christmas **196**
  Cory's **469**
  Flesh-footed **194**
  Great **469**
  Manx **196**
  Newell's **196**
  Pink-footed **194**
  Short-tailed **192**
  Sooty **192**
  Streaked **469**
  Wedge-tailed **192**
Shelduck
  Common **30**
  Ruddy 30, 31
Shorebirds, anatomy 11
Shoveler, Northern **32**
Shrike
  Brown **462**
  Loggerhead **292**
  Northern **292**
  Red-backed **470**
Sialia
  currucoides **342**
  mexicana **342**
  sialis **342**
Sibirionetta formosa **454**
Sicalis flaveola **452**
Silky-flycatcher, Gray **471**
Silverbill, African **354**
Siskin
  Eurasian **472**
  Pine **378**
Sitta
  canadensis **326**
  carolinensis **326**
  pygmaea **326**
Sittidae (family) 326–327
Skimmer, Black **172**
Skua, South Polar **134**
Skylark, Eurasian **308**
Smew **455**
Snipe
  Common **458**
  Jack **457**
  Pin-tailed **468**
  Solitary **468**
  Wilson's **128**
Snowcock, Himalayan **66**
Solitaire
  Brown-backed **471**
  Townsend's **344**
Somateria
  fischeri **42**
  mollissima **42**
  spectabilis **42**
Sora **96**
Sparrow
  American Tree **394**
  anatomy 10
  Baird's **402**
  Bell's **400**

Black-chinned **390**
Black-throated **390**
Botteri's **386**
Brewer's **392**
Cassin's **386**
Chipping **392**
Clay-colored **392**
Eurasian Tree **471**
Field **394**
Five-striped **388**
Fox **394**
Golden-crowned **398**
Grasshopper **388**
Harris's **398**
Henslow's **472**
House **356**
Java **354**
Lark 10, **390**
LeConte's **402**
Lincoln's **404**
Nelson's **402**
Rufous-crowned **406**
Rufous-winged **386**
Sagebrush **400**
Savannah **388**
Song **404**
Swamp **404**
Vesper **402**
White-crowned **398**
White-throated **400**
Worthen's **472**
Sparrowhawk, Chinese **470**
Spatula
   clypeata **32**
   cyanoptera **32**
   discors **32**
   querquedula **454**
Sphyrapicus
   nuchalis **248**
   ruber **248**
   thyroideus **250**
   varius **248**
Spinus
   lawrencei **378**
   pinus **378**
   psaltria **378**
   spinus **472**
   tristis **378**
Spiza americana **452**
Spizella
   atrogularis **390**
   breweri **392**
   pallida **392**
   passerina **392**
   pusilla **394**
   wortheni **472**
Spizelloides arborea **394**
Spoonbill, Roseate **460**
Starling, European **340**
Starthroat, Plain-capped **86**
Stelgidopteryx serripennis **312**
Stercorariidae (family) 134–137
Stercorarius
   longicaudus **136**
   maccormicki **134**
   parasiticus **136**
   pomarinus **134**
Sterna
   forsteri **170**

   hirundo **170**
   paradisaea **170**
Sternula
   albifrons **459**
   antillarum **168**
Stilt
   Black-necked **102**
   Black-winged **468**
Stint
   Little **457**
   Long-toed **457**
   Red-necked **122**
   Temminck's **457**
Stonechat, Asian **464**
Stork, Wood **198**
Storm-Petrel
   Ainley's **469**
   Ashy **184**
   Band-rumped **186**
   Black **184**
   Fork-tailed **184**
   Leach's **182**
   Least **184**
   Ringed **469**
   Swinhoe's **469**
   Townsend's **182**
   Tristram's **186**
   Wedge-rumped **459**
   Wilson's **182**
Streptopelia
   chinensis **74**
   decaocto **72**
   orientalis **455**
Streptoprocne zonaris **467**
Strigidae (family) 232–241
Strix
   nebulosa **238**
   occidentalis **238**
   varia **238**
Sturnella
   lilianae **412**
   magna **412**
   neglecta **412**
Sturnidae (family) 340–341
Sturnus vulgaris **340**
Sula
   dactylatra **200**
   granti **459**
   leucogaster **200**
   nebouxii **200**
   sula **200**
Sulidae (family) 200–201
Sungrebe **468**
Surfbird **114**
Surnia ulula **236**
Swallow
   Bank **312**
   Barn **314**
   Cave **314**
   Cliff **314**
   Northern Rough-winged **312**
   Tree **310**
   Violet-green **310**
Swan
   Black 28, 29
   Mute **28**
   Trumpeter **28**
   Tundra **28**
   Whooper **454**

Swift
   Black **82**
   Chimney **84**
   Common **467**
   Fork-tailed **456**
   Vaux's **84**
   White-collared **467**
   White-throated **84**
Swiftlet, Mariana **84**
Sylvia curruca **471**
Sylviidae (family) 318–319
Synthliboramphus
   antiquus **142**
   craveri **142**
   hypoleucus **142**
   scrippsi **142**

T

Tachybaptus dominicus **455**
Tachycineta
   bicolor **310**
   thalassina **310**
Tadorna tadorna **30**
Tanager
   Flame-colored **446**
   Hepatic **444**
   Scarlet **444**
   Summer **444**
   Western 7, **444**
Tarsiger cyanurus **463**
Tattler
   Gray-tailed **458**
   Wandering **128**
Taxonomy 7
Teal
   Baikal **454**
   Blue-winged **32**
   Cinnamon 16, **32**
   Green-winged **38**
Telespiza
   cantans **362**
   ultima **362**
Tern
   Aleutian **166**
   Arctic **170**
   Black **168**
   Bridled **468**
   Caspian **168**
   Common **170**
   Elegant **172**
   Forster's **170**
   Gray-backed **166**
   Great Crested **468**
   Gull-billed **168**
   Inca **468**
   Least **168**
   Little **459**
   Royal **172**
   Sandwich **468**
   Sooty **166**
   White **164**
   White-winged **459**
Tetraogallus himalayensis **66**
Thalassarche
   cauta **459**
   eremita **468**
   salvini **468**
Thalasseus
   bergii **468**

INDEX

elegans **172**
maximus **172**
sandvicensis **468**
Thrasher
  Bendire's **338**
  Brown **336**
  California **338**
  Crissal **338**
  Curve-billed 12, **336**
  LeConte's **338**
  Long-billed **471**
  Sage **340**
Thraupidae (family) 452–453
Threskiornithidae (family) 214–215
Thrush
  Aztec **463**
  Clay-colored **471**
  Dusky **348**
  Eyebrowed **348**
  Gray-cheeked **346**
  Hermit **346**
  Naumann's **471**
  Song **471**
  Swainson's **346**
  Varied 13, 17, **348**
  White-throated **471**
  Wood **463**
Thryomanes bewickii **332**
Thryophilus sinaloa **471**
Thryothorus ludovicianus **332**
Titmouse
  Black-crested **306**
  Bridled **306**
  Juniper **306**
  Oak **306**
Tityridae (family) 266–267
Towhee
  Abert's **406**
  California **406**
  Canyon **406**
  Eastern **465**
  Green-tailed **408**
  Spotted **408**
Toxostoma
  bendirei **338**
  crissale **338**
  curvirostre **336**
  lecontei **338**
  longirostre **471**
  redivivum **338**
  rufum **336**
Tringa
  brevipes **458**
  erythropus **458**
  flavipes **130**
  glareola **130**
  incana **128**
  melanoleuca **130**
  nebularia **458**
  ochropus **458**
  semipalmata **130**
  solitaria **128**
  stagnatilis **458**
Trochilidae (family) 86–95
Troglodytes
  aedon **330**
  hiemalis **330**
  pacificus **330**
Troglodytidae (family) 330–335

Trogon, Elegant **242**
Trogon elegans **242**
Trogonidae (family) 242–243
Tropicbird
  Red-billed **174**
  Red-tailed **174**
  White-tailed **174**
Turdidae (family) 342–349
Turdus
  assimilis **471**
  eunomus **348**
  grayi **471**
  iliacus **471**
  migratorius **348**
  naumanni **471**
  obscurus **348**
  philomelos **471**
  pilaris **471**
  rufopalliatus **348**
Turkey, Wild **56**
Turnstone
  Black **114**
  Ruddy **114**
Tympanuchus
  cupido **62**
  pallidicinctus **62**
  phasianellus **62**
Tyrannidae (family) 266–283
Tyrannulet, Northern Beardless- **266**
Tyrannus
  couchii **470**
  crassirostris **272**
  forficatus **272**
  melancholicus **270**
  savana **461**
  tyrannus **272**
  verticalis **270**
  vociferans **270**
Tyto alba **232**
Tytonidae (family) 232–233

U

Ula-ai-hawane **473**
Upupa epops **470**
Uria
  aalge **138**
  lomvia **138**
Urile
  pelagicus **202**
  penicil **202**
  urile **202**

V

Vanellus vanellus **468**
Veery **346**
Verdin **302**
Vermivora
  chrysoptera **465**
  cyanoptera **465**
Vidua macroura **352**
Viduidae (family) 352–353
Violetear, Mexican **467**
Vireo
  Bell's **284**
  Black-capped **470**
  Blue-headed 13, **286**
  Cassin's 13, **286**
  Gray **284**
  Hutton's **284**
  Philadelphia **288**
  Plumbeous 13, **286**
  Red-eyed **288**
  Warbling **288**
  White-eyed **461**
  Yellow-green **286**
  Yellow-throated **462**
Vireo
  atricapilla **470**
  bellii **284**
  cassinii **286**
  flavifrons **462**
  flavoviridis **286**
  gilvus **288**
  griseus **461**
  huttoni **284**
  olivaceus **288**
  philadelphicus **288**
  plumbeus **286**
  solitarius **286**
  vicinior **284**
Vireonidae (family) 284–289
Viridonia sagittirostris **473**
Volatinia jacarina **472**
Vulture
  Black **216**
  Turkey **216**

W

Wagtail
  Citrine **472**
  Eastern Yellow 13, **356**
  Gray **464**
  White **356**
Warbler
  Arctic **316**
  Bay-breasted **434**
  Black-and-white **424**
  Black-throated Blue **466**
  Black-throated Gray **440**
  Black-throated Green **440**
  Blackburnian **434**
  Blackpoll **436**
  Blue-winged **465**
  Blyth's Reed **470**
  Canada **442**
  Cape May **432**
  Cerulean **472**
  Chestnut-sided **436**
  Colima **426**
  Connecticut **430**
  Crescent-chested **465**
  Dusky **463**
  Fan-tailed **466**
  Golden-cheeked **472**
  Golden-crowned **472**
  Golden-winged **465**
  Grace's **438**
  Hermit **440**
  Hooded **432**
  Icterine **470**
  Kamchatka Leaf **463**
  Kentucky **466**
  Lanceolated **462**
  Lucy's **428**
  MacGillivray's **430**
  Magnolia **434**
  Middendorff's Grasshopper **462**
  Mourning **430**

Nashville **428**
Olive **352**
Orange-crowned **426**
Pallas's Grasshopper **470**
Pallas's Leaf **471**
Palm **438**
Pine **466**
Prairie **466**
Prothonotary **465**
Red-faced **442**
River **471**
Rufous-capped **442**
Sedge **470**
Swainson's **472**
Tennessee **426**
Thick-billed **470**
Townsend's **440**
Virginia's **428**
Willow **462**
Wilson's **442**
Wood **462**
Worm-eating **465**
Yellow **436**
Yellow-browed **463**
Yellow-rumped **438**
Yellow-throated **466**
Waterthrush
  Louisiana **465**
  Northern **424**
Waxbill
  Common **354**
  Lavender **471**
Waxwing
  Bohemian **324**
  Cedar **324**
Wheatear
  Northern 13, **350**
  Pied **471**
Whimbrel **110**
Whip-poor-will
  Eastern **467**
  Mexican **82**
Whistling-Duck
  Black-bellied **22**
  Fulvous **22**
White-eye
  Swinhoe's **320**
  Warbling **320**
Whitethroat, Lesser **471**
Whydah, Pin-tailed **352**
Wigeon
  American **34**
  Eurasian **34**
Willet **130**
Woodcock, American **457**
Woodpecker
  Acorn **244**
  American Three-toed **250**
  Arizona **254**
  Black-backed **250**
  Downy **252**
  Gila **246**
  Golden-fronted **246**
  Great Spotted **460**
  Hairy **254**
  Ladder-backed **252**
  Lewis's **244**
  Nuttall's **252**
  Pileated **256**
  Red-bellied **246**
  Red-headed **244**
  White-headed 12, **254**
Wren
  Bewick's **332**
  Cactus **334**
  Canyon **334**
  Carolina **332**
  House **330**
  Marsh **332**
  Pacific 13, **330**
  Rock **334**
  Sedge **332**
  Sinaloa **471**
  Winter **330**
Wrentit 12, **318**
Wryneck, Eurasian **470**

X
Xanthocephalus xanthocephalus **410**
Xema sabini **148**
Xenus cinereus **458**

Y
Yellowlegs
  Greater **130**
  Lesser **130**
Yellowthroat, Common **430**

Z
Zapornia
  palmeri 473
  sandwichensis 473
Zenaida
  asiatica **76**
  macroura **76**
Zonotrichia
  albicollis **400**
  atricapilla **398**
  leucophrys **398**
  querula **398**
Zosteropidae (family) 320–321
Zosterops
  japonicus **320**
  simplex **320**

Since 1888, the National Geographic Society has funded more than 14,000 research, conservation, education, and storytelling projects around the world. National Geographic Partners distributes a portion of the funds it receives from your purchase to National Geographic Society to support programs including the conservation of animals and their habitats.

National Geographic Partners, LLC
1145 17th Street NW
Washington, DC 20036-4688 USA

Get closer to National Geographic Explorers and photographers, and connect with our global community. Join us today at nationalgeographic.org/joinus

For rights or permissions inquiries, please contact National Geographic Books Subsidiary Rights: bookrights@natgeo.com

Copyright © 2025 National Geographic Partners, LLC. All rights reserved. Reproduction of the whole or any part of the contents without written permission from the publisher is prohibited.

NATIONAL GEOGRAPHIC and Yellow Border Design are trademarks of the National Geographic Society, used under license.

The Library of Congress cataloged the first edition as follows:

National Geographic field guide to the birds of western North America / edited by Jon L. Dunn and Jonathan Alderfer.
  p.cm.
  Includes index.
  ISBN 978-1-4262-0331-2 (trade pbk.)
  1. Birds--West (U.S.)--Identification.
  2. Birds--Canada, Western--Identification.
  I. Dunn, Jon, 1954-. II. Alderfer, Jonathan.
  QL683.W4.N38 2008
  598.097-dc22

2008023312

ISBN: 978-1-4262-2278-8

Printed in China

24/RRDH/1